T0257948

Nanoparticles: Biological Activities and Nano-Technology

Nanoparticles: Biological Activities and Nano-Technology

Edited by **Mindy Adams**

New York

Published by NY Research Press,
23 West, 55th Street, Suite 816,
New York, NY 10019, USA
www.nyresearchpress.com

Nanoparticles: Biological Activities and Nano-Technology
Edited by Mindy Adams

International Standard Book Number: 978-1-63238-338-9 (Hardback)

Contents

Preface

Every book is initially just a concept; it takes months of research and hard work to give it the final shape in which the readers receive it. In its early stages, this book also went through rigorous reviewing. The notable contributions made by experts from across the globe were first molded into patterned chapters and then arranged in a sensibly sequential manner to bring out the best results.

The study of nanoparticles is dynamic. This book compiles state-of-the-art information on nanoparticles. Nanoparticles are a universal challenge for today's technology and future researches. Nanoparticles encompass approximately all kinds of sciences and manufacturing technologies. The characteristics of these particles are overcoming current scientific hurdles and have passed the restrictions of conventional science. This book explores the field of nano-scale particles, grounding methods and the methods of reaching their objective. It deals with significant issues regarding biological actions and nano-technology. This book will be beneficial for students and experts interested in this field.

It has been my immense pleasure to be a part of this project and to contribute my years of learning in such a meaningful form. I would like to take this opportunity to thank all the people who have been associated with the completion of this book at any step.

Editor

Section 1

Biological Activities

Biological Activities of Carbon Nanotubes

Anurag Mishra[1,2], Yon Rojanasakul[2] and Liying Wang[1]
[1]National Institute for Occupational Safety and Health, HELD/PPRB, Morgantown, WV,
[2]West Virginia University, Department of Pharmaceutical Sciences, Morgantown, WV,
USA

1. Introduction

During the past several years, nanotechnology based on novel nanomaterials has gained considerable attention in various scientific disciplines such as biotechnology, medicine and material engineering (McCarthy and Weissleder 2008). According to British Standards Institute Report (2007), nanoparticles are those particles with at least one dimension of less than or equal to 100 nm (1 nm = 1 x 10^{-6} m) in size. Since particle size is directly related to surface area and associated surface energy, nanoscaled materials relatively exhibit unique physicochemical, optical and electrical properties than micron-sized particles. Nanomaterials have exceptional properties and are beneficial in a wide range of applications. Nanotechnology based on these novel nanomaterials is fueling the modern industrial revolution which is already a multi-billion dollar market capitalization. Among the different types of nanomaterials, carbon nanoparticles have gained much attention in recent years due to their exceptional physicochemical properties. Some of the most popular carbon-based nanomaterials are fullerene (C_{60}), carbon nanohorn, single wall carbon nanotubes (SWCNT), and multi wall carbon nanotubes (MWCNT).

Carbon nanotubes (CNT) are one of the most commonly used nanomaterials possessing unique physicochemical properties such as high aspect ratio and a diameter of less than 100 nm (Iijima 1991). Due to their exceptional characteristics, CNT, if incorporated will enhance the efficiency of a number of applications including electronics (Bandaru 2007), biosensors (Le Goff, Holzinger et al. 2011), drug and biomolecule carriers (Prato, Kostarelos et al. 2007). Other potential biomedical applications of CNT include bone scaffold, dental tissue support, and neuronal cell growth scaffold (Li, Fan et al. 2010). Increasing evidence has shown that certain CNT properties such as nano-sized dimension, high surface energy, and large reactive surface area are directly correlated to their biological activities (Oberdorster, Ferin et al. 1994; Oberdorster, Oberdorster et al. 2005). The bioactivity of nanoparticles differs from micron-size particles of the same material. Although the underlying mechanism remains to be understood, small size, high surface area and chemical composition of nanomaterials play an important role. Recent studies have shown that CNT could be harmful to human health. Fiber morphology and high surface energy of CNT raise health concerns among scientists due to their structural similarities with asbestos fibers (Donaldson, Murphy et al. 2010). The biological properties of nanoparticles are currently under intense investigations and are the subject of this review.

2. Routes of nanoparticle exposure and associated pathologies

Nanomaterials such as CNT have very low specific weight and can be easily aerosolized and come in contact with humans during manufacturing, transportation or usage of the CNT-based products. Apart from unintentional exposures, for certain biomedical applications such as drug delivery, artificial tissues and diagnostic agents, CNT need to be introduced into the human body. Therefore, it is important to consider the potential adverse effects of nanoparticles. Most recent studies have focused on the adverse effects of nanoparticle exposure on pulmonary or dermal tissues.

2.1 Pulmonary exposure

Lung is the major target organ for nanoparticle exposure. Because of their low density and small size, CNT can be aerosolized and inadvertently inhaled during their manufacturing or handling. Therefore, respiratory exposure of nanomaterials including CNT has been the focus of intense research. Lung exposure to solid particles has been linked to asthma (Bonner 2010), fibrosis (Shvedova, Kisin et al. 2005), mesothelioma (Sakamoto, Nakae et al. 2009), and other inflammatory diseases (Li, Muralikrishnan et al. 2010). The region of lung affected by accidently inhaled materials depends on the shape (fibrous, spherical), size (aerodynamic diameter), and other physical and chemical properties of the particles. Nano-sized particles are deposited deep inside the lung compared to micron-sized particles of similar chemical composition (Oberdorster, Ferin et al. 1994). As shown in Figure 1, a large fraction of inhaled particles with the size of less than 100 nm deposits mainly in the terminal alveolar region of the lung, whereas most of the micron-sized inhaled particles remain in the upper respiratory tract. Recent studies have shown that 80% of pulmonary exposed SWCNT reach the alveolar space of the lung in mice (Shvedova, Kisin et al. 2005; Mercer, Scabilloni et al. 2008). Pulmonary exposure of SWCNT and MWCNT has been shown to induce rapid interstitial lung fibrosis with non-persistent inflammatory response in rodents (Shvedova, Kisin et al. 2005; Porter, Hubbs et al. 2010). Inhaled CNT can also translocate to the surrounding regions of the lung such as pleural space (Mercer, Scabilloni et al. 2008; Wilson and Wynn 2009). In several *in vivo* studies, pulmonary exposure to CNT has been shown to induce granuloma formation in the terminal alveolar region of the lung (Lam, James et al. 2004; Warheit, Laurence et al. 2004; Shvedova, Kisin et al. 2005). Inflammatory granulomas are accumulation of epithelioidal macrophages engulfing persistent and non-biodegradable particles. Neutrophil infiltration was also observed in these studies.

Usually inhalation mimics pulmonary exposure of non-soluble particles in animal models, while intratracheal instillation or pharyngeal aspiration have shown similar results (Li, Li et al. 2007; Shvedova, Kisin et al. 2008). Upon alveolar deposition, an unexpectedly rapid translocation of dispersed SWCNT into the alveolar interstitium (1 day post-exposure) has been observed (Shvedova 2005). Subsequent development of lung fibrosis occurs as early as 1 week post-exposure and progresses through 60 days post-exposure without persistent inflammation. The mechanism of nanoparticle-induced lung fibrosis is still under investigation. Data suggest oxidative stress, cytotoxicity and apoptosis induction via DNA damage (Nam, Kang et al. 2011; Ravichandran, Baluchamy et al. 2011) or direct stimulation of lung fibroblasts (Wang, Mercer et al. 2010). *In vitro* studies provide detailed information on the mechanism of the unusual CNT-induced lung fibrosis. Several key lung cells have been selected to determine the specific CNT-lung cell interactions. For example, SWCNT directly stimulate fibroblasts to produce collagen (Wang, Mercer et al. 2010) or induce oxidative stress through ROS production in macrophages (Migliore, Saracino et al. 2010; Palomäki, Välimäki et al. 2011). A study using lung epithelial cells has shown that chronic exposure of low-dose SWCNT can induce cancer-like cell transformation (Wang, Luanpitpong et al. 2011). Several other studies have shown that CNT can induce toxicity of alveolar epithelial cells through a suppression of immune response or oxidative stress (Simon-Deckers, Gouget et al. 2008; Herzog, Byrne et al. 2009). Due to the large surface area and proximity to the circulatory system, inhaled nanoparticles can potentially translocate to other parts of the body and can cause toxic effects such as cardiovascular abnormality (Li, Hulderman et al. 2006).

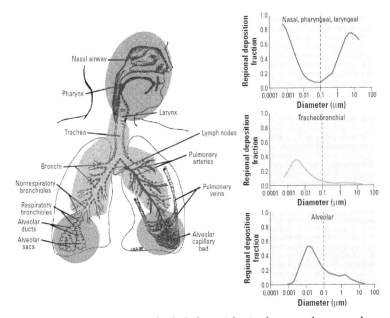

Fig. 1. Predicted fractional deposition of inhaled particles in the nasopharyngeal, tracheobronchial, and alveolar region of the human respiratory tract during nose breathing (Oberdorster, et al., 2005)

2.2 Skin exposure

Skin is the primary barrier preventing the entry of foreign particles into the body. Exposure of skin to nanoparticles can occur in the form of cosmetic formulations containing nanomaterials such as nano-sized titanium dioxide and accidental exposure of nanoparticles during manufacturing or handling. Recent *in vivo* and *in vitro* studies have shown that skin-exposed CNT can translocate to the deeper layers of skin such as dermis or subcutaneous layer. *In vivo* studies with dermal exposure of CNT have shown inflammation and inflammatory granuloma formation in dermal and subcutaneous tissues (Yokoyama, Sato et al. 2004; Koyama, Endo et al. 2006). Other studies have also shown that exposure to CNT can induce dermal granuloma formation, thickness of skin folding and neutrophil and macrophage-mediated inflammation (Sato, Yokoyama et al. 2005; Murray, Kisin et al. 2009).

A number of *in vitro* dermal studies has shown that SWCNT exposure can cause inflammation and oxidative stress in dermis and keratinocyte cells (Li, Hulderman et al. 2006; Poland, Duffin et al. 2008; Msiska, Pacurari et al. 2009). The inflammation and toxicity are less severe in skin exposure than lung exposure due to the stiffness and thick layer of the skin cutaneous tissue (Sato, Yokoyama et al. 2005).

2.3 Oral exposure

Oral exposure of nanoparticles could be through food packaging, contaminated food or water. In addition, ingestion of nano-drugs or nano-delivery systems can lead to gastrointestinal track (GIT) exposure to nanoparticles. Absorption from GIT depends on the physiochemical properties of nanomaterials. For example, Kolosnjaj-Tabi et al. (2010) reported that when SWCNT were administered orally (1000 mg/kg body weight), neither death nor growth or behavioral troubles were observed. However, intraperitoneal administration of SWCNT (50-1000 mg/kg body weight) can coalesce inside the body to form fiber-like structures. When SWCNT length exceeded 10 μm, they irremediably induced granuloma formation compared to smaller aggregates which did not induce granuloma but persisted inside the cells for up to 5 months post-administration. Individualized SWCNT (< 300 nm) can escape the reticuloendothelial system to be excreted through the kidney and bile ducts (Kolosnjaj-Tabi, Hartman et al. 2010). Oral exposure of corn oil-suspended CNT induced oxidative and genotoxic changes in the lung and liver of rats (Folkmann, Risom et al. 2009). However, the risk of CNT ingestion is not clearly understood because of the short history of usage of this new nanomaterial and the limited knowledge/studies in the field.

2.4 Cardiovascular exposure

One of the applications of CNT is as a carrier for drugs and biomolecules into the body. This requires these materials to be able to get into the systemic circulation. Therefore, implication of CNT exposure in blood vessels and heart muscles should be thoroughly investigated. Apart from directly injected into the blood for medical applications, nanoparticles can also enter blood via translocation from the lung or skin. Intravenous exposure in mice to SWCNT induces inflammatory reactions and an up-regulation of pro-inflammatory cytokines such as tumor necrosis factor-α (Li, Hulderman et al. 2007; Yang, Wang et al. 2008). Another study reported that CNT exposure to the heart tissue induces cellular mitochondrial damage, oxidative stress and apoptosis (Li, Hulderman et al. 2007). Additionally, vascular thrombosis and platelet aggregation was observed in both *in vivo* and *in vitro* exposures to CNT (Radomski, Jurasz et al. 2005). These studies suggest that cardiovascular exposure of nanomaterials could be hazardous to normal cardiac function. However, studies on the dose, time, physicochemical properties and pharmacokinetic parameters of the administrated CNT are limited.

2.5 Neuronal exposure

There are only a few studies on the effect of nanomaterial exposure on the nervous system. When exposed to neuroblast-glioma cells *in vitro*, SWCNT induced DNA damage and cytotoxicity (Belyanskaya, Weigel et al. 2009). Also SWCNT induced oxidative stress, cell membrane damage and DNA damage in neuronal pc12 cells (Zhang, Ali et al. 2010). *In vivo* systemic exposure studies did not find distribution or translocation of CNT to neuronal tissues or brain (Wang, Wang et al. 2004). However, another *in vivo* study showed the translocation of CNT across blood brain barrier (Yang, Guo et al. 2007)

3. Absorption, distribution and metabolism of carbon nanotubes *in vivo*

Biodegradation of nanoparticles after exposure depends on the nature of nanomaterial and its chemical constituents. It also depends on the bio-distribution and persistence of nanoparticles in the body. Much of the absorption of CNT depends on the translocation of CNT from the site of exposure (lung, skin, etc.) to the systemic circulation (Yang, Wang et al. 2008). Only a few studies have shown the translocation and distribution of systemically exposed CNT into organs such as liver, lung, spleen and kidney where a high level of CNT accumulation was observed (Wang, Wang et al. 2004; Rotoli, Bussolati et al. 2008). Usually biodegradable nanomaterials are broken down into constituent molecules. However, some nanomaterials including SWCNT and MWCNT are bio-persistent and difficult to excrete, which become a safety concern and deter their biomedical applications. Studies have shown that these CNT can stay in the lung tissue for months after exposure and lead to granuloma or progressive interstitial fibrosis (Shvedova, Kisin et al. 2005; Mercer, Hubbs et al. 2011). Bio-persistency of CNT increases their interaction with body's cells which can lead to harmful effects to specific tissues and organs. Contaminants, mainly metal ions which are used as catalysts of CNT synthesis, can induce reactive oxygen species (ROS) generation which can cause tissue inflammation, cell damage and even carcinogenesis.

4. Physicochemical properties of nanomaterials affecting their biological activities

Physicochemical properties of nanomaterials can have a great influence on the biological activities of the materials. Some key physicochemical properties affecting their bioactivities include size, shape, surface activity, dispersion status, and metal contaminants.

4.1 Size

Studies with micron- and nano-sized particles of the same material, e.g., silica and CNT, have shown that the nano-sized particles have a deeper lung deposition and are more toxic than large particles. As the particle size reduces, the surface atom increases. As the particles size reduces to less than 100 nm, the surface atom increases exponentially (Figure 2). Properties such as surface energy and electrical force also change accordingly (Garg and Sinnott 1998; Folkmann, Risom et al. 2009). These properties have been shown to influence the translocation and distribution of nanoparticles in the body.

4.2 Shape

High aspect-ratio fibers like asbestos are historically known to be cytotoxic (Jaurand, Renier et al. 2009). Compared to asbestos fibers, the aspect ratio of MWCNT is similar but larger. Recent animal studies have shown that the toxicity of CNT is several times higher than that of asbestos. Pulmonary exposed long fibers like CNT (> 20 μm) are difficult to be cleared by macrophages (Lam, James et al. 2004). Other studies have also reported that short CNT evade phagocytosis by macrophages when compared to larger micron-sized particles of similar composition (Muller, Huaux et al. 2005; Mercer, Scabilloni et al. 2008; Zhang, Bai et al. 2010; Palomäki, Välimäki et al. 2011)

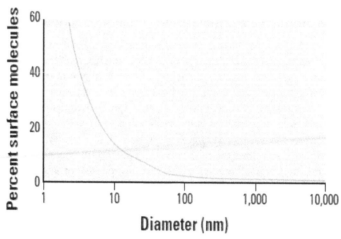

Fig. 2. Surface molecules as a function of particle size. Surface molecules increase exponentially when particle size decreases to less than 100 nm (Oberdorster et al. 2005).

4.3 Surface activity

The biological activity of CNT is directly related to its surface activity. CNT have a large surface area and high surface energy. Studies have shown that surface modifications can lead to changes in biological activity and toxicity. Surface modification with different chemical functionalities can affect the overall biological activity of nanoparticles. For example, acid treatment of CNT introduces carboxyl groups onto the surface of particles leading to an increase in dispersion and water solubility. Functionalization of CNT with COOH reduces the attractive electrical force between CNT surfaces and affects the biological response. COOH-functionalized MWCNT are more water soluble and dispersed than non-functionalized MWCNT (Upadhyayula and Gadhamshetty; Cao, Chen et al. 2011; Jacobs, Vickrey et al. 2011)

4.4 Dispersion status

Particle agglomeration and aggregation is a common phenomenon for CNT. High surface energy and surface area of nanoparticles result in Van der Waals interactions leading to agglomeration. For full exploitation of CNT, they should be used as well dispersed particles such as those observed with aerosolized particles. Dispersion results in an increase in particle number per unit mass and an associated increase in contact surface area with exposed cells, thus affecting their biological activities. Published data have shown that dispersed CNT exhibit more pronounced effects on cell proliferation and cytotoxicity than their non-dispersed counterpart. Several methods of nanoparticle dispersion have been investigated including the use of natural lung surfactants such as Survanta® (Wang et al. 2010), phospholipids such as dipalmitoylphosphatidylcholine (Sager, Porter et al. 2007), organic solvents such as acetone and dimethyl sulphoxide (Soto, Carrasco et al. 2005) and biomolecules such as single-stranded DNA, albumin, and cell culture serum (Cherukuri, Bachilo et al. 2004; Jia, Wang et al. 2005; Muller, Huaux et al. 2005)

4.4.1 Dispersion of CNT using natural lung surfactants

Survanta® is a surfactant replacement used in clinic which provides some advantages as a dispersing agent for CNT, i.e., it is commercially available, biocompatible and safe. Previous study has also shown that it is effective in dispersing CNT into fine particles comparable in size to that of aerosolized CNT (Figure 3). It has no apparent cytotoxic effect and does not mask the biological effect of CNT (Wang et al. 2010).

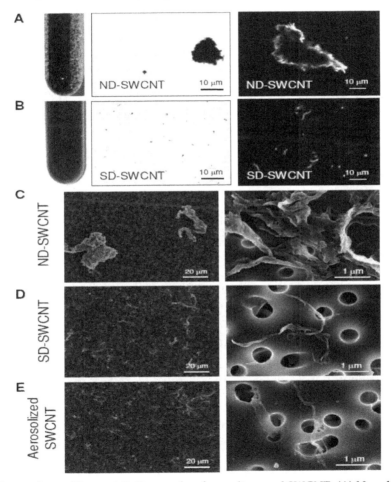

Fig. 3. Comparison of Survanta®-dispersed and non-dispersed SWCNT. (A) Non-dispersed SWCNT suspension showing visible clumping of SWCNT (left panel) with corresponding light microscopy (middle panel, 100x) and hyperspectral imagining of an individual clump (right panel, 400x). (B) Survanta®-dispersed SWCNT suspension showing much improved

dispersion with no visible large clumps (left panel) observed in the corresponding light microscopy and hyperspectral imaging. (C-E) Scanning electron microscopy of non-dispersed, Survanta®-dispersed and aerosolized SWCNT at low magnification (400x, left panel) and high magnification (30,000x, right panel)(Wang, Castranova et al. 2010)

4.4.2 Effect of dispersed CNT on fibrogenesis

Well dispersed SWCNT and MWCNT can deposit deep in the interstitial area of the alveoli, where it can enhance the fibrotic process by directly stimulating interstitial lung fibroblasts (Mercer, Scabilloni et al. 2008; Monteiro-Riviere, Inman et al. 2009). Dispersed SWCNT have been shown to evade engulfment by lung macrophages where 80% of pulmonary exposed SWCNT reach the alveolar interstitial space (Shvedova, Kisin et al. 2005; Mercer, Scabilloni et al. 2008). In case of exposure to micron-sized particles, most of these particles are engulfed by macrophages which induce a robust inflammatory response. Dispersed SWCNT bypass the inflammatory response which might be due to evading macrophage engulfment. Dispersed SWCNT can stimulate resident fibroblasts in the interstitial space to produce collagen. Ongoing research suggests that dispersed nano-sized SWCNT can induce lung fibrosis without persistent inflammation. Usually micron-sized particles induce robust inflammation followed by slow developing fibrosis in contrast to nano-sized particles like CNT which induce rapid fibrosis. The mechanism of this unusual fibrosis is unclear and is under investigation.

4.5 Metal contaminants

Metals are routinely used as catalysts in nanomaterial production. These metal ions are usually incorporated into the nanomaterials during the manufacturing process and contribute towards metal contaminants. Generally metal ions present in CNT as impurities are Fe, Ni, Co, etc. During the purification process, manufactured CNT are thoroughly washed in acid to remove these metal ions or any amorphous carbon. The metal ion concentration depends on the method used to manufacture CNT and the purification process. In some cases, metal ion contamination can reach 30% by weight in unpurified CNT. Metal impurities such as Fe in CNT are known to induce oxidative stress in cells (Warheit, Laurence et al. 2004; Le Goff, Holzinger et al. 2011). These ions can influence redox reactions by inducing ROS or inhibiting antioxidant enzymes. Formation of ROS leads to oxidative stress, inflammation and apoptosis.

5. Mechanisms of CNT toxicity

5.1 Oxidative stress

ROS are generated by distressed cells leading to oxidative stress and apoptosis. It is an imbalance between antioxidant proteins and ROS produced in cells and induced by toxic substances such as asbestos (Liu, Ernst et al. 2000). CNT have been found to induce oxidative stress and ROS in the lung (Manna, Sarkar et al. 2005; Shvedova, Kisin et al. 2005; Pacurari, Yin et al. 2008), skin (Murray, Kisin et al. 2009), and heart (Li, Hulderman et al. 2007). Metal contaminants such as Fe are the major source of ROS generation induced by CNT. Studies have shown that SWCNT containing up to 30% of Fe (%wt) are more toxic than SWCNT with 0.23% iron content (Shvedova, Castranova et al. 2003). Iron ion can initiate Fenton reaction with hydrogen peroxide in cells to produce highly oxidative species such as hydroxyl radicals.

$$Fe^{2+} + H_2O_2 \rightarrow Fe^{3+} + OH\cdot + OH^-$$

As mentioned above, metal ions are key contributor to nanomaterial-induced ROS generation. Since many nanoparticles are prepared by using metal ions, exposure to raw or as manufactured materials in occupational workers can exacerbate the toxic response due to these impurities. Oxidative stress further leads to inflammation and toxicity which ultimately results in cell death. Other studies have also suggested that oxidative stress plays an important role in nanoparticle-induced toxicities (Warheit, Laurence et al. 2004; Li, Muralikrishnan et al. 2010). The damage caused by nanoparticle-induced oxidative stress can be reduced by pretreatment with antioxidants (Shvedova, Kisin et al. 2007).

5.2 Inflammation

CNT have been shown to induce inflammatory response in a number of *in vivo* studies (Shvedova, Kisin et al. 2005; Poland, Duffin et al. 2008). Macrophages engulf inhaled particles and produce various inflammatory cytokines and chemokines, which attract and amplify inflammatory responses in the body. Pulmonary exposed CNT induce transient inflammation whereas micro-sized particles usually induce persistent inflammation. Previous study has shown that alveolar microphages ignore certain small-sized CNT (Shvedova, Kisin et al. 2005), which may explain the less inflammation induced by the nanoparticles. Some studies suggest that CNT suppress the immune response by reducing the inflammatory signal and preventing macrophage activation (Mitchell, Gao et al. 2007). *In vitro* studies using keratinocytes and macrophages showed that these cells secrete pro-inflammatory cytokines such as interlukin-8 and tumor necrosis factor-α in response to CNT stimulation (Monteiro-Riviere, Inman et al. 2005; Brown, Kinloch et al. 2007).

5.3 Fibrosis

Unlike micron-sized particles, nanoparticles induce an unusual rapid fibrosis. For example, inhaled CNT can quickly penetrate the alveolar epithelial barrier into the interstitial tissue to form a matrix which stimulates resident fibroblasts to produce collagen (Mercer, Scabilloni et al. 2008; Monteiro-Riviere, Inman et al. 2009). Persistent stimulation of fibroblasts has been shown to result in interstitial lung fibrosis *in vivo* and collagen production by lung fibroblasts *in vitro* (Wang, Mercer et al. 2010). Data have also shown that CNT induce fibrogenic cytokines and growth factors such as transforming growth factor-β1, matrix metalloproteinase-9, and fibroblast growth factor-2 in human lung cells both *in vitro* and *in vivo* (Shvedova, Kisin et al. 2008; Wang, Mercer et al. 2010).

5.4 DNA damage

CNT can interfere with the genetic constituents of the cells such as DNA and RNA (Zhu, Chang et al. 2007; Pacurari, Yin et al. 2008; Bonner 2010). These changes in nucleic acid structures can affect cell survival and genomic integrity. MWCNT induce clastogenic (DNA break) and aneuogenic response (chromosomal loss) raising the possibility of mutational changes in the genetic materials of the cell (Muller, Decordier et al. 2008). Mutational changes in K-ras have been observed in the lung of mice after CNT exposure (Shvedova, Kisin et al. 2008). A recent study showed that chronic exposure of SWCNT to lung epithelial cells causes malignant transformation of the cells and tumorigenesis in nude mice (Wang, Luanpitpong et al. 2011). Changes in p53 protein phosphorylation in embryonic stem cells after SWCNT exposure further support the potential genotoxicity and tumorigenicity of the nanomaterial (Zhu, Chang et al. 2007). Genotoxic damage due to CNT exposure has also reported in lung fibroblasts (Kisin, Murray et al. 2007).

6. Conclusion

The unique characteristics of nanomaterials offer potential novel applications as well as potential toxicities. A number of factors including the route of exposure and physico-chemical properties of nanoparticles can affect the biological activities of CNT and their toxic responses. Among the various exposure routes, the pulmonary route is the most common route of exposure to airborne nanoparticles, which have been shown to induce fibrotic and toxicological responses. Properties of nanoparticles such as dispersion status, size and shape, chemical composition and surface functionalization play an important role in the biological activities of nanoparticles. Mechanistic understanding of nanoparticle interactions with cells and tissues is still lacking. Most of the reported toxic effects of nanoparticles are caused by tissue penetration and induction of oxidative stress, DNA damage, inflammation and fibrosis. Careful evaluations of the toxic and fibrogenic effects of nanomaterials are critically needed for the safe and effective use of nanomaterials.

7. Disclaimer

The findings and conclusions in this report are those of the authors and do not necessarily represent the views of the National Institute for Occupational Safety and Health.

8. References

Bandaru, P. R. (2007). "Electrical Properties and Applications of Carbon Nanotube Structures." *Journal of Nanoscience and Nanotechnology* 7(4-5): 1239-1267.

Belyanskaya, L., S. Weigel, et al. (2009). "Effects of carbon nanotubes on primary neurons and glial cells." *Neurotoxicology* 30(4): 702-711.

Bonner, J. C. (2010). "Nanoparticles as a Potential Cause of Pleural and Interstitial Lung Disease." *Proc Am Thorac Soc* 7(2): 138-141.

Brown, D. M., I. A. Kinloch, et al. (2007). "An in vitro study of the potential of carbon nanotubes and nanofibres to induce inflammatory mediators and frustrated phagocytosis." *Carbon* 45(9): 1743-1756.

Cao, X., J. Chen, et al. (2011). "Effect of surface charge of polyethyleneimine-modified multiwalled carbon nanotubes on the improvement of polymerase chain reaction." *Nanoscale* 3(4): 1741-1747.

Cherukuri, P., S. M. Bachilo, et al. (2004). "Near-Infrared Fluorescence Microscopy of Single-Walled Carbon Nanotubes in Phagocytic Cells." *Journal of the American Chemical Society* 126(48): 15638-15639.

Donaldson, K., F. Murphy, et al. (2010). "Asbestos, carbon nanotubes and the pleural mesothelium: a review of the hypothesis regarding the role of long fibre retention in the parietal pleura, inflammation and mesothelioma." *Particle and Fibre Toxicology* 7(1): 5.

Folkmann, J. K., L. Risom, et al. (2009). "Oxidatively Damaged DNA in Rats Exposed by Oral Gavage to C(60) Fullerenes and Single-Walled Carbon Nanotubes." *Environmental Health Perspectives* 117(5): 703-708.

Garg, A. and S. B. Sinnott (1998). "Effect of chemical functionalization on the mechanical properties of carbon nanotubes." *Chemical Physics Letters* 295(4): 273-278.

Herzog, E., H. Byrne, et al. (2009). "Swcnt suppress inflammatory mediator responses in human lung epithelium in vitro." *Toxicol Appl Pharmacol* 234: 378 - 390.

Iijima, S. (1991). "Helical microtubules of graphitic carbon." *Nature* 354(6348): 56-58.

Jacobs, C. B., T. L. Vickrey, et al. (2011). "Functional groups modulate the sensitivity and electron transfer kinetics of neurochemicals at carbon nanotube modified microelectrodes." *Analyst* 136(17): 3557-3565.

Jaurand, M.-C., A. Renier, et al. (2009). "Mesothelioma: Do asbestos and carbon nanotubes pose the same health risk?" *Particle and Fibre Toxicology* 6(1): 16.

Jia, G., H. Wang, et al. (2005). "Cytotoxicity of Carbon Nanomaterials: Single-Wall Nanotube, Multi-Wall Nanotube, and Fullerene." *Environmental Science & Technology* 39(5): 1378-1383.

Kisin, E. R., A. R. Murray, et al. (2007). "Single-walled carbon nanotubes: Geno- and cytotoxic effects in lung fibroblast V79 cells." *Journal of Toxicology and Environmental Health-Part a-Current Issues* 70(24): 2071-2079.

Kolosnjaj-Tabi, J., K. B. Hartman, et al. (2010). "In Vivo Behavior of Large Doses of Ultrashort and Full-Length Single-Walled Carbon Nanotubes after Oral and Intraperitoneal Administration to Swiss Mice." *ACS Nano* 4(3): 1481-1492.

Koyama, S., M. Endo, et al. (2006). "Role of systemic T-cells and histopathological aspects after subcutaneous implantation of various carbon nanotubes in mice." *Carbon* 44(6): 1079-1092.

Lam, C. W., J. T. James, et al. (2004). "Pulmonary toxicity of single-wall carbon nanotubes in mice 7 and 90 days after intratracheal instillation." *Toxicol Sci* 77(1): 126 - 134.

Le Goff, A., M. Holzinger, et al. (2011). "Enzymatic biosensors based on SWCNT-conducting polymer electrodes." *Analyst* 136(7): 1279-1287.

Li, J.-G., W.-X. Li, et al. (2007). "Comparative study of pathological lesions induced by multiwalled carbon nanotubes in lungs of mice by intratracheal instillation and inhalation." *Environmental Toxicology* 22(4): 415-421.

Li, J. J. e., S. Muralikrishnan, et al. (2010). "Nanoparticle-induced pulmonary toxicity." *Experimental Biology and Medicine* 235(9): 1025-1033.

Li, X. and et al. (2010). "Current investigations into carbon nanotubes for biomedical application." *Biomedical Materials* 5(2): 022001.

Li, Z., T. Hulderman, et al. (2007). "Cardiovascular effects of pulmonary exposure to single-wall carbon nanotubes." *Environmental Health Perspectives* 115(3): 377-382.

Liu, W., J. Ernst, et al. (2000). "Phagocytosis of crocidolite asbestos induces oxidative stress, DNA damage, and apoptosis in mesothelial cells." *Am J Respir Cell Mol Biol* 23: 371 - 378.

Manna, S. K., S. Sarkar, et al. (2005). "Single-Walled Carbon Nanotube Induces Oxidative Stress and Activates Nuclear Transcription Factor-κB in Human Keratinocytes." *Nano Letters* 5(9): 1676-1684.

McCarthy, J. R. and R. Weissleder (2008). "Multifunctional magnetic nanoparticles for targeted imaging and therapy." *Advanced Drug Delivery Reviews* 60(11): 1241-1251.

Mercer, R. R., A. Hubbs, et al. (2011). "Pulmonary fibrotic resposnse to aspiration of multi-walled carbon nanotubes." *Particle and Fibre Toxicology*.

Mercer, R. R., J. Scabilloni, et al. (2008). "Alteration of deposition pattern and pulmonary response as a result of improved dispersion of aspirated single-walled carbon nanotubes in a mouse model." *Am J Physiol Lung Cell Mol Physiol* 294(1): L87-97.

Migliore, L., D. Saracino, et al. (2010). "Carbon nanotubes induce oxidative DNA damage in RAW 264.7 cells." *Environmental and Molecular Mutagenesis* 51(4): 294-303.

Mitchell, L. A., J. Gao, et al. (2007). "Pulmonary and Systemic Immune Response to Inhaled Multiwalled Carbon Nanotubes." *Toxicol. Sci.* 100(1): 203-214.

Monteiro-Riviere, N. A., A. O. Inman, et al. (2005). "Surfactant effects on carbon nanotube interactions with human keratinocytes." *Nanomedicine: Nanotechnology, Biology and Medicine* 1(4): 293-299.

Monteiro-Riviere, N. A., A. O. Inman, et al. (2009). "Limitations and relative utility of screening assays to assess engineered nanoparticle toxicity in a human cell line." *Toxicology and Applied Pharmacology* 234(2): 222-235.

Msiska, Z., M. Pacurari, et al. (2009). "DNA Double Strand Breaks by Asbestos, Silica and Titanium dioxide: Possible Biomarker of Carcinogenic Potential?" *Am. J. Respir. Cell Mol. Biol.*: 2009-0062OC.

Muller, J., I. Decordier, et al. (2008). "Clastogenic and aneugenic effects of multi-wall carbon nanotubes in epithelial cells." *Carcinogenesis* 29(2): 427-433.

Muller, J., F. Huaux, et al. (2005). "Respiratory toxicity of multi-wall carbon nanotubes." *Toxicology and Applied Pharmacology* 207(3): 221-231.

Murray, A. R., E. Kisin, et al. (2009). "Oxidative stress and inflammatory response in dermal toxicity of single-walled carbon nanotubes." *Toxicology* 257(3): 161-171.

Nam, C.-W., S.-J. Kang, et al. (2011). "Cell growth inhibition and apoptosis by SDS-solubilized single-walled carbon nanotubes in normal rat kidney epithelial cells." *Archives of Pharmacal Research* 34(4): 661-669.

Oberdorster, G., J. Ferin, et al. (1994). "Correlation between particle size, in vivo particle persistence, and lung injury." *Environ Health Perspect* 102(Suppl 5): 173 - 179.

Oberdorster, G., E. Oberdorster, et al. (2005). "Nanotoxicology: an emerging discipline evolving from studies of ultrafine particles." *Environ Health Perspect* 113: 823 - 839.

Pacurari, M., X. Yin, et al. (2008). "Raw single-wall carbon nanotubes induce oxidative stress and activate mapks, ap-1, nf-kappab, and akt in normal and malignant human mesothelial cells." *Environ Health Perspect* 116: 1211 - 1217.

Palomäki, J., E. Välimäki, et al. (2011). "Long, Needle-like Carbon Nanotubes and Asbestos Activate the NLRP3 Inflammasome through a Similar Mechanism." *ACS Nano* 5(9): 6861-6870.

Poland, C. A., R. Duffin, et al. (2008). "Carbon nanotubes introduced into the abdominal cavity of mice show asbestos-like pathogenicity in a pilot study." *Nature Nanotechnology* 3(7): 423-428.

Porter, D. W., A. F. Hubbs, et al. (2010). "Mouse pulmonary dose- and time course-responses induced by exposure to multi-walled carbon nanotubes." *Toxicology* 269(2-3): 136-147.

Prato, M., K. Kostarelos, et al. (2007). "Functionalized Carbon Nanotubes in Drug Design and Discovery." *Accounts of Chemical Research* 41(1): 60-68.

Radomski, A., P. Jurasz, et al. (2005). "Nanoparticle-induced platelet aggregation and vascular thrombosis." *British Journal of Pharmacology* 146(6): 882-893.

Ravichandran, P., S. Baluchamy, et al. (2011). "Pulmonary Biocompatibility Assessment of Inhaled Single-wall and Multiwall Carbon Nanotubes in BALB/c Mice." *Journal of Biological Chemistry* 286(34): 29725-29733.

Rotoli, B. M., O. Bussolati, et al. (2008). "Non-functionalized multi-walled carbon nanotubes alter the paracellular permeability of human airway epithelial cells." *Toxicology Letters* 178(2): 95-102.

Sager, T. M., D. W. Porter, et al. (2007). "Improved method to disperse nanoparticles for in vitro and in vivo investigation of toxicity." *Nanotoxicology* 1(2): 118-129.

Sakamoto, Y., D. Nakae, et al. (2009). "Induction of mesothelioma by a single intrascrotal administration of multi-wall carbon nanotube in intact male fischer 344 rats." *J Toxicol Sci* 34: 65 - 76.

Sato, Y., A. Yokoyama, et al. (2005). "Influence of length on cytotoxicity of multi-walled carbon nanotubes against human acute monocytic leukemia cell line THP-1 in vitro and subcutaneous tissue of rats in vivo." *Molecular BioSystems* 1(2): 176-182.

Shvedova, A., E. Kisin, et al. (2008). "Inhalation vs. Aspiration of single-walled carbon nanotubes in c57bl/6 mice: Inflammation, fibrosis, oxidative stress, and mutagenesis." *Am J Physiol Lung Cell Mol Physiol* 295: L552 - 565.

Shvedova, A. A., V. Castranova, et al. (2003). "Exposure to carbon nanotube material: Assessment of nanotube cytotoxicity using human keratinocyte cells. "*Journal of Toxicology and Environmental Health-Part A* 66(20): 1909-1926.

Shvedova, A. A., E. R. Kisin, et al. (2005). "Unusual inflammatory and fibrogenic pulmonary responses to single-walled carbon nanotubes in mice." *Am J Physiol Lung Cell Mol Physiol* 289(5): L698-708.

Shvedova, A. A., E. R. Kisin, et al. (2007). "Vitamin E deficiency enhances pulmonary inflammatory response and oxidative stress induced by single-walled carbon nanotubes in C57BL/6 mice." *Toxicology and Applied Pharmacology* 221(3): 339-348.

Simon-Deckers, A., B. Gouget, et al. (2008). "In vitro investigation of oxide nanoparticle and carbon nanotube toxicity and intracellular accumulation in A549 human pneumocytes." *Toxicology* 253(1-3): 137-146.

Soto, K. F., A. Carrasco, et al. (2005). "Comparative <i>in vitro</i> cytotoxicity assessment of some manufacturednanoparticulate materials characterized by transmissionelectron microscopy." *Journal of Nanoparticle Research* 7(2): 145-169.

Upadhyayula, V. K. K. and V. Gadhamshetty "Appreciating the role of carbon nanotube composites in preventing biofouling and promoting biofilms on material surfaces in environmental engineering: A review." *Biotechnology Advances* 28(6): 802-816.

Wang, H., J. Wang, et al. (2004). "Biodistribution of carbon single-wall carbon nanotubes in mice." *J Nanosci Nanotechnol* 4: 1019 - 1024.

Wang, L., V. Castranova, et al. (2010). "Dispersion of single-walled carbon nanotubes by a natural lung surfactant for pulmonary in vitro and in vivo toxicity studies." *Particle and Fibre Toxicology* 7(1): 31.

Wang, L., S. Luanpitpong, et al. (2011). "Carbon Nanotubes Induce Malignant Transformation and Tumorigenesis of Human Lung Epithelial Cells." *Nano Letters* 11(7): 2796-2803.

Wang, L., R. R. Mercer, et al. (2010). "Direct Fibrogenic Effects of Dispersed Single-Walled Carbon Nanotubes on Human Lung Fibroblasts." *Journal of Toxicology and Environmental Health, Part A* 73(5-6): 410-422.

Warheit, D. B., B. R. Laurence, et al. (2004). "Comparative pulmonary toxicity assessment of single-wall carbon nanotubes in rats." *Toxicological Sciences* 77(1): 117-125.

Wilson, M. S. and T. A. Wynn (2009). "Pulmonary fibrosis: pathogenesis, etiology and regulation." *Mucosal Immunol* 2(2): 103-121.

Yang, S.-t., W. Guo, et al. (2007). "Biodistribution of Pristine Single-Walled Carbon Nanotubes In Vivo†." *The Journal of Physical Chemistry C* 111(48): 17761-17764.

Yang, S.-T., X. Wang, et al. (2008). "Long-term accumulation and low toxicity of single-walled carbon nanotubes in intravenously exposed mice." *Toxicology Letters* 181(3): 182-189.

Yokoyama, A., Y. Sato, et al. (2004). "Biological Behavior of Hat-Stacked Carbon Nanofibers in the Subcutaneous Tissue in Rats." *Nano Letters* 5(1): 157-161.

Zhang, Y., S. F. Ali, et al. (2010). "Cytotoxicity Effects of Graphene and Single-Wall Carbon Nanotubes in Neural Phaeochromocytoma-Derived PC12 Cells." *ACS Nano* 4(6): 3181-3186.

Zhang, Y., Y. Bai, et al. (2010). "Functionalized carbon nanotubes for potential medicinal applications." *Drug Discovery Today* 15(11-12): 428-435.

Zhu, L., D. W. Chang, et al. (2007). "DNA Damage Induced by Multiwalled Carbon Nanotubes in Mouse Embryonic Stem Cells." *Nano Letters* 7(12): 3592-3597.

Nanoparticles and Nanostructures
for Biophotonic Applications

Enzo Di Fabrizio[1,2] et al.[*]
[1]Nanostructures Department, Italian Institute of Technology, Genova,
[2]BioNEM lab., Departement of Clinical and Experimental Medicine,
Magna Graecia University, Viale Europa, Catanzaro,
Italy

1. Introduction

The aim of this chapter is to expound on the theoretical analysis and experimental assessment of NanoParticles (NPs) for imaging, early detection and therapeutic applications. NPs are extremely small particulates with dimensions ranging from few micrometers down to few tens of nanometers. Their characteristics, including size, shape, physical and chemical properties, can be tailored during the fabrication/synthesis process and, on account of these, they would feature certain aspects that may be exploited for applications ranging from drug delivery to the enhancement of the local electric field, and thus the detection of few molecules. In particular, intravascularly injectable NPs (that are sometimes called nanovectors or nanocarriers) are probably the major class of nanotechnological devices of interest for use in cancer or, in general, for the treatment of diseases. On the other hand, aggregates of metallic NPs, either of silver or gold, represent extremely efficient SERS (Surface Enhanced Raman Scattering) substrates. In the following, after a brief description of NPs as a whole, a number of different applications will be discussed.

2. Nanoporous silicon nanoparticles: A drug delivery system

Intravascularly injectable NPs can be conveniently designed or engineered to release drug molecules or imaging tracers with a superior performance with respect to freely administrated agents (Ferrari, 2005; Whitesides, 2003; La Van et al., 2003); to this extent, they represent smart Drug Delivery Systems (DDS).

[*] Francesco Gentile[1,2], Michela Perrone Donnorso[1,3], Manohar Chirumamilla Chowdary[1],
Ermanno Miele[1], Maria Laura Coluccio[1,2], Rosanna La Rocca[1], Rosaria Brescia[4], Roman Krahne[1],
Gobind Das[1], Francesco De Angelis[1], Carlo Liberale[1], Andrea Toma[1], Luca Razzari[1],
Liberato Manna[4] and Remo Proietti Zaccaria[1]
[1]Nanostructures Department, Italian Institute of Technology, Genova, Italy
[2]BioNEM lab., Department of Clinical and Experimental Medicine, Magna Graecia University, Viale Europa, Catanzaro, Italy
[3]Nanophysics Department, Italian Institute of Technology, Genova, Italy
[4]Nanochemistry Department, Italian Institute of Technology, Genova, Italy

Fig. 1. Cartoon representing a nanoparticle featuring different coverings and payloads.

In other terms, on account on their geometrical, physical and chemical properties (**Fig.1**), the therapeutic or contrast agents would be targeted directly towards the site of interest (malignant cells) with a significant reduction of side effects, and a concurrent increase in the efficiency of delivery.

Liposomes are the simplest form of NPs (and, accordingly, they are sometimes referred to as first generation nanovectors) (Klibanov et al., 1991; Park, 2002; Crommelin & Schreier, 1994). Established for the treatment of Kaposi's sarcoma more than 10 years ago, nowadays certain liposomes based NPs are still being used for cancer treatment. Literature records a massive number of second generation NPs, (which, differently from liposomes, can be artificially produced using nanofabrication techniques), including polymer-based nanovectors, silicon and silica NPs and metal-based nanovectors like nanoshells (Kircher et al., 2003; Schellenberger et al., 2002; Zhang & Shang, 2004; Cohen et al., 2003; Hirsch et al., 2003; Langer, 1998; Duncan, 2003; Gilles & Frechet, 2002).

Despite this, and notwithstanding the merits that such an abundance of devices provides in terms of novel technological foundations and opportunities, only a small amount of these nanovectors would be really effective in delivering drugs or contrast agents (Ferrari, 2005).

A number of third generation NPs is currently under development which feature advanced properties and thus more efficacious delivery modalities, including nanoporous NPs with/without reduced silver for SERS analysis, mesoporous silicon particles as a multistage delivery systems, nanoporous NPs enhancing, via geometrical confinement of gadolinium-based contrast agents, T_1 contrast (Tasciotti et al., 2008; Jeyarama et al., 2010).

In general, an ideal nanoparticle should be able (i) to navigate into the circulatory system avoiding the immune system and recognizing the diseased cells (biological target) with high selectivity; (ii) to adhere firmly to them and (iii) to allow for endocytosis process. Targeting methods have been largely investigated and range from specific (covalently linked ligands decorate nanovectors and may recognize antibodies over-expressed on the cells of interest) to non specific (that are, mechanisms based on the size, shape and physical properties of the nanovector including density, porosity, surface charge) (Decuzzi, 2006a, 2006b, 2007).

Regardless the particular mechanisms of transport and adhesion that NPs can experience, it is clear that a thorough understanding of the physics behind these phenomena plays a fundamental role. Realistically, mathematical models provide an unprecedented tool for predicting the behaviour of NPs within the macro/microcirculation, thus also supplying a rationale for the best design of NPs (Gentile, 2007, 2008). In DDS the choice of bulk material constituent the nanocarrier is a key point since it must fulfill a tailored biological behavior (bioactivity, biocompatibility, biodegradability), it must improve payload capacity and be harmlessly eliminated from the body in a reasonable period of time after releasing the cargo and having carried out possible diagnostic function (Park et al., 2009). While many proposed nanocarriers do not meet these requirements, porous silicon (PSi), considered as silicon crystal having a series of voids, is a promising materials for its biocompatibility and biodegradability (Granitzer & Rumpf, 2010; Canham, 1997; O Farrel et al., 2006) which results in decomposition products not harmful for biological system. Porous silicon dissolves in body fluids into orthosilicic acid, commonly found in everyday foods and efficiently excreted from the body through the urine (Park et al., 2009). Thanks to the porous structures, silicon nanoporous nanoparticles show a great surface/volume ratio (200-800 m^2/cm^3) (Halimaoui, 1995), a very attractive characteristic since surface can be used to load drugs by physisoprtion process and modified with molecules that promote cell adhesion. Pore width, porosity and nanoparticles size can be tuned by adjustment of the parameters during the fabrication process. When the pore size is in the range of 2-5 nm, physisorbed drugs are efficiently entrapped and, due to the effect of the quantum confinement of the silicon structure (Godefroo et al., 2008), particles show emission at 620 nm (red-orange) at room temperature under UV illumination (wavelength of 365 nm), a very attractive feature for the realization of theranostic nanoparticles (Janib, 2010). Pores population can be divided in two different types: open pores that are connected to each other and to the external surface and closed pores isolated from the outside, that are less useful for drug delivery purpose.

Porosity is defined as the fraction of void in the porous structure. High porosity means great surface/volume ratio and it is preferable for drug delivery device, however, due to the surface tension, too high porosity silicon layer can undergo to collapse during the fabrication process and the obtained nanoparticles can disintegrate in water solution. Size and shape are two important features of nanoparticles used in drug delivery systems. Nanoporous silicon nanoparticles with size of 30-100 nm are suitable carriers for many anticancer agents (Petros & DeSimone, 2010; Mitragotri & Lahan, 2009).

2.1 Fabrication and characterization of NanoPorous Silicon Particles

The most widely used method to fabricate Nanoporous Silicon Nanoparticles (NPNPs) is a process that includes several steps, starting from the production of nanoporous silicon film by an electrochemical etching in ethanol/HF solution of Si wafer, in which silicon acts as an anode and a platinum electrode acts as a cathode. Since Si wafer surface is hydrophobic, ethanol increases the wettability of the substrate allowing the electrolyte penetrating into the pores and also helps in removing the H_2 bubble from the sample surface formed during the anodization process (Canham, 1997). A constant current density is applied to allow the formation of homogeneous porous layer. Samples are rinsed in de-ionized water, then in ethanol and pentane, and sonicated in water for a time required to remove all the nanoporous silicon film from the crystalline silicon substrate. The film is then fractured by ultrasonication and filtered through filtration membrane and centrifuged in order to select the desired particles size. It is possible to vary NPNPs morphology (pore size and distribution, pores interconnectivity, porosity and particles diameter) changing the fabrication parameters such as etching time, anodization current density, HF concentration, wafer type and ultrasonication time. Tab.1 shows the effect of some fabrication parameters on the formation of NPNPs as produced by the authors. Freshly anodized PSi has a remarkable surface hydrophobicity (contact angle ~110°, depending on porosity and surface roughness), a very attractive property that could be used in biomedical field to design nanostructure that enable plasma protein harvesting and concentration (Pujia et al., 2010). However hydrophobicity is a limitation in case of hydrophilic drugs loading, in this case porous silicon hydrophobic behavior has to be changed in hydrophilic one by thermal treatment just before porous silicon film sonication. To prevent particles aggregation and dissolution fabricated NPNPs are stored at 4°C until using.

Sample	Etching conditions (HF 25%/ethanol 1:2 v/v)	Ultrasonication time	Particles Size	Pore size	Gravimetric Porosity
p-type	$J = 10 \, mA/cm^2$ for 300 sec	10 min	101± 34 nm	< 20 nm	80%
p-type	$J = 10 \, mA/cm^2$ for 720 sec	10 min	75 ±24nm	< 10 nm	87%
p-type	$J = 10 \, mA/cm^2$ for 720 sec	20 min	30 ±15nm	< 10 nm	87%
p++	$J = 10 \, mA/cm^2$ for 720 sec	10 min	65±24 nm	< 5 nm	65%

Table 1. Characterization of p-type silicon nanoparticles varying the fabrication parameters.

The porosity of the porous silicon layer, is determined by gravimetric measurements, weighing the silicon substrate both before and after anodization (m1 and m2 respectively) and again after the complete dissolution of the porous layer by 25% KOH solution (m3). Porosity(P) is calculated by the relation $P = (m1-m2)/(m1-m3)$. Particles Dynamic Light Scattering (DLS), Transmission Electron Microscopy (TEM) and Scanning TEM (STEM) analyses are commonly used to determine hydrodynamic size and pores width of NPNPs. Porous silicon nanoparticles can be investigated using BET – BJH theory as well. BET (Brunauer – Emmet – Teller) adsorption isotherms allow for the calculation of surface / volume ratio (m^2/cm^3) and Barrett-Joyner-Halenda (BJH) method can be used in pores width (nm) and volume (cm^3/g) determination. **Fig.2** shows TEM, STEM and DLS analysis of nanoparticles as developed in the authors' laboratory. Nanoparticles chemical analysis, performed by X-ray energy dispersive spectroscopy (EDS), indicates that fluoride is a common contaminant that is residual from the electrolyte solution used during the fabrication and disappears during annealing at 300°C as shown in **Fig.3**.

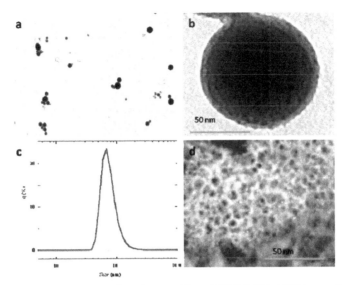

Fig. 2. TEM images (a, b), DLS size distribution (c) and STEM–HAADF image (d) of 75 nm porous silicon nanoparticles

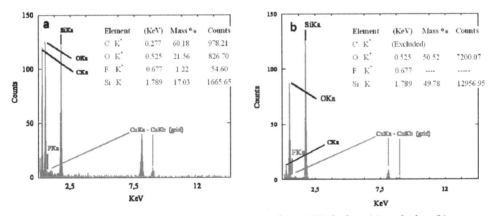

Fig. 3. Chemical analysis with EDS of nanoporous silicon NPs before (a) and after (b) annealing at 300°C.

In order to evaluate the cytotoxicity of NPNPs, apoptosis test has been performed by using iodure propidium agent and cytofluorimetric analysis. Human colon carcinoma cells (HCT116) and health monocytes (THP1) were incubated with 100 µg of two different types of NPNPs (before and after annealing) for 48 h in RPMI 1640 medium at 37°C at 5% CO_2 and cell viability detected and compared with control samples treated with medium only (**Fig.4**). No significant toxic effect was observed in the two cell lines incubated with NPNPs compared with controls. In order to evaluate the really usefulness for therapeutic applications, drug loading can be carried out by physisorption process, incubating NPNPs with drug in water at room temperature overnight. Loaded nanoparticles can be recovered from the solution by centrifugation at 12,000 rpm for 30 minutes. The amount of incorporated drug can be quantified through UV-Visible spectroscopy analysis of supernatant compared with drug standard curve. In the case of anti cancer Doxorubicin, loading test performed in the authors' laboratory, shows 5% (weight) payload for NPNPs obtained from silicon film annealed at 260°C for 4h, while a stronger thermal treatment (500°C for 12 h) can enhance physisorption process. Drug embedded NPNPs could be stored in DI water at 4°C for long time while under physiological conditions dissolve in 120 hours releasing the loaded drug (Pujia et al., 2010; De Angelis, 2010a).

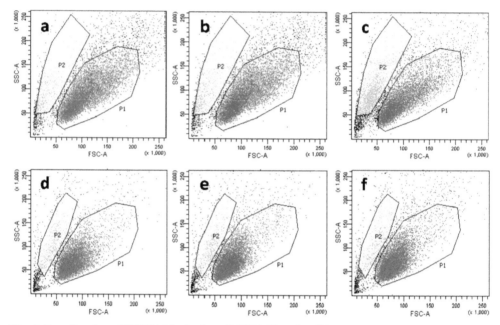

Fig. 4. Density plot of HCT116 (a, b, c) and THP1 (d, e, f) cell lines. Gate P1 shows cell viability %. a) HCT116 cells control without NPNPs, (P1 79%). b) HCT116 cells incubated with not annealed NPNPs (P1 78%). c) HCT116 cells incubated with annealed NPNPs (P1 68%). d) THP1 cells control without NPNPs (P1 83%). e) THP1 cells incubated with not annealed NPNPs (P1 82%). f) THP1 cells incubated with annealed NPNPs (P1 82%).

In conclusion, porous silicon nanoparticles represent a powerful and versatile tool in developing new drug delivery strategies, with high loading capacity, biocompatibility, cheap and scalable fabrication process. Porous silicon-based nanocarriers properties (size, shape and surface chemistry) can be easily tuned operating on the fabrication parameters to achieve a tailored biological behavior and improved bioavailability of transported drug.

3. Core-shell nanorods for light emitting applications

Inorganic semiconductor nanocrystals hold a great promise for emerging nanotechnologies, since they feature physical and chemical properties unique to the nanometer length scale, which are useful for diverse applications such as microelectronic and optical devices, as well as for sensing. These properties depend not only on the nanoparticle's composition, but also on its *size, shape,* and *mode of organization*. For this reason, a major effort has been dedicated in recent years to the development of synthetic methods providing control over morphological parameters of the nanocrystals and their assembly into organized structures. Colloidal synthesis, where judicious choice of protective ligands is employed to morphologically control crystal growth, has proven to be the most versatile approach for the synthesis of non-spherical nanoparticles, providing a variety of shapes, such as rods, prisms, cubes, disks, and others. In this section we will focus on *rod-shaped* nanocrystals, whose shape anisotropy is translated into polarized physical properties (Hu et al., 2001), and which can be harnessed for the creation of polarized light emitting materials (Kazes et al., 2002, 2004; Rizzo et al., 2009). For an extensive review on the physical properties of rod-shaped nanocrystals we refer the reader to reference (Krahne et al., 2011a). In particular, we will discuss a novel core-shell architecture for nanorods, in which a spherical core of a smaller band gap material is embedded in a rod-shaped shell of a high band gap semiconductor (Talapin et al., 2003; Carbone et al., 2007) as sketched in **Fig.5a**.

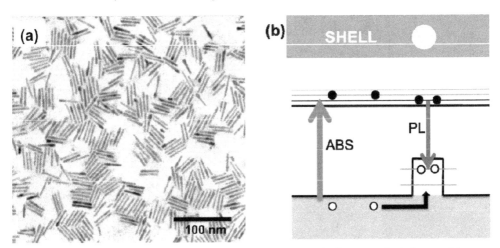

Fig. 5. (a) Transmission electron microscopy image of core shell nanorods. Scheme of the CdSe/CdS core-shell nanorod architecture and of the related band structure.

Such a system is often referred as "dot-in-a-rod", where the charge carriers can be excited in the UV-blue spectral region and then quickly (on the picosecond timescale) relax into the more long living core states from which the light emission occurs (Lupo et al., 2008). The core-shell nanorods have numerous advantages concerning their optical emission properties: (i) the stronger confinement in the core leads well defined energy levels for the optical transitions. (ii) The shell passivates the surface states of the emitting low band gap material. (iii) Due to their rod shape they have an enhanced absorption cross section with respect to spherical particles. In the specific case of CdSe/CdS core shell nanorods the lower energy levels of the holes are localized in the core, whereas the electrons are mostly delocalized over the rod volume, as illustrated in **Fig.5**. This particular electronic level structure proved to reduce non-radiative Auger recombination significantly (Zavelani-Rossi, 2010a) and therefore provides great advantages for lasing devices. Furthermore, the energy of the emitting light is determined by the quantum confinement in the core, which allows to tune their emission wavelength independently of the rod length, at least to a certain extent (Krahne, 2011b), as can be seen in **Fig.6**, which shows emission spectra of core shell nanorods with different core diameter and rod length.

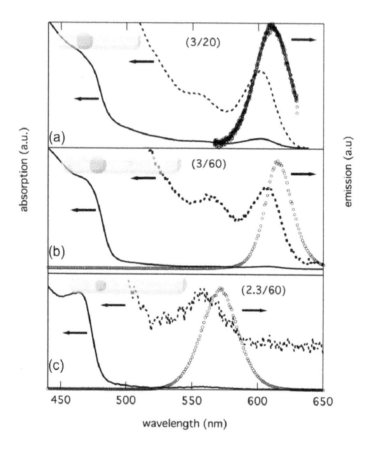

Fig. 6. Absorption and emission spectra of core-shell CdSe/CdS nanorods with different core sizes and different lengths. The core diameter and the rod length are specified by the numbers in brackets in units of nm. The emission wavelength is dominated by the core size. Taken with permission from (Krhane et al., 2011a).

Light emission from the nanorods can be obtained either by optical or via electrical pumping. In the latter case the charge carriers are injected via external electrodes that are in contact with the nanorods. Metal electrodes have the disadvantage that a Schottky barrier is formed at the interface with semiconductor material which significantly hinders the charge injection. Furthermore, the nanorod luminescence gets quenched by a direct contact of the nanorods with a metal, for example with gold. Instead, the implementation of a nanorod layer in a sandwich-like geometry, in between hole- and electron-injection organic layers, has proven to be a successful approach to fabricate light emitting diodes (LED) based on semiconductor nanorods as the active material (Rizzo et al., 2009). **Fig.7** illustrates this fabrication scheme where an ITO (indium-tin-oxide) substrate is used as a back electrode onto which *a N,N=*-bis(naphthalen-1-yl)-*N,N*=-bis(phenyl)benzidine (-NPD) hole injection layer (HIL) doped with 2,3,5,6-tetrafluoro-7,7,8,8-tetracyanoquinodimethane (F4-TCNQ) and a CBP hole transporting layer (HTL) were thermally evaporated

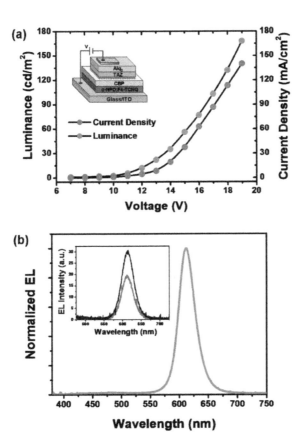

Fig. 7. (a) Current density and luminance from a LED based on an oriented layer of laterally aligned core-shell nanorods as illustrated by the scheme in the inset.
(b) Electroluminescence spectrum. The inset shows two spectra for orthogonal polarization directions. Taken with permission from (Rizzo et al., 2009).

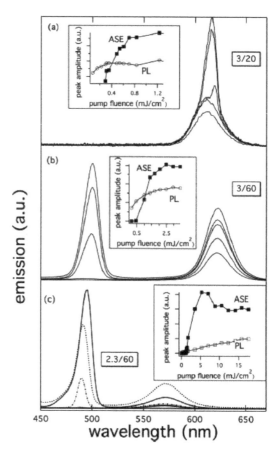

Fig. 8. Amplified spontaneous emission from a dense layers of core-shell nanorods with different core and shell size recorded at different values of pump fluence. Taken with permission from (Krhane et al., 2011a).

Then, a self-assembled oriented layer of laterally aligned nanords was transferred onto the CBP via a stamping technique. After the nanorod deposition the structure was over-coated with a 3-(4-biphenylyl)-4-phenyl-5-*t*-butylphenyl-1,2,4-triazole (TAZ) hole blocking layer (HBL), the tris(8-(hydroxyl-quinoline) aluminum (Alq3) electron transporting layer (ETL), and LiF/Al electrodes. In this type of device structure the excitation formation in the nanorod layer can occur either *via* charge trapping or *via* Forster energy transfer process from the organic material (Li et al, 2005; Anikeeva, 2007).

Amplified spontaneous emission (ASE) can be observed from dense aggregates of nanorods, for example in the form of nanorod layers fabricated by drop deposition from highly concentrated nanorod solutions onto planar surfaces. **Fig.8a** shows ASE from core states recorded from relatively short nanorods with large CdSe core. For rods with a length significantly larger than 25 nm ASE from the shell states, at 490 nm, was observed, while the photoluminescence from the core transitions was maintained, as can be seen in **Fig.8b-c**.

Lasing devices based on colloidal semiconductor nanorods have, to the best of our knowledge, only been obtained by optical pumping so far. In order to obtain lasing, an optical gain medium has to be positioned into a resonant cavity that provides sufficient feedback. Core-shell nanorods have demonstrated optical gain both from the core (Zavelani-Rossi, 2010b), and recently also from the shell emission (Krahne et al., 2011b). A conventional approach to obtain optical feedback is to embed the optical gain medium into an external resonator, for example a physical cavity consisting of a series of Bragg mirrors. However, this approach is not straightforward for self assembled layers of nanocrystals as gain medium because the roughness and thickness of this layer cannot be well controlled. An innovative solution to this problem was demonstrated by Zavelani et al. who used the nanorod layer itself as a resonant cavity. In this work ordered assemblies of nanorods were obtained via the coffee stain effect, i.e. the fluid dynamics in an evaporating droplet (Zavelani-Rossi, 2010b). Here the nanorods self assembled in large-scale ordered superstructures that are reminiscent of nematic/smectic liquid crystal phases. In particular, a dense and highly ordered region was obtained at the edge of the film that formed the outer ring (see **Fig.9a**). Within this outer ring the rods were well aligned and on average the long axis of the nanorods was oriented parallel to the ring edge. From such ordered regions of nanorods polarized emission (Carbone et al., 2007) and directionally dependent photoconductivity (Persano et al., 2010) have been observed. **Fig.9b** shows lasing spectra recorded from regions of the outer ring demonstrating that the lateral facets of the ring can function as a Fabry-Perot resonator.

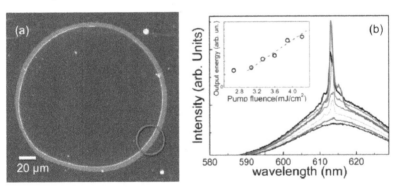

Fig. 9. Scanning electron microscopy image of a coffee stain ring that showed lasing. The red circle illustrates a spot from which lasing spectra were recorded. (b) Emission spectra above and below lasing threshold. The inset shows the characteristic input-output curve of a lasing device.

Such self-assembled micro-lasers provide new possibilities for the integration of narrow band emitters into device architectures such as lab-on-chip for point of care diagnostics, or for optical components in local area network datacom structures. Although the technology regarding light emitting devices based on colloidal nanorods is still in its infancy, the results described in this section are very encouraging and this bottom –up technology can be expected to take larger impact in the lighting industry in the near future.

4. Elongated nanoparticle arrays

The label-free ultrasensitive detection of biological molecules such as proteins, nucleic acids etc. is of utmost importance in the field of clinical medicine, especially with regards to the early diagnosis of diseases (Rosi & Mirkin, 2005). Vibrational spectroscopies (i.e. Raman Scattering and IR absorption) allow for direct detection of biological species (Colthup et al., 2010), but their cross-sections are extremely low in common experimental conditions. The possibility to excite localized surface plasmon resonances (LSPRs) in metallic nanoparticles has indicated a feasible and efficient way for enhancing the electromagnetic field on the local scale (Bohren & Huffman, 1998; Nie & Emory, 1997).

Therefore, the combination of plasmonic nanostructures with vibrational spectroscopies can be used to manipulate light-matter interactions, giving rise to fascinating perspective towards the production of novel biosensor devices (Anker et al., 2008; Das et al., 2009; De Angelis, 2008, 2010b). There is thus intensive activity oriented at the fabrication of tailored nanostructures endowed with the desired plasmonic properties. Metal nanoparticles present specific optical properties that depend on their size and geometry. While a symmetrical shape allows for polarization-independent plasmonic excitation, a dichroic absorption is expected in elongated NPs (Toma et al., 2008; Fazio et al., 2011). Beside this, the opto-plasmonic response can be additionally tailored by acting on the nanoparticle distribution and mutual coupling (Fischer & Martin, 2008).

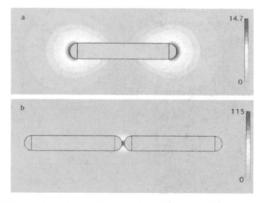

Fig. 10. (a) Calculated absolute value of the electric field distribution around an isolated nanostructure. The illuminating plane wave impinges perpendicularly on the array, and is polarized along the long axis of the elongated nanoparticle. The metallic structure length is 410 nm, while its width and height are set to 60 nm. The excitation wavelength is $\cong 1.9$ μm (b) Field distribution in the case of coupled nanostructures (gap width: 10 nm).

In order to elucidate this peculiar behavior, we have performed 3D numerical simulations using a commercial software based on a finite integration technique (CST, Computer Simulation Technology, Darmstadt, Germany). **Fig.10** shows the absolute value of the electric field around the nanostructures, on a plane that is perpendicular to the direction of the illuminating wave and cuts the nanostructure exactly at its half height. While an isolated particle concentrates the radiation at its extremities (**Fig.10a**), a dimer (i.e. a couple of closely spaced particles) creates a "hot spot" (Fischer & Martin, 2008; Stockman et al., 1994) in correspondence of the inter-particle nanocavity (**Fig.10b**).

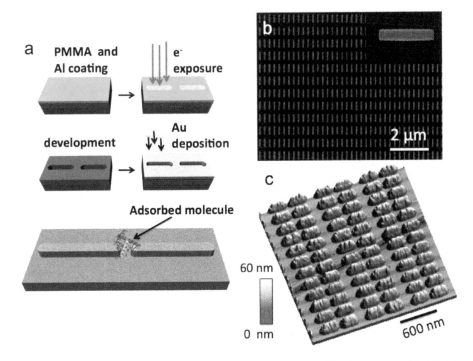

Fig. 11. (a) Schematic block diagram of the fabrication process. The last cartoon highlights the nano-biosensor concept: fabrication of a highly sensitive and specific device based on plasmonic signal enhancement. (b, c) Representative SEM and AFM images of two different nanoparticle arrays fabricated by EBL technique. The NPs are 410 nm (b) and 200 nm (c) long, while width and height are both set at 60 nm for all the structures.

Here we investigate these two specific examples, i.e. uncoupled and dimer nanoparticle arrays, with the aim to prove that their field enhancement and localization capabilities can be used for high-sensitivity Raman spectroscopy. A schematic diagram, summarized in **Fig.11a**, elucidates the main steps involved in the fabrication process. A 120 nm thick layer of PMMA (950K) was spin-coated on a CaF_2 (100) substrate. To prevent charging effects during the electron exposure, a 10 nm thick Al layer was thermally evaporated on the PMMA surface. Electron beam direct-writing of the nanoparticle patterns was carried out using a high resolution Raith150-Two e-beam writer at 15 keV beam energy and 25 pA beam current. After the Al removal in a KOH solution, the exposed resist was developed in a conventional solution of MIBK:IPA (1:3) for 30 s. Then, a 5 nm adhesion layer of Ti and a 60 nm Au film were evaporated, using a 0.3 Å/s deposition rate in a 10^{-7} mbar vacuum chamber (Kurt J Lesker PVD75). Finally, the unexposed resist was removed with acetone and rinsed out in IPA. Large scan overviews of different nanostructure arrangements are reported in Fig. 2b,c. The sample topography has been characterized recurring to scanning electron microscopy (SEM) and atomic force microscopy (AFM, Veeco MultiMode with NanoScope V controller) equipped with ultra-sharp Si probes (ACLA-SS, AppNano) and operating in tapping mode. The resulting arrays present, as shown in **Fig.11b,c**, a high degree of reproducibility with a surface RMS roughness value of around 1 nm.

The optical properties of the nanoparticle array were investigated by means of spectroscopic transmission of polarized light in the range between 450 and 900 nm (see **Fig.12a,b**). For this purpose, we used a fiber-optic spectrometer (AvaSpec-256, Avantes) while the light source was a combined deuterium-halogen lamp (AvaLight-DHc, Avantes). The polarization of the incident light was varied from parallel to perpendicular orientation with respect to the long axis of the nanostructure. **Fig.12a** shows the results concerning isolated nanoparticles while in **Fig.12b** we report transmission spectra for coupled nanostructures. In both cases the optical transmittance spectra present evidence of a clear anisotropic behavior; for perpendicular polarization a localized minimum around 620 nm is found. This is the typical behavior exhibited by sub-wavelength metal nanoparticles sustaining a localized surface plasmon resonance (Bohren & Huffman, 1998). For parallel polarization a red shifted extinction peak, centered around 700 nm, is observed. This can be attributed to a higher order plasmonic resonance (Aizpurua et al., 2005).

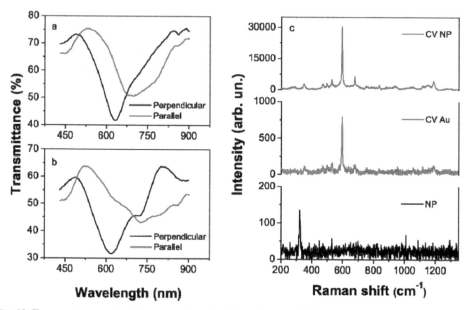

Fig. 12. Transmission optical spectra of single (a) and coupled (b) nanoparticle arrays. Red/black lines are for perpendicular and parallel polarization respectively. (c) SERS spectra of cresyl violet molecules deposited on NP sample (blue line) and flat Au film (red line). Background measurement performed on bare NP structure is also reported (black line).

Moving from the isolated nanoparticle array (**Fig.12a**) to the dimer case (**Fig.12b**) we can notice a broadening and a slightly shift of the resonance peaks due to near field coupling (Fischer & Martin, 2008).

In order to probe the enhancement behavior in Raman spectroscopy related to the plasmon excitation, the sample of **Fig.11b** was immersed in a cresyl violet (CV) solution (3.46 μ M in H_2O) for 15 min. The sample was gently rinsed in DI-water to remove molecules in excess not chemisorbed on the metal surface and then dried in nitrogen flow. SERS measurements were carried out by means of inVia microspectroscopy (Renishaw). The samples were excited by 633 nm laser wavelength (laser power = 0.14 mW and accumulation time = 50 s) through a 150X objective. The measurements were performed on bare nanoparticle arrays (sample NP), on CV molecules chemisorbed on a flat Au film (sample CV Au) and on CV deposited on the nanostructure arrays (sample CV NP). As shown in **Fig.12c** (NP line) no characteristic Raman band, except at around 320 cm^{-1} related to the Ca-F vibration from the substrate, is observed. The characteristic vibrational bands of CV are observed in the SERS spectrum (CV NP trace). Intense Raman bands centered at around 591, 882, 927 and 1189 cm^{-1} can be attributed to the N-H_2 rocking vibration, two benzene group bending, out-of phase N- H_2 rocking vibration, and combination of N-H_2 rocking and C-H_X rocking, respectively (Vogel, 2000; Sackmann et al., 2007). CV molecules chemisorbed over a flat Au film are also shown in **Fig.12c**. We can clearly observe that there is a giant enhancement in Raman signal for CV NP sample with respect to the CV Au one. The evaluated SERS enhancement for the fabricated device is around 10^7.

5. NPs based plasmonics devices for SERS applications

Metallic NPs are characterized from a great mobility of theirs electrons, resulting in a characteristic ability of sustaining coherent electronic oscillations, when light is impinged on them. The phenomenon, interesting generally the metallic surfaces, consists of the localized surface plasmons (LSP) formation (Raether, 1988; McCall et al., 1980; Haynes et al., 2005; Das et al., 2009; Kneipp et al., 2002) associated to the collective oscillation of the electrons moving in the small metal particles volume. The local electromagnetic field near the nanostructures surface results enormously enhanced. This property makes metallic NPs particularly interesting in spectroscopy. In Raman spectroscopy, this phenomenon is known as Surface Enhanced Raman Scattering (SERS) effect; it allows the detection or very diluted solutions, where very few molecules are present, overcoming the normal limit of Raman due to the great fluorescence which, in these cases, may cover molecules signals (Kneipp et al., 1997; De Angelis, 2008, 2010b). In fact, when a molecule is very close to a metallic nanostructure the LSP formation, under the laser effect, generates a giant enhancement of Raman signal (Das et al., 2009; Nie & Emory, 1997; Creighton et al., 1979).

It is clear that LSPR and, consequently, SERS intensity is strongly influenced from size, shape, inter–particle spacing of the nanoparticles and the dielectric environment of material. For particles diameters d « λ (λ = wave length of excitation laser), electrons move in phase on the excitation wave plan with the consequently formation of polarization charges and, then, of a dipolar field on particles surface. The estimated enhancement for Au and Ag isolated particles is around to 10^6 –10^7 (Otto, 1984; Kneipp & Kneipp, 2006).

For this reason the roughness of a surface obtained from a NPs deposition is of fundamental importance and overall the fabrication of specific metal nanostructures, in which size, shape, inter–particle spacing of the nanoparticles may be chosen carefully, has become an important factor for research on plasmonic devices (Gunnarsson et al., 2001). Traditionally, as already mentioned above, NPs for SERS applications were prepared in chemical way, producing Ag colloids to whom attacking bio-molecules for the detection (Xu et al., 1999; Kneipp et al., 2004; Hao & Schatz, 2004) with good results regarding SERS effect, but with some limits in the reproducibility of the systems. Afterwards, other kinds of SERS substrates were realized, as metallic films (Constantino et al., 2002; Garoff et al., 1983) and nanostructures (NPs) with characteristic length scale in the nanometer range, as nanovoids, nanoshells, nanorods, nanorings and nanocubes (Le Ru et al., 2008; Grzelczak et al., 2010; Tao et al., 2008). Recently development of nanofabrication techniques, as electron beam lithography (EBL) or focalized ion beam (FIB), has improved plasmonic nanostructures control and consequently has improved the efficiency of these nano-devices, focalizing the attention on substrate-bound nanostructure fabrication (Zhang et al., 2006).

Here, the fabrication of silver/gold nano-aggregates into well-defined lithographic structures is described. Micro- and nano- structured patterns are realized by EBL. The procedure of fabrication consists in three steps: 1)spin-coating of the silicon wafer by a resist, ZEP for microstructures and PMMA-A2 for the nano-structures; 2)the spinned resist is exposed to high energy electrons to design the pattern; 3)the exposed resist is removed by the appropriate solvent (developer). The exposition phase is regulated by means of typical EBL parameters, as electron beam current and exposition time, regulated on the basis of resist characteristics, on its thickness and on pattern size.

The metallic nanoparticles (especially gold and silver) may be fabricated into the lithographic pattern by physical techniques (evaporation), chemical techniques (reduction from metallic ions) or self-assembly techniques (Jensen et al., 1999; Felidj et al., 2002).

Among these, electroless deposition is a very fast, simple and economic technique, based on the auto-catalytic reduction of metal salts. The redox reaction consists in an electrons exchange between the metallic ions of an opportune solution where the substrate is immersed, and a reducing agent, which may be in the same solution or may be the substrate itself. In literature, several studies report the deposition of metals, as silver, gold, copper, nickel or their alloy, by this technique, obtaining thin films with higher roughness, dendritic structures, submicrometric metallic structures or NPs (Qiu & Chu, 2008; Gao et al., 2005; Yang et al., 2008; Peng & Zhu, 2004; Goia & Matijevic, 1998).

Here, the attention is centered on deposition of silver and gold nanoparticles for SERS application by electroless technique in which the reducing agent is the substrate itself (Coluccio et al., 2009). After the resist developing (the 3° steps of lithographic process) the pattern on Si wafer consists in holes where silicon remains uncovered. The patterned substrate is then dept into a solution of silver nitrate (AgNO3) or chloroauric acid (HAuCl$_4$) prepared in fluoridric acid. The standard solution concentration is around 1 mM, but it is possible to modulate it according to the morphology that we want to obtain. The solution enters the cavity and the reaction between silver/gold ions and the silicon wafer occurs. The metal deposition increases with the increase of temperature and reaction time.

The electroless process may be divided in two steps:

1. the patterned silicon reacts with fluoridric acid (HF) determines the removing of the superficial silicon oxide (SiO$_2$) layer and the formation of a hydrogenated surface, inert

to reactions with O_2, CO_2, CO, etc., while presenting a good reactivity with silver ions (Palermo & Jones, 2001);

2. the metal nano grains growth into the holes obtained onto the substrate, according the following reactions:

$$4\,Ag^+ + Si(s) + 6\,HF \rightarrow 4Ag^0 + H_2SiF_6 + 4H^+ \tag{1}$$

which can be separated into two half-cell reactions (2) and (3):

Anode

$$Si + 2H_2O \rightarrow SiO_2 + 4H^+ + 4e^- (silicium\ oxidation) \tag{2}$$

Cathode

$$Ag^+ + e^- \rightarrow Ag^0\ (silver\ reduction). \tag{3}$$

The oxidation of silicon, produced the electrons, is necessary for the silver reduction (Goia & Matijevic, 1998; Coluccio et al., 2009; Palermo & Jones, 2001). The mechanism of formation of the nanoparticles begins with few silver ions that react directly with the substrate forming metallic nuclei. These Ag nuclei are strongly electronegative, thereby, attracting other electrons from the silicon bulk and consequently getting negatively charged; then they attract silver ions which reduced to Ag^0 thus inducing the growth of the original Ag nuclei (Qiu & Chu, 2008). The gold reactions are very similar, the ion reacting obviously is Au^{3+}. Gold forms smaller NPs then silver and the reaction kinetic is slower.

The morphology of nano-aggregates depends on the lithographic pattern and, in particular, on the size: in nano sized structures, the grains grow compact and more mild reaction conditions are necessary for avoiding an excessive nano grains formation, in particular, concentrations under 1 mM and temperatures under 50°C may be used. In micro sized structures, for obtaining a uniform surface covering is important to use concentration up to 1mM and temperature around 50°C. For every SERS structures, the optimal reaction conditions must be searched, modulating the parameter for obtaining the correct particles sizes and density.

Both gold and silver nano aggregates give a good SERS effect, even if silver has the disadvantage that oxidizes with time, giving a decreasing of SERS efficiency. To overcome this problem bimetallic substrates may be fabricated depositing in a first step silver nano grains and over them gold nano grains. Gold assures the SERS device protection from time depending oxidation and maintains a good efficiency of the substrate. SEM images of silver and gold SERS devices are shown in **Fig.13**.

As mentioned above the fabrication of regular array of metallic NPs, may increase the efficiency and the reproducibility of the SERS device, and their optical characteristics may be selected varying size and mutual distance between particles. NPs arrays examples are reported in **Fig.14**.

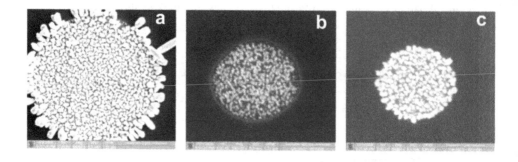

Fig. 13. SEM images of nano grains assembled on microstructures, fabricated using e-beam lithography and electroless deposition: (a) silver nano grains, (b) gold nano grains, (c) silver/gold nano grains.

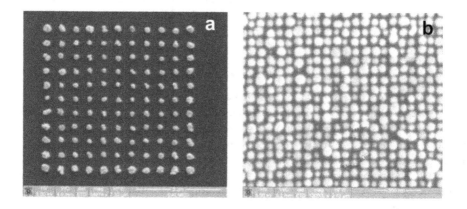

Fig. 14. SEM images of nano grains assembled on nanostructures, fabricated using e-beam lithography and electroless deposition.

The SERS effect of gold, silver or of bimetallic (silver/gold) microstructures is investigated using rhodamine-6G as probe molecules. The substrates are dipped into rhodamine-6G water solution with different concentrations, rinsed with water dried in N_2 and then used for the spectroscopic investigation. SERS experiments show spectra with well-defined peaks at very low molecular concentration, for all the metallic/bimetallic substrates (**Fig.15**). The highest SERS efficiency is found for SERS device based on gold over silver surface.

a

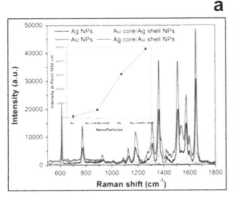

b

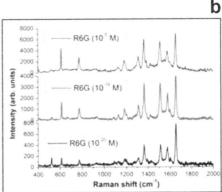

Fig. 15. a) R6G [10^{-12}M], absorbed on the silver, gold, Ag/Au and Au/Ag structures. Raman signal increases from Au, Au core/Ag shell, Ag and Ag core/Au shell; b) 3. SERS spectra acquired from 10^{-5}, 10^{-16} and 10^{-20} M R6G, absorbed on the silver nanostructures.

6. References

Aizpurua, J.; Bryant, G. W.; Richter, L. J.; Garcia de Abajo, F. J.; Kelley, B. K. & Mallouk, T. (2005). Optical properties of coupled metallic nanorods for field-enhanced spectroscopy. *Phys. Rev. B*, Vol. 71, pp. 235420.

Anikeeva, P.O., et al., Electroluminescence from a mixed red-green-blue colloidal quantum dot monolayer. Nano Letters, 2007. 7(8): p. 2196-2200.

Anker, J. N.; Hall, W. P.; Lyandres, O.; Shah, N. C.; Zhao, J. & Van Duyne, R. P. (2008). Biosensing with plasmonic nanosensors. *Nature Materials*, Vol. 7, pp. 442-453.

Bohren, C. F. & Huffman, D. R., (1998). *Absorption and Scattering of light by small particles* (Wiley, New York).

Canham L., Properties of Porous Silicon. INSPEC - The Institution of Electrical Engineers London - United Kingdom, 1997.

Carbone, L., et al., Synthesis and micrometer-scale assembly of colloidal CdSe/CdS nanorods prepared by a seeded growth approach. Nano Letters, 2007. 7(10): p. 2942-2950.

Cohen M. H., Melnik K., Boiasrki A., Ferrari M. & Martin F. J., Microfabrication of silicon-based nanoporous particulates for medical applications, *Biomed. Microdevices* (5), 253–259, 2003.

Colthup, N. B., Daly, L. H. & Wiberley, S. E., (2010). *Introduction to Infrared and Raman Spectroscopy* (Academic Press, USA).

Coluccio M.L., Das G., Mecarini F., Gentile F., Pujia A., Bava L., Tallerico R., Candeloro P., Liberale C., De Angelis F., Di Fabrizio E., Silver-based surface enhanced Raman scattering (SERS) substrate fabrication using nanolithography and site selective electroless deposition, Microelectronic Engineering 86 (2009) 1085–1088.

Constantino C.J.L., Lemma T., Antunes P.A., Aroca R., Single molecular detection of a perylene dye dispersed in a Langmuir–Blodgett fatty acid monolayer using surface-enhanced resonance Raman scattering, Spectro chimica Acta Part A 58 (2002) 403–409.

Creighton J. A., Blatchford C. G., Albrecht M. G., Plasma Resonance Enhancement of Raman-Scattering by Pyridine Adsorbed on Silver or Gold Sol Particles of Size Comparable to the Excitation Wavelength, J. Chem. Soc., Faraday Trans II 75 (1979) 790-798.

Crommelin D.J.A., Schreier H., Liposomes, in: *Colloidal Drug Delivery Systems*, J. Kreuter, editor. Marcel Dekker, Inc. New York, 1994.

Das G., Mecarini F., Gentile F., De Angelis F., Kumar M., Candeloro P., Liberale C., Cuda G., Di Fabrizio E., Nano-patterned SERS substrate: Application for proteinanalysis vs. temperature, Biosens. Bioelectron. 24 (2009) 1693–1699.

De Angelis, F.; Patrini, M.; Das, G.; Maksymov, I.; Galli, M.; Businaro, L.; Andreani, L. C. & Di Fabrizio, E. (2008). A Hybrid Plasmonic Photonic Nanodevice for Label-Free Detection of a Few Molecules. *Nano Letters*, Vol. 8, No. 8, pp. 2321–2327

De Angelis F., Pujia A., Falcone C., Iaccino E., Palmieri C., Liberale C., Mecarini F., Candeloro P., Luberto L., de Laurentiis A., Das G., Scala G., and Di Fabrizio E., Water soluble nanoporous nanoparticles for in vivo targeted drug delivery and controlled release in b cells tumor context. Nanoscale, 2:2230-2236, 2010.

De Angelis, F.; Das, G.; Candeloro, P.; Patrini, M.; Galli, M.; Bek. A.; Lazzarino, M.; Maksymov, I.; Liberale, C.; Andreani, L. C. & Di Fabrizio, E. (2010). Nanoscale chemical mapping using three-dimensional adiabatic compression of surface plasmon polaritons. *Nature Nanotech.*, Vol. 5, pp. 67-72

Decuzzi P. and Ferrari M., The Adhesive Strength of Non-Spherical Particles Mediated by Specific Interactions, *Biomaterials*, 27(30):5307-1534, 2006.

Decuzzi P., Causa F., Ferrari M. and Netti P. A., The Effective Dispersion of Nanovectors Within the Tumor Microvasculature, *Ann Biomed Eng.*, 34(4):633-641, 2006.

Decuzzi P. and Ferrari M., The Role of Specific and Non-Specific Interactions in Receptor-Mediated Endocytosis of Nanoparticles, *Biomaterials*, 28(18):2915-2922, 2007.

Duncan R., The dawning era of polymer therapeutics, *Nature Rev. Drug Discov.* (2), 347–360, 2003.

Fazio B.; D'Andrea, C.; Bonaccorso, F.; Irrera, A.; Calogero, G.; Vasi, C.; Gucciardi, P. G.; Allegrini, M.; Toma, A.; Chiappe, D.; Martella, C. & Buatier De Mongeot F. (2011). Re-radiation Enhancement in Polarized Surface-Enhanced Resonant Raman Scattering of Randomly Oriented Molecules on Self-Organized Gold Nanowires. *ACS Nano*, Vol. 5 No. 7, pp. 5945-5956.

Felidj N., Aubard J., Levi G., Krenn J.R., Salerno, Schider G., Lamprecht B., Leitner A., Aussenegg F.R., Controlling the optical response of regular arrays of gold particles for surface enhanced Raman scattering, Phys. Rev. B Condens. Matter Mater. Phys. 65 (2002) 075419/1–075419/9.

Ferrari M., Cancer Nanotechnology: opportunities and challenges, *Nature Reviews Cancer*, 2005; (5):161-171.

Fischer, H. & Martin, O. J. F. (2008). Engineering the optical response of plasmonic nanoantennas. *Optics Express*, Vol. 16 No.12, pp. 9144-9154.

Gao J., Tang F., Ren J., Electroless nickel deposition on amino-functionalized silica spheres, Surface & Coatings Technology 200 (2005) 2249–2252.

Garoff S., Weitz D.A., Alvarez M.S. and Chung J.C., Electromagnetically Induced Changes in Intensities, Spectra and Temporal Behavior of Light Scattering from Molecules on Silver Island Films, J. Phys. Colloq. 44, C10 (1983) 345-348.

Gentile F., Ferrari M., Decuzzi P. (2007), Transient Diffusion of Nanovectors in Permeable Capillaries, J. Serbian Soc. Comput. Mech., 1(1), 1-19.

Gentile F., Ferrari M. and Decuzzi P., `The Transport of Nanoparticles in Blood Vessels: The Effect of Vessel Permeability and Blood Rheology`, Annals of Biomedical Engineering, 2008:2(36); 254-261.

Gilles E. M. & Frechet J. M. J., Designing macromolecules for therapeutic applications: Polyester dendrimerpolyethylene oxide 'bow-tie' hybrids with tunable molecular weights and architecture, J. Am. Chem. Soc. (124), 14137–14146, 2002.

Godefroo S., Hayne M., Jivanescu M., Stesmans A., Zacharias M., Lebedev O.I., van Tendeloo G., and Moschchalkov V. V., Classification and control of the origin of photoluminescence from si nanocrystals. Nature Nanotechnology, 3:174-178, 2008.

Granitzer P. and Rumpf K., Porous silicon - a versatile host material. Materials, 3:943-998, 2010.

Grzelczak M., Vermant J., Furst E. M., Liz-Marza´n L. M., Directed Self-Assembly of Nanoparticles, ACSNano VOL. 4 ▪ NO. 7 ▪ 3591–3605 ▪ 2010;

Gunnarsson L., Bjerneld E.J., Xu H., Petronis S., Kasemo B. and Käll M., Interparticle-coupling effects in Nanofabricated Substrates for Surface Enhanced Raman Scattering, Applied Physics Letters 78, 802-804 (2001).

Halimaoui A., Porous silicon: material processing, properties and applications, in JC Vial and J. Derrien (editors), Porous silicon science and technology, Springer-Verlag (1995).

Hao E. and Schatz G. C., Electromagnetic fields around silver nanoparticles and dimmers, J. Chem. Phys. 120 (2004) 357-367.

Haynes C.L., McFarland A.D., VanDuyne R.P., Surface-Enhanced Raman Spectroscopy, Anal. Chem. 77 (2005) 338A-346A.

Hirsch, L. R., Halas, N. J. & West, J. L. Nanoshell-mediated near-infrared thermal therapy of tumors under magnetic resonance guidance, Proc. Natl Acad. Sci. USA (100), 13549–13554, 2003.

Hu, J.T., et al., Linearly polarized emission from colloidal semiconductor quantum rods. Science, 2001. 292(5524): p. 2060-2063.

Jensen T.R., Schatz G.C., Van Duyne R.P., Nanosphere Lithography: Surface plasmon resonance spectrum of a periodic array of silver nanoparticles by UV-vis extinction spectroscopy and electrodynamic modeling, J. Phys. Chem. B 103 (1999) 2394–2401 .

Jeyarama S. Ananta, Biana Godin, Richa Sethi, Loick Moriggi, Xuewu Liu, Rita E. Serda, Ramkumar Krishnamurthy, Raja Muthupillai, Robert D. Bolskar, Lothar Helm, Mauro Ferrari, Lon J. Wilson, Paolo Decuzzi, Geometrical confinement of gadolinium-based contrast agents in nanoporous particles enhances T_1 contrast, Nature Nanotechnology 5, 815–821 (2010).

Ji-Ho Park, Luo Gu, Geo rey von Maltzahn, Erkki Ruoslahti, Sangeeta N. Bhatia, and Michael J. Sailor. Biodegradable luminescent porous silicon nanoparticles for in vivo applications. Nature Materials, 8:331, 336, 2009.

Kazes, M., et al., Lasing from semiconductor quantum rods in a cylindrical microcavity. Advanced Materials, 2002. 14(4): p. 317-321.

Kazes, M., et al., Lasing from CdSe/ZnS quantum rods in a cylindrical microcavity. Quantum Dots, Nanoparticles and Nanowires, 2004. 789: p. 11-16429.

Kircher M. F., Mahmood U., King, R. S., Weissleder R. & Josephson L., A multimodal nanoparticle for preoperative magnetic resonance imaging and intraoperative optical brain tumor delineation, *Cancer Res.* (63), 8122–8125, 2003.

Klibanov A. L. et al., Activity of amphipathic PEG 5000 toprolong the circulation time of liposomes depends on the liposome size and is unfavourable for immunoliposome binding to target, *Biochem. Biophys. Acta* (1062), 142–148, 1991.

Kneipp K., Wang Y., Kneipp H., Perelman L.T., Itzkan I., Dasari R. R., and Feld M. S., Single Molecule Detection using Surface-Enhanced Raman Scattering, Phys. Rev. Lett. 78 (1997) 1667-1670.

Kneipp K., Kneipp H., Itzkan I., Dasari R. R., Feld M.S., Kneipp H., Itzkan I., Dasari R. R., and Feld M. S., Surface-enhanced Raman scattering and biophysics, J. Phys. Condens. Matter 14 (2002) R597-R624.

Kneipp, K., Kneipp, H., Abdali, S. et al., Single Molecule Raman Detection of Enkephalin on Silver Colloidal Particles. Spectroscopy 18 (2004) 433–440.

Kneipp K., Kneipp H., Kneipp J., Surface-Enhanced Raman Scattering in Local Optical Fields of Silver and Gold Nanoaggregatess From Single-Molecule Raman Spectroscopy to Ultrasensitive Probing in Live Cells, Acc. Chem. Res. 39 (2006) 443-450.

Krahne, R., et al., Amplified Spontaneous Emission from Core and Shell Transitions in CdSe/CdS Nanorods fabricated by Seeded Growth. Applied Physics Letters, 2011. 98(6): p. 063105.

Krahne, R., et al., Physical properties of elongated inorganic nanoparticles. Physics Reports, 2011. 501(3-5): p. 75-221.

La Van D. A., McGuire T. & Langer R., Small-scale systems for *in vivo* drug delivery, *Nature Biotechnol.*, (21), 1184–1191, 2003.

Langer R., Drug delivery and targeting, *Nature* (392), 5–10, 1998.

Le Ru E.C., Etchegoin P.G., Grand J., Félidj N., Aubard J., Lévi G., Hohenau A., Krenn J.R., Surface enhanced Raman spectroscopy on nanolithography-prepared substrates, Current Applied Physics 8 (2008) 467–470.

Li, Y.Q., et al., White organic light-emitting devices with CdSe/ZnS quantum dots as a red emitter. Journal Of Applied Physics, 2005. 97(11): p. art. n. 113501.

Lupo, M.G., et al., Ultrafast Electron-Hole Dynamics in Core/Shell CdSe/CdS Dot/Rod Nanocrystals. Nano Letters, 2008. 8(12): p. 4582-4587.

McCall S.L., Platzman P.M., Wolf P.A., Surface enhanced Raman scattering, Phys. Lett. 77A (1980) 381-383.

Mitragotri Samir and Lahan Joerg, Physical approaches to biomaterial design. Nature Materials, 8:15-23, 2009.

Nie S. and Emory S. R., Probing Single Molecules and Single Nanoparticles by Surface-Enhanced Raman Scattering, Science 275 (1997) 1102-1106.

Nobile, C., et al., Self-assembly of highly fluorescent semiconductor nanorods into large scale smectic liquid crystal structures by coffee stain evaporation dynamics. Journal Of Physics-Condensed Matter, 2009. 21(26): p. 264013.

O Farrel N., Houlton A., B. R. Horrocks, Silicon nanoparticle application in cell biology and medicine. International Journal of Nanomedicine, 1(4):451-472, 2006.

Otto A., In Light scattering in solids IV. Electronic scattering, spin effects, SERS and morphic effects; M. Cardona, G. Guntherodt, Eds.; Springer-Verlag: Berlin, Germany, 1984; Vol. 1984, pp 289- 418.

Palermo, D. Jones, Morphological changes of the Si [100] surface after treatment with concentrated and diluted HF, Materials Science in Semiconductor Processing, 4 (2001) 437–441.

Park J. W., Liposome-based drug delivery in breast cancer treatment, *Breast Cancer Res.* (4), 95–99, 2002.

Peng K., Zhu J., Morphological selection of electroless metal deposits on silicon in aqueous fluoride solution, ElectrochimicaActa 49 (2004) 2563–2568.

Persano, A., et al., Photoconduction Properties in Aligned Assemblies of Colloidal CdSe/CdS Nanorods. Acs Nano, 2010. 4(3): p. 1646-1652.

Petros Robby A. and DeSimone Joseph M., Strategies in the design of nanoparticles for therapeutic applications. Nature Reviews Drug Discovery, 9:615-627, 2010.

Pujia A., De Angelis F., Scumaci D., Gaspari M., Liberale C., Candeloro P., Cuda G., Di Fabrizio E., Highly efficient human serum filtration with water-soluble nanoparticles. International Journal of Nanomedicine, 5:1005-1015, 2010.

Qiu T., Chu P.K., Self-selective electroless plating: An approach for fabrication of functional 1D nanomaterials, Mater. Sci. Eng. R 61 (2008) 59-77.

Raether H., Surface Plasmonson Smooth and Rough Surfaces and on Gratings (Springer, Berlin, 1988).

Rizzo, A., et al., Polarized Light Emitting Diode by Long-Range Nanorod Self-Assembling on a Water Surface. Acs Nano, 2009. 3(6): p. 1506-1512.

Rosi, N. L. & Mirkin, C. A., (2005). Nanostructures in Biodiagnostics. *Chem. Rev.,* Vol. 105, pp. 1547-1562.

Sackmann, M.; Bom, S.; Balster, T. & Materny, A. (2007). Nanostructured gold surfaces as reproducible substrates for surface-enhanced Raman spectroscopy. *J. Raman Spect.,* Vol. 38, pp. 277-282.

Schellenberger E. A. et al., Annexin V-CLIO: a nanoparticle for detecting apoptosis by MRI, *Mol. Imaging* (1), 102–107, 2002.

Siti M. Janib, Ara S. Moses, and J. Andrew MacKay. Imaging and drug delivery using theranostic nanoparticles. Advanced Drug Delivery Reviews, 62:1052-1063, 2010.

Stockman, M. I.; Pandey, L. N.; Muratov, L. S. & George, T. F. (1994). Giant fluctuations of local optical fields in fractal clusters. *Phys. Rev. Lett.,* Vol. 72, pp. 2486-2489.

Talapin, D.V., et al., Highly emissive colloidal CdSe/CdS heterostructures of mixed dimensionality. Nano Letters, 2003. 3(12): p. 1677-1681.

Tao A. R., Habas S., Yang P., Shape Control of Colloidal Metal Nanocrystals , Small 4 No 3 (2008), 310 – 325.

Tasciotti E, Liu X, Bhavane R, Plant K, Leonard AD, Price BK, Cheng MM, Decuzzi P, Tour JM, Robertson F, Ferrari M. Mesoporous silicon particles as a multistage delivery system for imaging and therapeutic applications, Nature Nanotechnol. 2008 3(3):151-157.

Toma, A.; Chiappe, D.; Massabò, D.; Boragno, C. & Buatier de Mongeot, F. (2008). Self-organized metal nanowire arrays with tunable optical anisotropy. *Appl. Phys. Lett.,* Vol. 93, pp. 163104.

V. Dan G., Matijevic E., Preparation of monodispersed metal particles, New J. Chem. 22 (1998) 1203–1215.

Vogel, E.; Gbureck, A. & Kiefer, W. (2000). Vibrational spectroscopic studies on the dyes cresyl violet and coumarin 152*, *J. Mol. Struct.*, Vol. 550, pp. 177-190.

Whitesides G. M., The 'right' size in nanotechnology, *Nature Biotechnol.*, (21), 1161–1165, 2003.

Xu H., Bjerneld E.J., Kall M., et al..Spectroscopy of Single Hemoglobin Molecules by Surface Enhanced Raman Scattering. Phys. Rev. Lett. 83 (1999) 4357–4360.

Yang Y., Shi J., Kawamura G., Nogami M., Preparation of Au-Ag, Ag-Au core-shell bimetallic nanoparticles for surface-enhanced Raman scattering, Scr. Mater. 58 (2008) 862-865.

Zavelani-Rossi, M., et al., Lasing in self-assembled microcavities of CdSe/CdS core/shell colloidal quantum rods. Nanoscale, 2010. 2(6): p. 931-935.

Zavelani-Rossi, M., et al., Suppression of Biexciton Auger Recombination in CdSe/CdS Dot/Rods: Role of the Electronic Structure in the Carrier Dynamics. Nano Letters, 2010. 10(8): p. 3142-3150.

Zhang Y. & Shang M., Self-assembled coatings on individual monodisperse magnetite nanoparticles for efficient intracellular uptake, *Biomed. Microdevices* (6), 33–40, 2004.

Zhang, X., Yonzon, C.R., Duyne, R.P.V. 2006. Nanosphere Lithography Fabricated Plasmonic Materials and Their Applications. *J. Mater. Res.* 21: 1083–1092.

3

Rapid Nanoparticle Characterization

Rajasekhar Anumolu[1] and Leonard F. Pease III[1,2,3]
[1]Department of Chemical Engineering, University of Utah, Salt Lake City, UT
[2]Department of Internal Medicine, Division of Gastroenterology, Hepatology,
and Nutrition, University of Utah, Salt Lake City, UT
[3]Department of Pharmaceutics and Pharmaceutical Chemistry,
University of Utah, Salt Lake City, UT
USA

1. Introduction

Nanoparticles are the focus of intense scientific and engineering attention, due to their unique properties and wide array of potential biomedical, optical, and electronic applications. Their unique properties bridge those of bulk materials and atomic or molecular structures. Whereas bulk materials display constant physical properties regardless of size, nanomaterial properties may be size-dependent, such as quantum confinement in semiconductor particles, surface plasmon resonance in many metal particles and superparamagnetism in magnetic materials. However, design, synthesis, and fabrication of nanoparticles for specific applications or fundamental inquiry remains immature without accurate and well resolved characterization of nanoparticle size and structure. To date, nanoparticles may be characterized using a variety of traditional techniques including electron microscopy (EM), atomic force microscopy (AFM), dynamic and static light scattering (DLS and SLS), x-ray photoelectron spectroscopy (XPS), powder X-ray diffraction (XRD), Fourier transform infrared spectroscopy (FTIR), size exclusion chromatography (SEC), asymmetric flow field flow fractionation (AFFFF), X-ray crystallography, small angle neutron scattering (SANS), matrix-assisted laser desorption/ionization time-of-flight mass spectrometry (MALDI-TOF), ultraviolet-visible spectroscopy, dual polarization interferometry, and nuclear magnetic resonance (NMR) [Attri & Minton, 2005; Bondos, 2006; Casper & Clug, 1962; Chang et al., 1992; Colter & Ellem, 1961; Dai et al., 2008; Kim et al., 2010; Knapman et al., 2010; Pease et al., 2009; Russel et al., 1989; Shoemaker et al., 2010; Siuzdak et al., 1996; Swann et al., 2004; Umbach et al., 1998; Wang, 2005]. However, techniques such as electron microscopy and X-ray crystallography are expensive, time consuming, and require extensive computational resources. Other techniques such as SANS also suffer from limited availability [Kuzmanovic et al., 2008]. Despite the wide array of available techniques, there remains a need for rapid, label free, and statistically powerful characterization techniques to resolve dynamic multimodal distributions within approximately an hour or less.

Electrospray differential mobility analysis (ES-DMA) is a rapid technique (analysis time scales on the order of 1-100 min) with sub-nanometer resolution. ES-DMA can detect particles from 0.7 nm – 700 nm [Hollertz et al., 2011; Tsai et al., 2008]. At its smallest, ES-DMA has sub-angstrom resolution on the molecular diameter. ES-DMA can be used to determine the concentration of nanoparticles and also deposit the nanoparticles for additional characterization. ES-DMA is particularly attractive because it operates at atmospheric pressure and requires neither fluorescent tagging nor calibration curves. It separates particles based on their charge-to-size ratio, similar to mass spectrometry or capillary electrophoresis but without the expensive equipment such as turbo pumps [Pease et al., 2008]. While ES-DMA does not possess the atomic scale resolution of X-ray crystallography, EM, or SANS, its simplicity gives it decided advantages in speed, cost, and statistical significance [Pease et al., 2011]. The "coarse-grain" structures it resolves provide significant information regarding the structure of several biological particles such as assembling viruses, virus-like particles, and vaccines for biomedical applications and the structure of nanoparticle clusters and aggregates [Cole et al., 2009; Hogan et al., 2006; Lute et al., 2008; Thomas et al., 2004; Wick et al., 2005].

Here we review this label-free, quantitative, and rapid technique that provides full multimodal size distributions with sub-nanometer resolution. We first describe the operation and physics underlying this instrument. We then describe exemplary applications of this instrument to nanoparticle characterization. We finally conclude by comparing ES-DMA to several of the techniques listed above and provide an outlook for future growth of the technique.

1.1 ES-DMA – operation and physics

In ES-DMA, also referred to as a scanning mobility particle sizing (SMPS) or gas-phase electrophoretic mobility molecular analysis (GEMMA) [Bacher et al., 2001; Saucy et al., 2004], a particle suspension is first conveyed into the gas phase (Figure 1 shows ES-DMA system components). This is achieved by electrospray ionization, which produces a narrow distribution of droplets, typically 150-400 nm in initial diameter.

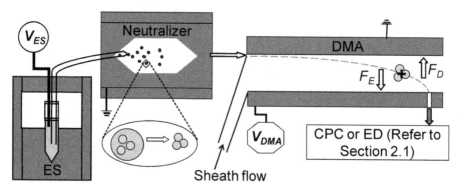

Fig. 1. Schematic of ES-DMA nanoparticle characterization system. ES-DMA produces either a size distribution using an ultrafine condensation particle counter (CPC) or deposits nanoparticles using an electrostatic deposition (ED) chamber as discussed in Section 2.1. Here, F_E is the electrostatic force and F_D is the radial component of drag force acting on the nanoparticle system.

The electrospray system uses pressure (3-5 psig) to drive flow through a 25-40 micron inner diameter capillary. At the capillary tip, a Taylor cone forms due to an applied electric field (1-3.7 kV across ~0.3 cm). Droplets emitted from the cone encapsulate one or multiple discrete nanoparticles (e.g. DNA coated gold nanoparticles or recombinant polymer strands) and are entrained in a mixed stream of air (1.0 L/min) and carbon dioxide (0.2 L/min) at atmospheric pressure. The droplets quickly evaporate leaving a dry nanoparticle. For example, Figure 1 shows three gold nanoparticles forming into a trimeric cluster. As these drying particles pass through a neutralizing chamber, collisions with charged ions reduce the charge on the nanoparticle received in the electrospray to a modified Boltzmann distribution [Kaddis et al., 2007; Loo et al., 2005; Wiedensohler, 1988]. Consequently, the positively charged particles analyzed in the DMA carry predominantly a single net positive charge. Within the annular DMA analysis chamber, a potentiated ($\leq$-10 kV) center electrode attracts charged particles dragged towards the exit of the DMA by a carrier gas (nitrogen). The electrical force acting on the particles carrying n_e electron charges of magnitude e (=1.602 x 10^{-19} C) through an electric field, E_s, is given by $F_E = n_e e E_s$. The mobility of the particle, Z_p, is then defined as the ratio of the particle's velocity v to the force giving rise to that velocity [Anumolu et al., 2011; Knutson & Whitby, 1975]. This force gives rise to a resulting radial drag force acting on the particle, $F_D = 3\pi\mu_g v d_p/C_c$, where d_p is the spherically equivalent particle mobility diameter and μ_g is the gas viscosity. This form of Stokes' Law corrects for moderate to high Knudsen number flow using the Cunningham slip correction factor, C_c, because many particle sizes are equal to or less than the mean free path of air, λ (=66 nm) [Radar, 1990]. This correction factor is given by $C_c = 1 + Kn [a + \beta \, Exp(-\gamma/Kn)]$, where $Kn = 2\lambda/d_p$, $a = 1.257$, $\beta = 0.40$, and $\gamma = 1.110$ [Allen & Raabe, 1985; Radar, 1990]. When the particles are not spherical, several authors show that the drag force in the free molecular regime (where $Kn>1$) depends on the projected area [Epstein, 1924; Hollertz et al., 2011; Pease et al., 2011]. Particles quickly reach their terminal velocity in a small fraction of the particle's DMA residence time allowing us to equate the radial component of the two forces to obtain the particle's electrical mobility as

$$Z_p = \frac{n_e e C_c}{3\pi\mu_g d_p} . \tag{1}$$

The instrument also has a unique mobility, which holds for any laminar velocity profile, given by [Knutson & Whitby, 1975],

$$Z_p^* = \frac{(q_a - q_s)/2 + q_{sh}}{2\pi V L} \ln\frac{r_2}{r_1} , \tag{2}$$

where, V is the average voltage on the center electrode, q_{sh} is the sheath flow, q_s is the sampling or monodispersed flow out of the DMA, q_a is the aerosol flow out of the ES into the DMA, L is the length between polydisperse aerosol inlet and exit slit (4.987 cm), r_1 is the inner radius of annular space of the DMA (0.937 cm), and r_2 is the outer radius of annular space of the DMA (1.905 cm). Here the values of L, r_1, and r_2 represent the dimensions of nano-DMA. When the particle and instrument mobilities are equal, the particle passes through a collection slit at the distal end of the center electrode to be counted or further analyzed. Combining particle and instrument mobilities determines the mobility diameter of the particle to be

$$\frac{d_p}{C_c} = \frac{2n_e eVL}{3\mu_g q_{sh} \ln\left(\frac{r_2}{r_1}\right)}. \tag{3}$$

Here C_c is grouped with the diameter d_p, because C_c is diameter dependent. Stepping through a series of voltages (or sizes) while counting with a condensation particle counter (CPC), yields a complete multimodal distribution. The continuous and high throughput (millions of particles per run) design of the instrument ensures statistical significance of both mean and tails of ES-DMA size distributions. The distributions are also highly repeatable with a standard deviation on the number-average diameter of only ± 0.1 nm for nominally 10 nm gold nanoparticles [Pease et al., 2007].

By collecting data for several seconds and averaging over time the mean or number–average diameter may be calculated with $d_{ave} = \sum_i N_i d_i / \sum_i N_i$, where N_i is the number of particles counted by the CPC of size d_i [Fissan et al., 1983]. Because we apply a negative bias to ions within the DMA, only particles that acquire a positive charge are detected. A modified expression for the Boltzmann distribution [Weidensohler, 1988] is used to correct for this effect, transforming the distribution of positively charged particles into the complete distribution of all particles regardless of charge. The fraction of singly and doubly charged particles was determined by Weidensohler,

$$f(j) = 10^{\sum_{i=0}^{5} a_i^j [\log(d_p / d_o)]^i}, \tag{4}$$

where d_p represents the mobility diameter of the particle, d_o = 1 nm, j is the number of charges on a particle, and a_0^1 through a_5^1 are -2.3484, 0.6044, 0.4800, 0.0013, -0.1553, and 0.0320, while a_0^2 through a_5^2 are -44.4756, 79.3772, -62.8900, 26.4492, -5.7480, and 0.5049, respectively, based on the probability of charging provided by Fuchs [Fuchs, 1963]. To remove the influence of doubly counted particles from the size distribution, the charge corrected count is multiplied by the overlap factor, f_{no}, which becomes unity for d_p < 5 nm and $10.94 / d_p - 29.94 / d_p^2$ for d_p > 5 nm as described by Pease, et al. [Pease et al., 2010a].

Alternatively, monodispersed nanoparticles or nanoparticle clusters from the DMA (or polydispersed nanoparticles bypassing the DMA) may be directed into an electrostatic deposition (ED) chamber. Within the chamber an electrode beneath a collecting substrate exerts an electrostatic field (≤-10 kV over ~2.5 cm) to attract entering particles. Electrostatic deposition is particularly effective at assembling nanoparticles onto the substrates, in contrast to impaction, because nanoparticle electrical mobilities are high (Eq. 1) and their inertia is relatively low (i.e., the Stokes number of nanoparticles remains much less than unity) such that they follow gas streamlines in the absence of an external field [Hinds, 1999]. Several factors affect the collection efficiency with smaller particles, higher particle charging, higher applied voltage, and lower gas flow rates increasing deposition. Figure 2 shows that smaller nanoparticles indeed have higher deposition efficiency (~100% when d < 30 nm) [Dixkens & Fissan, 1999]. The electrode features also affect the deposition profiles. For example, smaller electrodes lead to more focused deposition.

Fabrication of nanoparticle-based devices requires addressing nanoparticles to specific locations, which may be accomplished using electrostatic forces. For instance, several authors have demonstrated that charged nanoparticles may be directed to specific substrate locations by tuning electric fields near surfaces using charge patterning [Barry et al., 2003a, 2003b; Fissan et al., 2003; Jacobs et al., 2002; Krinke et al., 2002, 2003]. Similarly, Tsai, et al., used planar p-n junction patterned substrates to generate an array of tunable electric fields [Park & Phaneuf, 2003; Tsai et al., 2005]. Several other substrates are used for electrostatic deposition such as TEM grids (~3 mm in diameter) with thin holey carbon films [Anumolu et al., 2011], silicon wafers, and glass substrates pretreated in KOH and UVO-cleaner [Kang et al., 2011, 2012]. The optimum deposition time onto holey carbon TEM grids requires the product of the aerosol number density and time to exceed 3000 particle·hr/cm³, while ~100 particle·hr/cm³ is optimal to prepare glass rounds for precision optical measurements. Pease, et al., used freshly cleaved mica surfaces for deposition of carbon nanotubes for AFM observation [Pease et al., 2009b]. More recently, Saffari, et al., deposited particles into live DU145 prostate cancer cells in culture stored in Petri dishes [Hedieh et al., 2012].

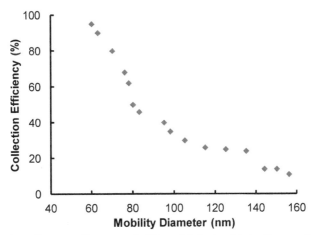

Fig. 2. Nanoparticle collection efficiency via ED at 1 L/min of inlet flow and 10 kV [Adapted from Dixkens & Fissan, 1999].

1.1.1 Predicting molecular size and structure

Bacher, et al., correlated empirically the mobility diameter (d_p in nm) of a variety of large biomolecules including proteins, antibodies, and viral proteins with their molecular weight, M_w (in kDa), assuming proteins to be globular spheres of constant density. Their composite empirical expression, M_w = -22.033 + 9.830d_P - 1.247d_P^2 + 0.228 d_P^3 (which when inverted for d_P becomes d_P = 1.832$M_W^{0.3256}$ ≈ (6M_w/$\pi\rho$)$^{1/3}$ [Pease et al., 2001], where ρ is the density of protein), predicted individual protein and protein aggregates sizes [Bacher et al., 2001]. Pease, et al., demonstrated how to convert the projected areas, A_i, of DNA coated gold particles into d_p, accounting for Brownian motion in the DMA [Pease et al., 2007] with

$$d_p = \left(\frac{\pi^{1/2}}{6} \sum_{i=1}^{3} A_i^{-1/2} \right)^{-1} . \tag{5}$$

This equation may be applied to a wide variety of biomolecules using coordinates from the protein databank to obtain the projected areas. For example, Pease, et al., used protein databank coordinates to predict the ES-DMA measured size of IgG antibodies and insulin oligomers within 1 nm [Pease et al., 2008, 2010c]. Additionally, this formula may be used to characterize the icosahedral structure of viruses by converting the ES-DMA measured size into the edge-to-edge lengths of regular icosahedra [Pease et al., 2011]. The equation also accommodates more challenging geometries such as nanorods, clusters composed of heterogeneously sized nanospheres, and clusters composed of both nanospheres and nanorods [Pease et al., 2010a, 2010b]. Pease further extended this formulation to determine the selectivity of specific cluster compositions (dimers, trimers, etc.), as will be discussed subsequently [Pease, 2011a].

2. Materials characterized by ES-DMA

A wide variety of materials have been characterized by ES-DMA including gold nanoparticles, nanotubes (e.g. single wall carbon nanotubes), nanorods (e.g. gold nanorods), bionanoparticles (polymers, proteins, viruses, etc.), functionalized nanoparticles (e.g. with DNA), quantum dots, aggregated and conjugated nanoparticles, etc. Here we review the contributions of ES-DMA to the characterization of these materials.

2.1 Gold and metallic nanoparticles

Gold nanoparticles remain among the most extensively analyzed materials by ES-DMA because they are very stable, inert and (by many definitions) biocompatible. Figure 3 shows the size distribution of nominally 15 nm gold nanoparticles and the TEM image of size-separated gold nanoparticles using ES-DMA. The National Institute of Standards and Technology (NIST) and the Nanotechnology Characterization Lab (NCL) issued a Joint Assay Protocol, "PCC-10: Analysis of Gold Nanoparticles by Electrospray Differential Mobility Analysis (ES-DMA)," which details a protocol for size analysis of liquid borne gold nanoparticles via ES-DMA [Pease et al., 2010d]. The use of ES-DMA to measure the size distribution of colloidal nanoparticles was first demonstrated well over a decade ago [Juan & Fernandez de la Mora, 1996]. Since then gold and silver nanoparticles have been analyzed using ES-DMA [Elzey & Grassian, 2010; Lenggoro et al., 2002, 2007] by comparing with colloids of known sizes, such as polystyrene particles. However, polystyrene particles are not monodispersed. In 2007, NIST issued 10 nm, 30 nm, and 60 nm gold nanoparticles as official NIST reference materials, which are very monodispersed and can replace polystyrene particles for DMA calibration. Various techniques including DLS, EM, and SAXS were used to characterize these gold nanoparticles, and the diameter of gold nanoparticles measured by each of the different techniques was in reasonable agreement with ES-DMA (see Section 3). These reference materials were issued at the behest of the National Cancer Institute (NCI) to evaluate and qualify the methodology and instrument performance related to the physical characterization of nanoparticle systems.

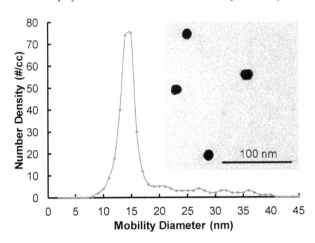

Fig. 3. ES-DMA size distribution and the TEM image (inset) following electrostatic deposition of gold nanoparticles nominally 15 nm in diameter. Number density is the number of particles per cubic centimeter of gas flow through the CPC at a rate of 1.5 L/min.

2.2 Nanotubes and nanorods

Characterization of non-spherical particles is of immense interest, as the shape of a particle along with its size greatly influences particle properties (optical, mechanical, etc.). Though several techniques are available to quantify the length distributions of nanorods and nanotubes, ES-DMA remains competitive (see Section 3 for a detailed comparison). Several authors have reported the characterization of rod- or tube-like particles, such as, gold nanorods and multi-walled carbon nanotubes [Baron et al., 1994; Chen et al., 1993; Deye et al., 1999; Kim & Zachariah, 2005, 2006, 2007a; Moisala et al., 2005; Song et al., 2005]. Song, et al., used a shape factor analysis to convert the average mobility diameter into the average length of gold nanorods. Kim, et al., used Dahneke's theory to convert ES-DMA mobility size distributions into length distributions [Kim et al., 2007b]. Their adaptation includes the orientation of the nanowire as a key factor. Below 70 nm Brownian motion randomizes the orientation of the particles, whereas longer particles align with the electric field in DMA. Pease, et al., used ES-DMA to characterize rapidly the length distribution of single-walled carbon nanotubes from liquid suspensions [Pease et al., 2009b]. Their model, also based on Dahneke's theory, converts the mobility diameter distribution to a length distribution but also accounts for thin salt layers present on the nanotubes that form during electrospray. They found d_{nt}, the diameter of an individual nanotube, and d_s, the diameter of a spherical salt particle, to be key parameters in their model. Figure 4 shows the conversion of mobility size into length distribution based on nanotube diameter and salt layer thickness. Figure 4 also suggests that neglecting to correct for the salt layer overestimates the nanotube length by 42 nm to 56 nm, whereas a small increase in d_{nt} (from 1.4 nm to 2 nm) for a 20 nm mobility diameter results in a decrease in carbon nanotube length of 71 nm [Pease et al., 2009b].

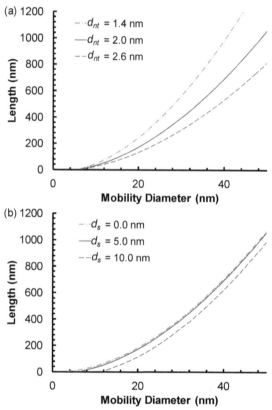

Fig. 4. Length of carbon nanotubes versus mobility diameter as a function of nanotube diameter, d_{nt}, and salt particle size, d_s, for (a) d_{nt} values of 1.4 nm (dash dot), 2.0 nm (solid), and 2.6 nm (long dash) with d_s = 5.0 nm; (b) for d_s = 0.0 nm (dash dot), 5.0 nm (solid), and 10.0 nm (long dash) with d_{nt} = 2.0 nm. Reprinted with permission from [Pease et al., 2009b]. Copyright (2009) Wiley-Interscience.

2.3 Bionanoparticles

Biological systems are of immense interest and ES-DMA is perfectly suited to analyze their soft components. Several biological particles such as viruses, proteins, and protein polymers have soft structures for which ES-DMA is sufficiently gentle to preserve their structure.

2.3.1 Viral nanoparticles

Despite the importance of viral structure, very few methods quantify or validate it. ES-DMA has been demonstrated to be gentle for both enveloped viruses and protein complexes [Wick et al., 2005]. For instance, Wick, et al., used ES-DMA to measure the enveloped alpha virus to be 70 nm ± 3 nm in good agreement with that reported in the structural databases of viruses (~70 nm), indicating electrospray and neutralizer to be sufficiently gentle to preserve the lipid envelop despite shear forces present in the Taylor cone at exit of the electrospray capillary. Furthermore, Hogan, et al., and Thomas, et al., have shown that icosahedral viruses remain infectious following electrospray [Hogan et al., 2006; Thomas et al., 2004].

Pease, et al., demonstrated that the ES-DMA technique can be used to quantify the dimensions of icosahedral viruses [Pease et al., 2011]. A recent review highlights the use of ES-DMA to analyze virus particles [Pease, 2012]. Previous ES-DMA studies of viruses report only the mobility diameter, neglecting the inherent geometry of the virus [Bacher et al., 2001; Hogan et al., 2006; Lute et al., 2008; Thomas et al., 2004]. This left the connection between mobility and actual dimensions of the virus unclear and poorly defined. Pease, et al., converted the mobility diameter, d_p, into the icosahedral geometry expected of *Tectiviridae* viruses using the projected area formulation introduced previously [Pease et al., 2010a]. Figure 5 shows the three orthogonal projections for icosahedra and also gives the corresponding projected areas, where a represents the length of an edge between neighboring vertices.

Fig. 5. Three orthogonal projections and projected areas for an icosahedron. Here, $A_1 = 5a^2/(2(5-\sqrt{5}))^{1/2}$, $A_2 = (2+\sqrt{5})a^2/2$, $A_3 = a^2((25+11\sqrt{5})/10)^{1/2}$. Reprinted with permission from [Pease et al., 2011]. Copyright (2011) American Chemical Society.

Figure 6 shows serial ES-DMA size distributions of a PR772 virus sample tracking the temporal disintegration of the capsid. The primary capsid peak in the size distribution was identified at 61.4 nm, allowing determination of the symmetry and the number of major capsid proteins per capsid. Capsomers and capsomer assemblies were identified by estimating their mobility diameter via Eq. 5 using coordinates from the protein data bank [Pease et al., 2010a]. Figure 6 also provides insight into the mechanism of degradation. Initial degradation products take the form of mostly individual capsomers though some larger assemblies (peak at 18.6 nm) are observed. Continued loss of smaller pieces leaves partially degraded capsids (peak at 51.6 nm). The small but nonzero number density between 20 and 40 nm suggests a continuum of degradation products. However, only after many weeks do the individual capsomers begin to breakdown as seen by the appearance of peaks <7 nm at t = 20 weeks.

2.3.2 Polymeric nanoparticles

Nanoparticles hold potential for a variety of biomedical applications including targeted gene and drug delivery. However, most nanoparticles with refined size (*i.e.*, where the ratio of standard deviation in the diameter to the mean diameter, also called coefficient of variation, is <0.15) are metallic with potential *in vivo* toxicity issues (e.g. quantum dots, silver particles, *etc.*) [Elzey & Grassian, 2010; Pease, 2011a]. Anumolu, et al., fabricated nanoparticles using ES-DMA from recombinant silk elastin-like protein polymers (SELPs) by encapsulating multiple polymer strands in evaporating electrospray droplets. SELPs were selected because they showed promise in clinical trials for localized gene delivery (i.e. by direct injection into tumor containing tissue) [Gustafson & Ghandehari, 2010].

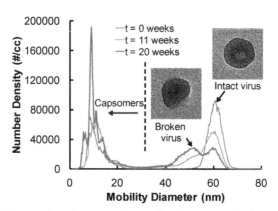

Fig. 6. Serial ES-DMA size distributions of a single PR772 sample showing the disintegration of the capsid (61.4 nm) at t = 0 weeks (short dash), 11 weeks (solid), and 20 weeks (long dash) after storage at room temperature. The number density was corrected with a modified Boltzmann distribution [Wiedensohler, 1988] to reconstruct the full size distribution. Insets are TEM images of representative capsids electrostatically collected at (a) 61.4 nm and (b) 50.0 nm without fixing or stain. Reprinted with permission from (Pease et al., 2011). Copyright (2011) American Chemical Society.

A key feature of their work is the use of the DMA to purify the particles. Figure 7a shows that prior to separation in the DMA, the particles are heterogeneous in size, but after separation the distributions narrow dramatically. Indeed, Figure 7b shows two histograms of SELP particles nominally 24.0 nm and 36.0 nm in diameter, each assembled from nearly 200 nanoparticle TEM images. Statistical compilation shows the standard deviation on the diameter of these purified particles to be 1.2 nm and 1.4 nm for the two sizes, respectively, leading to coefficients of variation of <5% [Anumolu et al., 2011]. This manufacturing precision meets or exceeds that of metallic nanoparticles and rivals that of biologically assembled particles such as viruses [Cole et al., 2009; Lute et al., 2008; Pease, 2011a; Pease et al., 2007]. These results provide the first compelling evidence that ES-DMA can both *generate* and *purify* polymeric nanoparticles with high dimensional uniformity without the addition of hazardous solvents.

These highly uniform nanoparticles may be developed into carriers of therapeutic agents [Anumolu et al., 2011]. Simply including the therapeutic agent in the polymer solution to be electrosprayed, leads to incorporation within the nanoparticle. For example, SELP-815K was mixed with plasmid DNA and fluorescein isothiocyanate (FITC). In both cases new peaks arise 7-8 nm from the primary peak and the distribution of all particles is wider, confirming incorporation of these model agents of gene and drug delivery.

Combining precise control over nanoparticle size with precise control over polymer structure enabled by recombinant techniques presents a unique opportunity to precisely tune the payload and rate of release of the therapeutic agents as well as their biological fate.

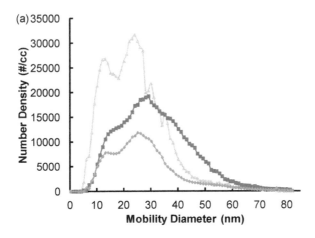

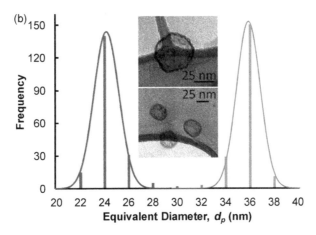

Fig. 7. (a) Size distributions of nanoparticles fabricated from polymers SELP-815K (▲), SELP-415K (■), and SELP-47K (♦) at a polymer weight fractions and buffer concentrations of 0.00133 and 2 mM, respectively. The designation 815K indicates that the strands are assembled from multiple consecutive repeats of 8 silk units, 15 elastin units, and one lysine modified elastin. (b) Histograms representing the diameter of SELP-815K nanoparticles as determined from TEM following electrostatic deposition of nominally 24.0 nm and 36.0 nm. The mean and standard deviation of the size distribution of these particles are 24.2 ± 1.2 nm and 35.8 ± 1.4 nm, respectively. The insets show micrographs of SELP-415K nanoparticles electrostatically collected on TEM grids. Reprinted with permission from [Anumolu et al., 2011]. Copyright (2011) American Chemical Society.

2.4 Surface-functionalized nanoparticles

Determining the surface density of ligands attached to nanoparticle surfaces is a challenging problem in nanoscience and nanotechnology and a major barrier to commercial development. Pease, et al., used ES-DMA to determine the surface density of thiol terminated single-stranded DNA (ssDNA) tethered to 20 nm gold nanoparticles. Comparing the diameter of coated and bare particles (Figure 8a) shows that ES-DMA measured sizes are sensitive to *both* hard and soft components of complex nanoparticles. Pease, et al., investigated the dependence of the coating thickness on the number of deoxythymine (dT) nucleotides or bases, N_b, within a ssDNA strand. The dependence of the coating thickness on N_b is related to the spatial configuration of the bases within the strand in the dry state. If the strands pack together tightly in a brush structure, similar to alkanethiol self-assembled monolayers, the coating thickness should scale linearly on the length of the ssDNA backbone [Tsai et al., 2008]. However, if packing allows for sufficient space between the strands, the bases may adopt a random coil configuration to maximize entropy (appropriate for dried strands) such that the coating thickness is proportional to the linear end-to-end distance to the ½ power. Other exponents are available for hydrated or collapsed configurations due to interactions with solvent, if present. Figure 8b shows that the data follow square root curve fits, indicating that the strands adopt a random coil configuration on the nanoparticle surface [Adamuți-Trache et al., 1996; Netz & Andelman, 2003; Russel et al., 1989]. Knowing the configuration of the strands enables estimation of the surface density because the drag force experienced by the coated particle in the DMA depends on the diameter of the particle, the projected area of the coiled strands, and the number of those strands on the surface. The "lumpy sphere" model combines these variables by approximating each strand as a hemispherical cap, enabling direct determination of the surface density [Mansfield, 2007]. The reported densities are in reasonably agreement with those for "brushes" prepared under similar conditions [Demers et al., 2000; Liao & Roberts, 2006; Petrovykh et al., 2006; Xu & Craig, 2007]. Tsai, et al., used a similar strategy to determine the size, thermal stability, and surface density of alkanethiol self-assembled monolayers (SAMs) on gold nanoparticles. They measured the coating thickness and binding energy of SAMs with excellent precision [Tsai et al., 2008]. These results indicate the potential of ES-DMA to quantify the surface density, configuration and binding energy of biological molecules and organic coatings on nanoparticles.

2.5 Quantum dots and nanoparticle conjugates

Quantum dots (QDs) are semiconducting nanoparticles or nanocrystals that exhibit quantum confinement and are useful in nanophotonics, advanced lighting and displays, as the next generation of photovoltaics (i.e. solar cells), and as dye replacements for molecular biology. Incident photons elevate electrons from the valence to the conduction band, leaving an excited electron-hole pair called an exciton. When the exciton is confined within a nanoparticle that is smaller than its Bohr radius, quantum confinement leads to increased separation between the valance and conduction bands with corresponding increases in the band gap and energy of the photon emitted upon recombination. Therefore, QD size is an essential feature because small changes in QD diameter lead to large changes in the energy of the emitted photon.

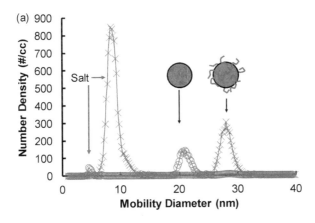

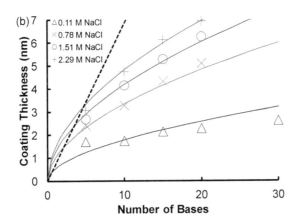

Fig. 8. (a) Size distributions of nominally 20 nm Au nanoparticles, one bare (O) and the other coated (x) with tethered single stranded DNA $(dT)_{20}$-SH. The difference between the two particle size distributions determines the apparent coating thickness. (b) Apparent coating thickness, H, versus number of dT nucleotides per strand, N_b, for a variety of salt concentrations, n_s. The dashed and solid lines respectively represent fits for a contour length model for fully stretched out DNA ($H \sim N_b$) versus that of a square root dependence ($H \sim N_b^{1/2}$) characteristic of strands coiled into low-grafting density layers (see text above). Reprinted with permission from [Pease et al., 2007]. Copyright (2011) American Chemical Society.

Figure 9a shows a typical ES-DMA size distribution of carboxylic acid terminated QDs, with peaks at 15.0 and 18.0 nm representing individual QDs and dimeric QD clusters. This size varies from that anticipated from optical measurements between 5 nm and 7 nm for several reasons. First, these rod-shaped particles (see inset of Figure 9a) are coated with 40 kDa polyethylene glycol (PEG) and salt layers; the equivalent external size of an individual quantum dot without salt and polymer coatings is ~12.4 nm. Second, optical measurements (e.g., Figure 9b-d) reflect the size of the core for QDs with type I band alignment rather than the external size measured by ES-DMA. These differences indicate that optical, TEM, and ES-DMA measurements are mutually complementary for core-shell QDs: optical measurements ascertain the core, TEM gives the combined core and shell measurements, and ES-DMA provides core, shell and coating thicknesses.

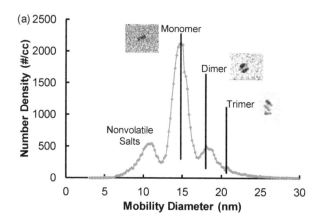

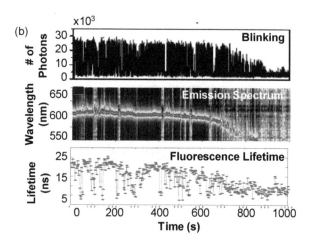

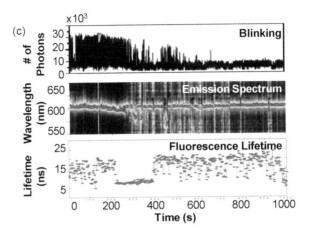

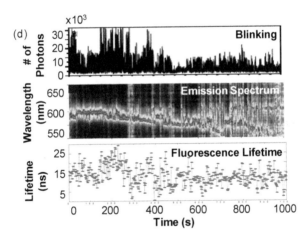

Fig. 9. (a) Raw ES-DMA size distribution of quantum dots at 0.4 µmol/L ($2.4 \cdot 10^{14}$ particles/mL) in 11 mmol/L acetic acid with insets representing TEM images of monomer, dimer and trimer. The first peak at ~11 nm is due to nonvolatile salts present in the original quantum dot solution. The intensity, spectrum, and lifetime of QD (b) monomeric, (c) dimeric, and (d) trimeric clusters

The ability to distinguish the soft and hard components indicates the potential to determine macromolecular conjugation. For example, Figure 10a compares QDs having a ZnS shell coated with mercaptoethanol (ME), ssDNA and double stranded DNA (dsDNA). ES-DMA reported an increase in the mobility diameter of the QDs complexes in the order of $d_{dsDNA} > d_{ssDNA} > d_{ME}$. This order is expected because mercaptoethanol is a shorter molecule than ssDNA, and dsDNA (persistence length ~50 nm) has a more extended conformation than ssDNA (persistence length ~1-3 nm). The extended conformation provides additional drag to the nanoparticle complex resulting in the larger apparent size measured. Similar detection of macromolecular binding is seen in Figure 10b where dextran coated particles are conjugated with the lectins concanavalin (ConA) and wheat germ agglutinin (WGA). The conjugation clearly increased the apparent size of the nanoparticle complex.

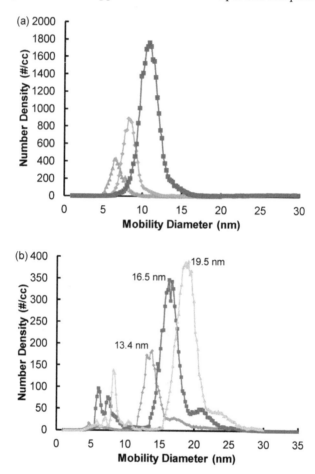

Fig. 10. ES-DMA size distributions of (a) QDs coated with mercaptoethanol (▲), ssDNA (♦), and dsDNA (■) and (b) dextran decorated bare nanoparticles bare (♦) or conjugated with WGA (■) or ConA (▲).

A further highlight of the use of ES-DMA to evaluate macromolecular conjugation is the effort of Yim, et al., to determine the number of QDs attached to genetically modified lambda phage [Yim et al., 2009]. Their biotinylated lambda phage present a rapid means of detecting harmful E. Coli bacteria because bacteriophages replicate much more rapidly than bacteria. Streptavidin coated QDs then bind to the biotinylated phage to provide colorimetric detection. The amount of bacteria is proportional to the amount of phage produced, and the signal from the assay is proportional to the average number of QDs bound to each phage. By comparing the conjugate size in ES-DMA size distributions with that of individual QDs and the lambda phage, the number of QDs attached to the bacteriophage was determined using Eq. 5 by modeling the QD-phage conjugates as clusters of spheres (i.e., the phage) and rods (the streptaviding coated QDs) [Pease, 2011a]. More recently Tsai, et al., studied viral neutralization by quantifying antibody-virus binding [Tsai et al., 2011]. They compared the size of bare MS2 virus particles to those conjugated with anti-MS2 antibodies and reported an increase in the apparent size. They found that antibodies were responsible for formation of multiple virus complexes by acting as connectors between two virus particles (no dimerization was observed with the initial virus sample), indicating that ES-DMA is able to quantify viral aggregation as a form of virus neutralization. They then used their model to determine the stoichiometric ratio of bound antibodies to virus particles to find that infectivity strongly depends on binding stoichiometry not the type of antibody. Monoclonal and polyclonal antibodies both fell on a single infectivity versus stoichiometry curve.

2.6 Nanoparticles aggregates – Packing structure and kinetics

The development of structured materials for nanotechnology and nanobiotechnology requires readily available, rapid analytical techniques to determine the composition of nanostructured particles and clusters. ES-DMA is uniquely qualified to determine the packing and aerodynamic size of colloidal clusters in the nanometer range. For instance, aggregation of metallic nanoparticles may be detected via coupled plasmon resonances using UV-vis, however, this approach does not provide direct information on aggregate size and structure [Weisbecker et al., 1996]. Figure 11a shows an ES-DMA size distribution of multimers of gold nanoparticles with each multimer (e.g. dimers or trimers) appearing as a distinct peak separated by 2-3 nm. These multimers were prepared at the limit of colloidal stability by adding ammonium acetate buffer to a solution of gold nanoparticles to suppress electrostatic screening forces between like charged particles. Figure 11b shows how the ammonium acetate concentration induces aggregation in solution. Tsai, et al., followed the temporal kinetics of the nanoparticle aggregation in the liquid-phase using ES-DMA, as shown in Figure 11c [Tsai et al., 2008]. The monomer concentration decreases with the appearance of each higher order cluster (dimer, trimer, etc.) as anticipated from simple Brownian flocculation/aggregation. By fitting the data to a first principles kinetic model of the aggregation process, the surface charge on particles was extracted from the kinetic rate constants and found to be in good agreement with values reported in the literature (see Figure 11d).

ES-DMA can also be used to monitor protein aggregation. For example, Pease, et al., demonstrated that protein aggregates can be individually resolved and determined an equilibrium constant for the formation of dimers. This equilibrium constant was evaluated based on the number density, which can be obtained from the ES-DMA distribution [Pease et al., 2008]. Pease, et al., also demonstrated that ES-DMA can be used to characterize the size and packing structure of small clusters of identical nanoparticles in colloidal suspension [Pease et al., 2010a]. Essential to their analysis was the projected area model (Eq. 5) to calculate the cluster size, distinguishing collinear from close packed sphere arrangements without resorting to TEM. Several others used ES-DMA to analyze aggregates. For example, Lall, et al., used ES-DMA to determine the ultrafine aggregate surface area and volume distribution of silver aggregates [Lall et al., 2006a, 2006b]. This model is particularly apt for very large aggregates comprising >10 particles/aggregate. Agglomeration of silver nanoparticles was also reported by Elzey & Grassian using ES-DMA [Elzey & Grassian, 2010]. They investigated agglomeration as a function of pH, which determines the dissolution of silver ions in aqueous environments.

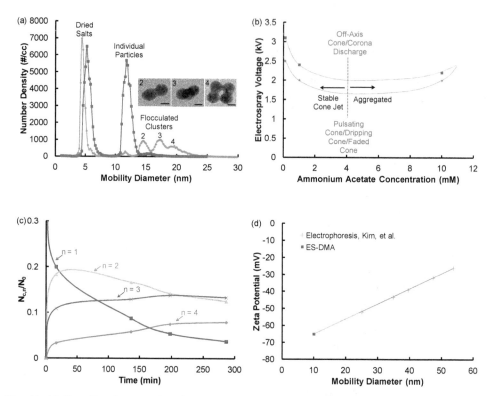

Fig. 11. (a) Size distribution of multimers of gold nanoparticles. The inset shows the TEM images of dimers, trimers, etc. (b) Onset of gold nanoparticle aggregation as a function of ammonium acetate buffer concentration. (c) Normalized concentration of aggregates vs reaction time at an ammonium acetate concentration of 7.89 mmol/L. (d) Relation between surface potential and particle diameter for gold nanoparticles as measured by ES-DMA (■) and reported in the literature (+). The absolute value from ES-DMA was 65 mV. Panel c reprinted with permission from [Tsai et al., 2008]. Copyright (2008) American Chemical Society.

2.7 Optimizing cluster production and purity

Though most ES-DMA research centers on analytical scale characterization, this technique also holds potential as a preparative scale purification tool. For example, ES-DMA separates two particle clusters (dimers) from three particle clusters (trimers), and Kang, et al., have shown that cluster composition may affect cluster photonic properties such as the wavelength evolution, intensity, blinking and radiative lifetime [Kang et al., 2012] (see Figures 9b, 9c, and 9d). To enhance the yield and purity (i.e., how many nominally three-particle clusters actually contain three particles), Pease modeled the ES-DMA technique as a function of nanoparticle properties (size, concentration, etc.) using Monte Carlo simulations [Pease, 2011a]. Figure 12a shows that tuning the initial solution concentration of nanoparticles optimizes the yield of each kind of cluster. The concentration of dimers and subsequently trimers begins to increase as the fraction of monomers declines. Figure 12a also shows that at a concentration of ~$7.0 \cdot 10^{14}$ particles/mL, only 5% of the particles/strands in the size distribution are not monomers providing a sensitive metric for the onset of aggregation. Around 10^{15} particles/mL the concentrations of monomer, dimer, trimer, and higher order clusters all approach parity.

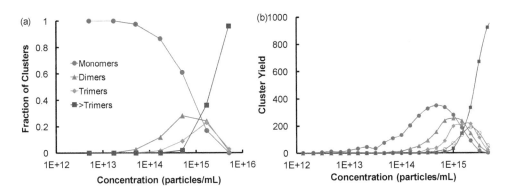

Fig. 12. (a) Monte Carlo simulation using 1000 droplets each exactly 150 nm in diameter for solution concentrations ranging from 10^{12} to 10^{16} particles/mL. (b) The yield for monomers (filled circles), dimers (filled triangles), trimers (filled diamonds), tetramers (open circles), and clusters composed of five or more particles, as a function of initial solution concentration.

To determine the cluster purity, the size distribution of heterogeneously sized particles must be calculated. There are two options for determining the average mobility diameter, d_{ave}. The first option as outlined in Pease, et al., calculates the three projected areas, A_i, orthogonal to the cluster's principle axes (see Figure 13) as given in Eqs. 5 and 6 [Pease et al., 2007]. The area of each projection may be determining from the center-to-center distances, h_i, between the circles in that projection as given by

$$A_1 = \pi/4(\, d_1^2 + d_2^2 + d_3^2 \,)$$
$$A_2 = \pi/4(\, d_1^2 + d_3^2 \,)\text{-}A_{ov}(h_1)$$
$$A_3 = A_1\text{-}A_{ov}(h_2)\text{-}A_{ov}(h_3), \tag{6}$$

where d_1, d_2, and d_3 are the diameters of 3 clusters and $d_1 \geq d_2 \geq d_3$ [Pease, 2011a].

The second option is to employ the proportionality constants determined previously by Pease, et al., for homogeneous clusters [Pease et al., 2010a]. They show that $d_{ave} = kd_p$, where k is the proportionality with values of 1.526, 1.724, and 1.875 for close packed trimers, tetramers, and pentamers, respectively. For heterogeneously sized particles, $d_{ave} = \sum_{i=1}^{n} d_i / n$. Overlaps in the size distribution (see Figure 12b) increase the probability that particles of neighboring size will contaminate each other, decreasing the purity of selected clusters. Pease found that and when the coefficient of variation is <13%, this overlap in the size distributions is not significant. In more recent work, Li, et al., also investigated the formation of clusters during the electrospray process [Li et al., 2011]. They developed a statistical model to determine the extent of ES droplet induced aggregation and how it alters the results.

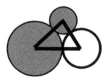 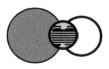

Fig. 13. Depiction of the three projected areas, A_i, used to determine the mobility of a close packed trimeric cluster. Crosshatching in the second and third geometries represent overlap area, A_{ov}.

3. Comparison of ES-DMA to similar techniques

Each of the above examples highlights the potential of ES-DMA to analyze nanoparticles. The remainder of this section compares this powerful technique to generally available classes of techniques commonly used to characterize nanomaterials such as electron microscopy and light scattering.

3.1 Electron microscopy

EM techniques such as transmission electron microscopy (TEM) and scanning electron microscopy (SEM) are direct methods to characterize nanoparticles with subnanometer resolution. EM methods provide compelling images that provide unbiased insight into the *actual* geometry and structure of electro-opaque particles, especially for the hard portions of bionanoparticles as described above. Crystallinity of the particles can also be investigated by generating diffraction patterns using EM. Despite providing very high resolution and beautiful images, EM methods remain expensive to install and operate. To rigorously determine the average nanoparticle size, several hundred to thousands of nanoparticles must be measured [Tsai et al., 2008a, 2008b], which is time consuming (of the order of days to weeks) without automation that is not widely available. For less homogeneous systems or soft materials, multimodal distributions and electron beam damage present further challenges to rigorous TEM measurement. Pease, et al., compared TEM with ES-DMA for quantitative characterization of virus particles and concluded that ES-DMA has a greater rapidity (millions of particles per hr compared to thousands of particles per hour) and statistical significance than TEM [Pease et al., 2009a, 2010b, 2011].

3.2 Light scattering techniques

Light scattering techniques such as multiangle light scattering (MALS) and dynamic light scattering (DLS) measure the intensity fluctuations of scattered light in solution. As light is focused using a laser on the particles moving due to Brownian motion, measured photon intensity fluctuates, from which the z-average hydrodynamic diameter may be estimated. The z-average molecular weight typically exceeds the mass average molecular weight for the most common distributions. In contrast, ES-DMA measures the number average diameter. Otherwise, light scattering and ES-DMA techniques are similar in that they are rapid, inexpensive, and produce statistical significant averages. Though the suspending medium of the sample influences the DLS measurements, it may be preferable for dirty samples minimizing the preparation time required for sample clean up. When characterizing structures with high aspect ratios (e.g., carbon nanotubes), MALS and DLS produce ensemble-average lengths of broad distributions using root-mean square length metrics assuming the nanotubes to have monomodal length distributions [Bauer et al., 2008; Chun et al., 2008; Hinds, 1999]. In contrast ES-DMA measures the size of each nanotube individually to assemble the full multimodal length distributions [Pease et al., 2009b]. Also, DLS does not directly correlate the size or number of individual particles with the size of an aggregate and cannot resolve individual aggregate concentrations within multimodal distributions. Clearly distinguishing clusters differing by only a single particle remains challenging, especially for particles less than 30 nm [Nguyent & Flagan, 1991; Nie et al., 2006].

3.3 Statistical comparison with different techniques

Perhaps the most comprehensive comparison of the primary techniques available to analyze nanoparticle size was performed at NIST on gold nanoparticles. Table 1 shows the statistical data comparison of mean diameters of gold nanoparticles using at least 7 different techniques. Though all the values are in reasonable agreement, some uncertainty is to be expected even from well calibrated systems. Several other authors have compared techniques including MALS, DLS, AFFFF, EM, and ES-DMA using organic (e.g., viruses and virus like particles) and inorganic (carbon nanotubes) nanomaterials [Lute et al., 2008; Pease et al., 2008, 2009a, 2009b]. For example, Lute, et al., and Pease, et al., both found ES-DMA measurements of viruses and virus like particles to be smaller than those of AFFFF-MALS. Reasons for the difference include the following. ES-DMA measures the external size of the virus particle, whereas AFFFF-MALS reports the radius of gyration that depends on the mass distribution of the particle. Also, ES-DMA measures the dry particle size in contrast to the light scattering techniques and AFFFF that measure particles in solution [Pease et al., 2008]. Organic particles are more likely to experience favorable enthaplic interactions in water rich environments that increase their hydrodynamic radius. In dry environments TEM and ES-DMA report similar averages for sufficiently large sample sizes [Lute et al., 2008; Pease, 2012]. Pease, et al., also report analysis of the length distribution of carbon nanotubes using several techniques including AFM, MALS, DLS, and FFF. Here hydrophilic interactions exert little to no influence on the size of the particle and each of these techniques reports similar results [Pease et al., 2009b]. Most recently, Kapelios, et al., characterized the size and molecular mass of proteins and complexes, comparing ES-DMA with light scattering techniques, multi-angle laser light scattering (MALLS) and quasi-elastic light scattering (QELS) [Kapelios et al., 2011]. Similar to Pease, et al., they found that ES-DMA measurements are smaller than those of MALLS and QELS, as expected for proteinacious materials.

Technique	Analyte form	Size of nominally 10 nm particles (nm)	Size of nominally 30 nm particles (nm)	Size of nominally 60 nm particles (nm)
ES-DMA	Dry, aerosol	11.3±0.1	28.4±1.1	56.3±1.5
TEM	Dry, on substrate	8.9±0.1	27.6±2.1	56.0±0.5
DLS (a)	Diluted liquid suspension	13.5±0.1	28.6±0.9	56.6±1.4
DLS (b)	Diluted liquid suspension	13.5±0.1	26.5±3.6	55.3±8.3
AFM	Dry, on substrate	8.5±0.3	24.9±1.1	55.4±0.3
SEM	Dry, on substrate	9.9±0.1	26.9±0.1	54.9±0.4
SAXS	Native liquid suspension	9.1±1.8	24.9±1.2	53.2±5.3

Table 1. Mean diameter and expanded uncertainty average particle diameter of gold nanoparticles [adapted from, NIST certificate, 2007]. DLS results: (a) 173° scattering angle, (b) 90° scattering angle. The uncertainties represent repeatability not the width of the particle size distributions.

4. Future of ES-DMA

In summary, ES-DMA is a powerful tool to characterize and fabricate nanoparticles. Several nanoparticle systems have been characterized using ES-DMA, including quantum dots, linear polymers, and complex colloidal nanomaterials. However, a major growth area for this technique is likely to be in the analysis of soft complex nanomaterials. For example. ES-DMA has significant potential to characterize biochemical reactions including those that assay for nano-therapeutics and bio-pharmaceuticals. Indeed, real-time detection of biological reactions is one of the most promising emerging applications of this instrument. ES-DMA is well suited for this future given its dynamic operating range from 150 Da up to 10^{11} Da. The lower limit of detection is still under refinement with significant improvements underway using high precision home-built DMAs that employ funnel shaped flow injectors to delay the onset of the turbulence. These systems boast sub-angtrom precision on the molecule's mobility diameter [Eichler et al., 1998; Fernandez de la Mora, 2011; Hollertz et al., 2011]. ES-DMA is also well suited to evaluate the role of aggregation, which may be particularly advantageous for detecting protein aggregates in biomanufacturing environments. However, to realize its full potential, further exploration and excellent coordination between biologists, chemists and engineers is required to expand the boundaries of ES-DMA.

5. Acknowledgements

We express appreciation for helpful conversations with De-Hao Tsai, Michael R. Zachariah and Michael J. Tarlov. We sincerely appreciate Hedieh Saffari, Joshua Gustafson, and Hamid Ghandehari for their help with biological particle systems. We thank Silvia De Paoli Lacerda, HyeongGon Kang, Peter Yim, Jeremy I. Feldblyum, Jeeseong Hwang, Matthew L. Clarke, and Phillip Deshong for the nanoparticle samples and conjugate data.

6. References

Adamuti-Trache, M.; McMullen, W. E. & Douglas, J. F. (1996). Segmental Concentration Profiles of End-Tethered Polymers with Excluded-Volume and Surface Interactions. *J. Chem. Phys.* Vol. 105, No. 11 (September 1996), pp. 4798-4811.

Allen, M. D. & Raabe, O. G. (1985). Slip Correction Measurements of Spherical Solid Aerosol Particles in an Improved Millikan Apparatus. *Aerosol Sci. Technol..* Vol. 4, No. 3 (1985), pp. 269-286.

Anumolu, R.; Gustafson, J. A.; Magda, J. J.; Cappello, J.; Ghandehari, H. & Pease III, L. F. (2011) Fabrication of Highly Uniform Nanoparticles from Recombinant Silk-Elastinlike Protein Polymers for Therapeutic Agent Delivery. *ACS Nano.* Vol. 5, No. 7, (July 2011), pp. 5374-5382.

Attri, A. K. & Minton, A. P. (2005). New Methods for Measuring Macromolecular Interactions in Solution via Static Light Scattering: Basic Methodology, and Application to Non-Associating and Self-Associating Proteins. *Anal. Biochem.* Vol. 337, No. 1 (February 2005), pp. 103–110.

Bacher, G.; Szymanski, W. W.; Kaufman, S. L.; Zollner, P.; Blaas, D. & Allmaier, G. (2001). Charge-Reduced Nano Electrospray Ionization Combined with Differential Mobility Analysis of Peptides, Proteins, Glycoproteins, Noncovalent Protein Complexes and Viruses. *J. Mass Spectrom.* Vol. 36, No. 9, (September 2001), pp. 1038-1052.

Baron, P.A.; Deye, G. J. & Fernback, J. (1994). Length Separation of Fibers. *Aerosol Sci. Technol.* Vol. 21, No. 2, (1994), pp. 179–192.

Barry, C. R.; Steward, M. G.; Lwin, N. Z. & Jacobs, H. O. (2003). Printing Nanoparticles from the Liquid and Gas Phases Using Nanoxerography. *Nanotechnology.* Vol. 14, No. 10, (August 2003), pp. 1057-1063.

Barry, C. R.; Lwin, N. Z.; Zheng, W. & Jacobs, H. O. (2003). Printing Nanoparticle Building Blocks from the Gas Phase Using Nanoxerography. *Appl. Phys. Lett.* Vol. 83, No. 26, (October 2003), pp. 5527-5529.

Bauer, B. J.; Fagan, J. A.; Hobbie, E. K.; Chun, J. & Bajpai, V. (2008). Chromatographic Fractionation of SWNT/DNA Dispersions with On-Line Multi-Angle Light Scattering. *J. Phys. Chem. C.* Vol. 112, No. 6, (January 2008), pp. 1842-1850.

Benson, S. D.; Bamford, J. K.; Bamford, D. H. & Burnett., R. M. (2002). The X-ray Crystal Structure of P3, the Major Coat Protein of the Lipid-Containing Bacteriophage PRD1, at 1.65 A° Resolution. *Acta Crystallogr., Sect D: Biol. Crystallogr.* Vol. 58, No. 1, (January 2002), pp. 39–59.

Bondos, S. E. (2006). Methods for Measuring Protein Aggregation. *Curr. Anal. Chem.* Vol. 2, No. 2, (April 2006), pp. 157–170, ISSN: 1573-4110.

Casper, L. D. & Klug, A. (1962) Physical Principles in the Construction of Regular Viruses. *Cold Spring Harbor Symp. Quant. Biol.* Vol. 27, (1962), pp. 1–24.

Chen, B. T.; Yeh, H. C. & Hobbs, C. H. (1993). Size Classification of Carbon Fiber Aerosols. *Aerosol Sci. Technol.*, Vol. 19, No. 2, (August 1993), pp. 109–120.

Cole, K. D.; Pease III, L. F.; Tsai, D. H.; Singh, T.; Lute, S.; Brorson, K. A. & Wang, L. (2009). Particle Concentration Measurement of Virus Samples Using Electrospray Differential Mobility Analysis and Quantitative Amino Acid Analysis. *J. Chromatogr. A.* Vol. 1216, No. 30, (July 2009), pp. 5715–5722.

Chang, T. H.; Cheng, C. P. & Yeh, C. T. (1992). Deuterium Nuclear Magnetic Resonance Characterization of Particle Size Effect in Supported Rhodium Catalysts. *J. Catal.* Vol. 138, No. 2, (December 1992), pp. 457-462.

Chun, J.; Fagan, J. A.; Hobbie, E. K. & Bauer, B. (2008). Size Separation of Single-Wall Carbon Nanotubes by Flow-Field Flow Fractionation. *J. Anal. Chem.* Vol. 80, No. 7, (February 2008), pp. 2514-2523.

Colter, J. S. & Ellem, K. A. O. (1961). Structure of Viruses. *Annu. Rev. Microbiol.* Vol. 15, (October 1961), pp. 219–244.

Dai, Q.; Liu, X.; Coutts, J.; Austin, L. & Huo, Q. (2008). A One-Step Highly Sensitive Method for DNA Detection Using Dynamic Light Scattering. *J. Am. Chem. Soc.* Vol. 130, No. 26, (June 2008), pp. 8138-8139.

Deye, G. J.; Gao, P.; Baron, P. A. & Fernback, J. (1999). Performance Evaluation of a Fiber Length Classifier. *Aerosol Sci. Technol.* Vol. 30, No. 5, (1999), pp. 420–437.

Dixkens, J. & Fissan, H. (1999). Development of an Electrostatic Precipitator for Off-Line Particle Analysis. *Aerosol Sci. Technol.* Vol. 30, No. 5, (1999), pp. 438-453.

Eichler, T.; Juan, L. D. & Fernandez de la Mora, J. (1998). Improvement of the Resolution of TSI's 3071 DMA via Redesigned Sheath air and aerosol inlets. *Aerosol Sci. Technol.* Vol. 29, No. 1, (1998), pp. 39–49.

Elzey, S. & Grassian, V. H. (2010). Agglomeration, Isolation and Dissolution of Commercially Manufactured Silver Nanoparticles in Aqueous Environments. *J. Nanopart. Res.* Vol. 12, No. 5, (2010), pp. 1945-1958.

Epstein, P. S. (1924). On the Resistance Experienced by Spheres in their Motion through Gases. *Phys. Rev.* Vol. 23, No. 6, (1924), pp. 710-733.

Fernandez de la Mora, J. (2011). Electrical Classification and Condensation Detection of Sub-3 nm Aerosols, In: *Aerosol Measurement: Principles, Techniques, and Applications* (3rd edition), Baron, P.A.; Kulkarni, P. & Willeke, K., pp. 607-721, John Wiley & Sons, Inc., ISBN 978-0-470-38741-2, New York, USA.

Fissan, H. J.; Helsper, C. & Thielen, H. J. (1983). Determination of Particle-Size Distributions by Means of an Electrostatic Classifier. *J. Aerosol Sci.* Vol. 14, No. 3, (1983), pp. 354-357.

Fissan, H.; Kennedy, M. K.; Krinke, T. J. & Kruis, F. E. (2003). Nanoparticles from the Gas phase as Building Blocks for Electrical Devices. *J. Nanopart Res.* Vol. 5, No. 3-4, (August 2003), pp. 299-310.

Fuchs, N. A. (1963). On the Stationary Charge Distribution on Aerosol Particles in a Bipolar Ionic Atmosphere. *Pure Appl. Geophys.* Vol. 56, No. 1, (1963), pp. 185-193.

Hinds, W. C. (January 1999). *Aerosol Technology: Properties, Behavior, and Measurement of Airborne Particles* (2nd edition), John-Wiley & Sons, Inc., ISBN 978-0-471-19410-1, New York, USA.

Hogan, C. J.; Kettleson, E. M.; Ramaswami, B.; Chen, D. R. & Biswas, P. (2006). Charge Reduced Electrospray Size Spectrometry of Mega- and Gigadalton Complexes: Whole Viruses and Virus Fragments. *Anal. Chem.* Vol. 78, No. 3, (February 2006), pp. 844–852.

Hogan, C. J.; Yun, K. M.; Chen, D. R.; Lenggoro, I. W.; Biswas, P.; Okuyama, K. (2007) Controlled Size Polymer Particle Production via Electrohydrodynamic Atomization. *Colloids Surf., A: Physicochem. Eng. Aspects.* Vol. 311, No. 1-3, (December 2007), pp. 67-76.

Hogan, C. J. & Biswas, P. (2008). Monte Carlo Simulation of Macromolecular Ionization by Nanoelectrospray. *J. Am. Soc. Mass Spectrom.* Vol. 19, No. 8, (August 2008), pp. 1098–1107.

Hollertz, M.; Elliott, J. T.; Lewis, J.; Mansfield, E. R.; Whetten, W. D.; Knotts IV, T. A.; Tarlov, M. J.; Zachariah, M. R. & Pease III, L. F. (2011). Structural Differentiation of Oxytocin in Cyclical and Linear Conformations Using High Resolution Differential Mobility Analysis. *Anal. Chem.* (2011), submitted.

Hung, L. H. & Lee, A. P. (2007). Microfluidic Devices for the Synthesis of Nanoparticles and Biomaterials. *J. Med. Biol. Eng.* Vol. 27, No. 1, (2007), pp. 1-6.

Jacobs, H. O.; Campbell, S. A. & Steward, M. G. (2002). Approaching Nanoxerography: The Use of Electrostatic Forces to Position Nanoparticles with 100 nm Scale Resolution. *Adv. Mater.* 2002, Vol. 14, No. 21, (November 2002), pp. 1553-1557.

Johnson, B. K. & Prud'homme, R. K. (2003). Mechanism for Rapid Self-Assembly of Block Copolymer Nanoparticles. *Phys. Rev. Lett.* Vol. 91, No. 11, (2003), pp. 118302-1-118302-4.

Juan, L. D. & Fernandez de la Mora, J. (March 1996). On-line Sizing of Colloidal Nanoparticles via Electrospray and Aerosol Techniques, In: *Nanotechnology: Molecularly Designed Materials*, (Vol. 622), Chow, G. M. & Gonsalves, K. E., pp. 20-41, ACS Publications, ISBN13: 9780841233928

Kang, H.; Clarke, M. L.; Pease III, L. F.; DePaoli Lacaerda, S. H.; Karim, A. & Hwang, J. (2011). Analysis of the Optical Properties of Clustered Colloidal Quantum Dots by the Chi-Square Distribution of the Fluorescence Lifetime Curves, *ASC Nano* (2011) manuscript.

Kang, H.; Clarke, M. L.; DePaoli Lacaerda, S. H.; Pease III, L. F. & Hwang, J. (2012). Multimodal Optical Studies of Single and Clustered Colloidal Quantum Dots Towards the Long-term Performance Evaluation of Optical Properties of Quantum Dot-included Molecular Imaging Phantoms. *Biomed. Opt. Express* (2012), submitted.

Kapellios, S.; Karamanou, M. F.; Sardis, M.; Aivaliotis, A.; Economou, S. & Pergantis, A. (2011). Using Nanoelectrospray Ion Mobility Spectrometry (GEMMA) to Determine the Size and Relative Molecular Mass of Proteins and Protein Assemblies: A

Comparison with MALLS and QELS. *Anal. Bioanal. Chem.* Vol. 399, No. 7, (March 2011), pp. 2421-2433.

Kim, S. H. & Zachariah, M. R. (2005). In-Flight Size Classification of Carbon Nanotubes by Gas Phase Electrophoresis. *Nanotechnology.* Vol. 16, No. 10, (August 2005), pp. 2149–2152.

Kim, S. H. & Zachariah, M. R. (2006). In-Flight Kinetic Measurements of the Aerosol Growth of Carbon Nanotubes by Electrical Mobility Classification. *J. Phys. Chem. B.* Vol. 110, No. 10, (February 2006), pp. 4555–4562.

Kim, S. H. & Zachariah, M. R. (2007a). Gas-Phase Growth of Diameter-Controlled Carbon Nanotubes. *Mater. Lett.* Vol. 61, No. 10, (April 2007), pp. 2079-2083.

Kim, S. H. & Zachariah, M. R. (2007b). Understanding Ion-Mobility and Transport Properties of Aerosol Nanowires. *J. Aerosol Sci.* Vol. 38, No. 8, (August 2007), pp. 823-842.

Kim, S. K.; Ha, T. & Schermann, J. P. (2010). Advances in Mass Spectrometry for Biological Science. *Phys. Chem. Chem. Phys.* Vol. 12, No. 41, (November 2010), pp. 13366–13367.

Knapman, T. W.; Morton, V. L.; Stonehouse, N. J.; Stockley, P. G. & Ashcroft., A. E. (2010). Determining the Topology of Virus Assembly Intermediates Using Ion Mobility Spectrometry-Mass Spectrometry. *Rapid Commun. Mass Spectrom.* Vol. 24, No. 20, (October 2010), pp. 3033–42.

Knutson, E. O. & Whitby, K. T. (1975). Aerosol Classification by Electric Mobility: Apparatus, Theory, and Applications. *J. Aerosol Sci.* Vol. 6, No.6, (November 1975), pp. 443-451.

Krinke, T. J.; Deppert, K.; Magnusson, M. H. & Fissan, H. (2002). Nanostructured Deposition of Nanoparticles from the Gas Phase. *Part. Part. Syst. Char.* Vol. 19, No. 5, (November 2002), pp. 321-326.

Krinke, T. J.; Fissan, H. & Deppert, K. (2003). Deposition of Aerosol Nanoparticles on Flat Substrate Surfaces. *Phase Transitions.* Vol. 76, No. 4-5, (May 2003), pp. 333-345.

Kuzmanovic, D. A.; Elashvili, I.; O'Connell, C. & Krueger, S. (2008). A Novel Application of Small-Angle Scattering Techniques: Quality Assurance Testing of Virus Quantification Technology. *Radiat. Phys. Chem.* Vol. 77, No. 3, (March 2008), pp. 215–224.

Lall, A. A. & Friedlander, S. K. (2006). On-Line Measurement of Ultrafine Aggregate Surface Area and Volume Distributions by Electrical Mobility Analysis: I. Theoretical Analysis. *J. Aerosol Sci.* Vol. 37, No. 3, (March 2006), pp. 260-271.

Lall, A. A. & Friedlander, S. K. (2006). On-Line Measurement of Ultrafine Aggregate Surface Area and Volume Distributions by Electrical Mobility Analysis: II. Comparison of Measurements and Theory. *J. Aerosol Sci.* Vol. 37, No. 3, (March 2006), pp. 272-282.

Lenggoro, I. W.; Xia, B. & Okuyama, K. (2002). Sizing of Colloidal Nanoparticles by Electrospray and Differential Mobility Analyzer Methods. *Langmuir.* Vol. 18, No. 12, (May 2002), pp. 4584–4591.

Lenggoro, I. W.; Widiyandari, H.; Hogan Jr., C. J.; Biswas, P. & Okuyama, K. (2007). Colloidal Nanoparticle Analysis by Nanoelectrospray Size Spectrometry with a Heated Flow. *Anal. Chim. Acta.* Vol. 585, No. 2, (March 2007), pp. 193-201.

Li, M.; Guha, S.; Zangmeister, R. A.; Tarlov, M. J. & Zachariah, M. R. (2011). Quantification and Compensation of Nonspecific Analyte Aggregation in Electrospray Sampling. *Aerosol Sci. Technol.*. Vol. 45, No.7, (March 2011), pp. 849-860.

Loo, J. A.; Berhane, B.; Kaddis, C. S.; Wooding, K. M.; Xie, Y. M.; Kaufman, S. L. & Chernushevich, I. V. (2005). Electrospray Ionization Mass Spectrometry and Ion Mobility Analysis of the 20S Proteasome Complex. *J. Am. Soc. Mass Spectrom.* Vol. 16, No. 7, (July 2005), pp. 998-1008.

Lute, S.; Riordan, W.; Pease III, L. F.; Tsai, D. H.; Levy, R.; Haque, M.; Martin, J.; Moroe, I.; Sato, T.; Morgan, M.; Krishnan, M.; Campbell, J.; Genest, P.; Dolan, S.; Tarrach, K.; Meyer, A.; the PDA Virus Filter Task Force; Zachariah, M. R.; Tarlov, M. J.; Etzel, M. & Brorson, K. (2008). A Consensus Rating Method for Small Virus-Retentive Filters. I. Method Development. *PDA J. Pharm. Sci. Technol.* Vol. 62, No. 5, (October 2008), pp. 318–333.

Moisala, A.; Nasibulin, A. G.; Shandakov, S. D.; Jiang, H. & Kauppinen, E. I. (2005). On-line Detection of Single-Walled Carbon Nanotube Formation During Aerosol Synthesis Methods. *Carbon.* Vol. 43, No. 10, (August 2005), pp. 2066-2074.

Netz, R. R. & Andelman, D. (2003). Neutral and Charged Polymers at Interfaces. *Phys. Rep.* Vol. 380, No. 1-2, (June 2003), pp. 1-95.

Nguyent, H. V. & Flagan, R. C. (1991). Particle Formation and Growth in Single-Stage Aerosol Reactors. *Langmuir.* Vol. 7, No. 8, (August 1991), pp. 1807-1814.

Nie, Z.; Xu, S.; Seo, M.; Lewis, P. C. & Kumacheva, E. (2005). Polymer Particles with Various Shapes and Morphologies Produced in Continuous Microfluidic Reactors. *J. Am. Chem. Soc.* Vol. 127, No. 22, (May 2005), pp. 8058-8063.

Nie, Z. X.; Tzeng, Y. K.; Chang, H. C.; Chiu, C. C.; Chang, C. Y.; Chang, C. M. & Tao, M. H. (2006). Microscopy-Based Mass Measurement of a Single Whole Virus in a Cylindrical Ion Trap. *Angew. Chem. Int. Ed. Engl.* 2006, Vol. 45, No. 48, (December 2006), pp. 8131-8134.

NIST Certificate (2007). *Nanoparticle Standards at NIST: Gold Nanoparticle Reference Materials.*http://www.mel.nist.gov/tripdf/NIST/05_Hackley%20TNW%20AuRM %20talk%2002.05.08.VAH_final.pdf (2007).

Park, J. Y. & Phaneuf, R. J. (2003) Investigation of the Direct Electromigration Term for Al Nanodots within the Depletion Zone of a pn Junction. *J. Appl. Phys.* Vol. 94, No. 10, (2003), pp. 6883-6886.

Pease III, L. F.; Tsai, D. H.; Zangmeister, R. A.; Zachariah, M. R. & Tarlov, M. J. (2007). Quantifying the Surface Coverage of Conjugated Molecules on Functionalized Nanoparticles. *J. Phys. Chem. C.* Vol. 111, No. 46, (November 2007), pp. 17155-17157.

Pease III, L.F.; Elliott, J. T.; Tsai, D. H.; Zachariah, M. R. & Tarlov, M. J. (2008) Determination of Protein Aggregation with Differential Mobility Analysis: Application to IgG Antibody. *Biotechnol. Bioeng.* Vol. 101, No. 6, (December 2008), pp. 1214–1222.

Pease III, L. F.; Lipin, D. I.; Tsai, D. H.; Zachariah, M. R.; Lua, L. H. L.; Tarlov, M. J. & Middelberg, A. P. J. (2009a). Quantitative Characterization of Virus-like Particles by Asymmetrical Flow Field Flow Fractionation, Electrospray Differential Mobility Analysis, and Transmission Electron Microscopy. *Biotechnol. Bioeng.* Vol. 102, No. 3, (February 2009), pp. 845-855.

Pease III, L.F.; Tsai, D. H.; Fagan, J. A.; Bauer, B. J.; Zangmeister, R. A.; Tarlov, M. J. & Zachariah, M. R. (2009b). Length Distributions of Single Wall Carbon Nanotubes in Aqueous Suspensions Measured by Electrospray-Differential Mobility Analysis. *Small.* Vol. 5, No. 24, (December 2009), pp. 2894-2901.

Pease III, L. F.; Tsai, D.-H.; Zangmeister, R. A.; Hertz, J. L.; Zachariah, M. R. & Tarlov, M. J. (2010a). Packing and Size Determination of Colloidal Nanoclusters. *Langmuir.* Vol. 26, No. 13, (May 2010), pp. 11384–11390.

Pease III, L. F.; Feldblyum, J. I.; DePaoli Lacerda, S. H.; Liu, Y.; Hight-Walker, A.; Anumolu, R.; Yim, P. B.; Clarke, M. L.; Kang, H. G. & Hwang, J. (2010b). Structural Analysis of Soft Multicomponent Nanoparticle Clusters. *ACS Nano.* Vol. 4, No. 11, (November 2010), pp. 6982-6988.

Pease III, L. F.; Sorci, M.; Guha, S.; Tsai, D. H.; Zachariah, M. R.; Tarlov, M. J. & Belfort, G. (2010c). Probing the Nucleus Model for Oligomer Formation During Insulin Amyloid Fibrillogenesis. *Biophys. J.* Vol. 99, No. 12, (December 2010), pp. 3979-85.

Pease III, L. F.; Tsai, D. H.; Zangmeister, R. A.; Zachariah, M. R. & Tarlov, M. J. (2010d). Analysis of Gold Nanoparticles by Electrospray Differential Mobility Analysis (ES-DMA). NIST-NCL Joint Assay Protocol, PCC-10, Version 1.1.

Pease III, L. F.; Tsai, D. H.; Brorson, K. A.; Guha, S.; Zachariah, M. R.; & Tarlov, M. J. (2011). Physical Characterization of Viral Ultra Structure, Stability, and Integrity. *Anal. Chem.* Vol. 83, No. 5, (February 2011), pp. 1753-1759.

Pease III, L. F. (2011a). Optimizing the Yield and Selectivity of High Purity Nanoparticle Clusters. *J. Nanopart. Res.* Vol. 13, No. 5, (2011), 2157-2172.

Pease III, L.F. (2012). Physical Analysis of Virus Particles. *Trends Biotechnol.* (2012), in press, DOI: 10.1016/j.tibtech.2011.11.004.

Rader, D. J. (1990). Momentum Slip Correction Factor for Small Particles in Nine Common Gases. *J. Aerosol Sci.* Vol. 21, No. 2, (1990), pp. 161-168.

Regenmortel, M. H. V.; Fauquet, C. M. & Bishop, D. H. L. (2000). Virus Taxonomy: Classification and Nomenclature of Viruses: Seventh Report of the International Committee on Taxonomy of Viruses. (October 2000), Academic Press, San Diego, USA.

Russel, W. B.; Saville, D. A. & Schowalter, W. R. (1989). *Colloidal Dispersions* (1st edition). Cambridge University Press, ISBN 9780521426008, New York, USA.

Saffari, H.; Malugin, A.; Ghandehari, H. & Pease III, L.F. (2012). Electrostatic Deposition of Nanoparticles into Live Cell Culture Using an Electrospray Differential Mobility Analyzer (ES-DMA). *J. Aerosol Sci.* (2012), in press.

Saucy, D. A.; Ude, S.; Lenggoro, I. W. & Fernandez de la Mora, J. (2004). Mass Analysis of Water-Soluble Polymers by Mobility Measurement of Charge-Reduced Ions Generated by Electrosprays. *Anal. Chem.* Vol. 76, No. 4, (January 2004), pp. 1045-1053.

Shoemaker, G. K.; Duijn, E. V.; Crawford, S. E.; Uetrecht, C.; Baclayon, M.; Roos, W. H.; Wuite, G. J. L.; Estes, M. K.; Prasad, B. V. & Heck, A. J. R. (2010). Norwalk Virus Assembly and Stability Monitored by Mass Spectrometry. *J. Mol. Cell Proteomics.* Vol. 9, No. 8, (April 2010), pp. 1742–51.

Siuzdak, G.; Bothner, B.; Yeager, M.; Brugidou, C.; Fauquet, C. M.; Hoey, K. & Change, C.-M. (1996). Mass Spectrometry and Viral Analysis. *Chem. Biol.* Vol. 3, No. 1, (January 1996), pp. 45–48.

Song, D. K.; Lenggoro, I. W.; Hayashi, Y.; Okuyama, K. & Kim, S. S. (2005). Changes in the Shape and Mobility of Colloidal Gold Nanorods with Electrospray and Differential Mobility Analyzer Methods. *Langmuir.* Vol. 21, No. 23, (November 2005), pp. 10375-10382.

Swann, M. J.; Peel, L. L.; Carrington, S. & Freeman, N. (2004). Dual-Polarization Interferometry: An Analytical Technique to Measure Changes in Protein Structure in Real Time, to Determine the Stoichiometry of Binding Events, and to Differentiate between Specific and Nonspecific Interactions. *J. Anal. Biochem.* Vol. 329, No. 2, (June 2004), pp. 190-198.

Thomas, J. J.; Bothner, B.; Traina, J.; Benner, W. H. & Siuzdak, G. (2004). Electrospray Ion Mobility Spectrometry of Intact Viruses. *Spectroscopy.* Vol. 18, No. 1, (January 2004), pp. 31–36.

Tsai, D. H.; Kim, S. H.; Corrigan, T. D.; Phaneuf, R. J. & Zachariah, M. R. (2005). Electrostatic-Directed Deposition of Nanoparticles on a Field Generating Substrate. *Nanotechnology.* Vol. 16, No. 9, (July 2005), pp. 1856-1862.

Tsai, D. H.; Zangmeister, R. A.; Pease III, L. F.; Zachariah, M. R. & Tarlov, M. J. (2008). Gas-phase Ion-mobility Characterization of SAM functionalized Au Nanoparticles. *Langmuir.* Vol. 24, No. 16, (July 2008), pp. 8483-8490.

Tsai, D. H.; Pease III, L. F.; Zachariah, M. R. & Tarlov, M. J. Aggregation Kinetics of Colloidal Particles Measured by Gas-phase Differential Mobility Analysis. *Langmuir.* Vol. 25, No. 1, (December 2008), pp. 140-146.

Tsai, D.H.; Lipin, D. I.; Guha, S.; Feldblyum, J. I.; Cole, K. D.; Brorson, K. A.; Zachariah, M. R.; Tarlov, M. J.; Middelberg, A. P. J. & Pease III, L. F. (2011). Process Analytical Technology for Recombinant Pandemic Flu Vaccines: Viral Ultrastructure, Aggregation, and Binding. *AIChE Annual Meeting.* CD-ROM. (2011), New York.

Umbach, P.; Georgalis, Y. & Saenger, W. (1998). Time-Resolved Small-Angle Static Light Scattering on Lysozyme During Nucleation and Growth. *J. Am. Chem. Soc.* Vol. 120, No. 10, (March 1998), pp. 2382-2390.

Wang, W. (2005). Protein Aggregation and its Inhibition in Biopharmaceutics. *Int. J. Pharm.* Vol. 289, No. 1–2, (January 2005), pp. 1–30.

Weisbecker, C. S.; Merritt, M. V. & Whitesides, G. M. (1996). Molecular Self-Assembly of Aliphatic Thiols on Gold Colloids. *Langmuir.* Vol. 12, No. 16, (August 1996), pp. 3763–3772.

Wick, C. H.; McCubbin, P. E. & Birenzvige, A. (2006). Detection and Identification of Viruses using the Integrated Virus Detection System (IVDS); ECBC Technical Report: 2006, ECBC-TR-463.

Wiedensohler, A. (1988). An Approximation of the Bipolar Charge-Distribution for Particles in the Sub-Micron Size Range. *J. Aerosol Sci.* Vol. 19, No. 3, (1988), pp. 387–389.

Yim, P. B.; Clarke, M. L.; McKinstry, M.; De Paoli Lacerda, S. H.; Pease III, L. F.; Dobrovolskaia, M. A.; Kang, H. G.; Read, T. D.; Sozhamannan, S. & Hwang, J. (2009). Quantitative Characterization of Quantum Dot-Labeled Lambda Phage for

Escherichia Coli Detection. *Biotechnol. Bioeng.* Vol. 104, No. 6, (December 2009), pp. 1059-1067.

Zhu, Z.; Anacker, J. L.; Ji, S.; Hoye, T. R.; Macosko, C. W. & Prud'homme, R. K. (2007). Formation of Block Copolymer-Protected Nanoparticles via Reactive Impingement Mixing. *Langmuir.* Vol. 23, No. 21, (September 2007), pp. 10499-10504.

4

Water-Soluble Single-Nano Carbon Particles: Fullerenol and Its Derivatives

Ken Kokubo

Division of Applied Chemistry, Graduate School of Engineering, Osaka University
Japan

1. Introduction

Since its discovery in 1985, fullerene has been extensively investigated as a unique, "dissolvable," and "modifiable" nanocarbon material. The most representative fullerene, C_{60}, is a perfectly spherical molecule with a diameter of ca. 1 nm (0.7 nm when the distance between the furthest C–C bond is considered, and 1 nm when the π-orbitals are included). It has many interesting electronic and biological properties owing to its spherical π-conjugation. While fullerenes satisfactorily dissolve in aromatic solvents such as toluene and o-dichlorobenzene as well as in carbon disulfide, they dissolve poorly in most common solvents such as hexane, chloroform, diethyl ether, ethyl acetate, tetrahydrofuran (THF), acetone, acetonitrile, ethanol, and even in benzene. This limitation has been one of the important issues hindering their practical application, especially in the field of life sciences.

Although the single-crystal X-ray structural analysis of C_{60} has been successful, it is generally difficult to grow the crystal of fullerene derivatives. Such a poor crystallinity is because of the lack of molecular orientation and intermolecular interaction that are crucial for determining the molecular alignment in a crystal. The lack of molecular orientation and interaction is attributed to the unidirectional spherical shape of these derivatives, which is different from the shapes of other organic molecules such as cubic- or plate-shaped ones. For the same reason, the solid form of C_{60} is known to easily afford its nanoparticles with a top-down approach, in which the solid is reduced to small particles (as small as 20 nm) by applying mechanical forces; such particles can even be obtained by hand-grinding (Deguchi et al., 2006; Deguchi et al., 2010). These small nanocarbon particles, so-called nC_{60} (Oberdörster, 2004; Brant et al., 2005) or nano-C_{60} (Fortner et al., 2005), can be dispersed even in neutral water, and they remain dispersed for a long time, especially in the presence of humic acid (Chen & Elimelech, 2007; Isaacson & Bouchard, 2010). On the other hand, the aggregate of fullerene can be easily obtained by a bottom-up approach in many solvents such as benzene (Ying et al., 1994), benzonitrile (Nath et al., 1998), N-methylpyrrolidone (Yevlampieva et al., 2002; Kyzyma et al., 2010), and o-dichlorobenzene (Gun'kin & Loginova, 2006). The nanoparticle formation and aggregation behavior are among the outstanding features of fullerene in terms of both its practical application and the safe use of nanomaterials.

Polyhydroxylated fullerenes, so-called fullerenols or fullerols, are a class of fullerenes that has many hydroxyl groups, formed by the chemical modification of covalent C–O bonds, on their spherical surfaces. The chemical formula $C_{60}(OH)_n$ represents an average structure that consists of a mixture of fullerenols having different number of hydroxyl groups, each with its own regioisomer. The solubility of a fullerene molecule is dependent on the number of hydroxyl groups introduced; i.e., the low-degree hydroxylated fullerenols $C_{60}(OH)_{10-12}$ (Chiang et al., 1994) can dissolve in some polar solvents, e.g., THF, dimethylformamide (DMF), and dimethyl sulfoxide (DMSO), and the medium-degree fullerenols $C_{60}(OH)_{16}$ (Wang et al., 2005) and $C_{60}(OH)_{20-24}$ (Li et al., 1993) are reported to dissolve even in water. However, these later fullerenols may be contaminated with Na salt because of the reagents used in synthesis, resulting in compositions with the formula $Na^+{}_n[C_{60}Ox(OH)_y]^{n-}$ and exhibiting high water solubility in spite of their small number of hydroxyl groups (Husebo et al., 2004). In contrast, the high-degree hydroxylated fullerenols $C_{60}(OH)_{36}$ and $C_{60}(OH)_{44}$, which are synthesized without using any Na salt, are completely water soluble by as much as 17 and 65 mg/mL, respectively (Kokubo et al., 2008; Kokubo et al., 2010). Particle size analysis revealed that the high-degree fullerenols exhibited high dispersion properties at a molecular level (ca. 1 nm, which is as large as the molecular diameter). The behavior of water-soluble carbon particles in the single-nano region (1–10 nm) is less well understood in terms of their chemical and physical properties.

This chapter focuses on the methods of synthesizing fullerenols, provides examples of applications, and describes the particle-size measurements of the high-degree fullerenols as water-soluble single-nano carbon particles.

2. Synthesis of fullerenols

2.1 Hydroxylation of fullerene

Various types of fullerenols having different number of hydroxyl groups have been synthesized so far because of their promising water solubility and the expected bioactivities. In general, the structure of fullerenols is qualitatively identified by infrared spectroscopy as having a characteristic broad $vO-H$ band, along with three broad peaks assigned for $vC = C$, $\delta_s C-O-H$, and $vC-O$. The number of hydroxyl groups introduced is quantitatively determined by either elemental analysis or X-ray photoelectron spectroscopy (XPS). Such a quantitative analysis is founded on the hypothesis that the addend of a fullerene is composed of only hydroxyl group.

One of the most well-known fullerenols, $C_{60}(OH)_{12}$, was synthesized by L. Y. Chiang using oleum (H_2SO_4 SO_3), followed by the hydrolysis of the intermediate cyclosulfated fullerene (Scheme 1a) (Chiang et al., 1994). The compound is soluble in an alkaline solution and some polar solvents such as THF and DMSO, but it is not soluble in neutral water because it has few hydroxyl groups. L. Y. Chiang also studied its antioxidant activity toward a superoxide, a reactive oxygen species (Chiang et al., 1995). Some other related methods of accessing the lower-degree fullerenols have been reported (Zhang et al., 2010).

The most frequently used method for synthesizing the medium-degree fullerenol $C_{60}(OH)_{22-26}$ is the one reported by J. Li et al. (Scheme 1b) (Li et al., 1993). This method employs sodium hydroxide (NaOH) as a hydroxylation reagent, and molecular oxygen is required to neutralize the intermediate fullerenyl anion, which is formed by the attack of ⁻

OH to C_{60}, in order to induce the successive attack of -OH. The fullerenols $C_{60}(OH)_{22-26}$, as well as $C_{60}(OH)_{16}$ synthesized by a similar method using NaOH and H_2O_2 (Scheme 1c) (Wang et al., 2005), exhibits sufficient solubility in neutral water. The fullerenol has also been known to have many bioactivities, including antioxidant activity similar to pristine C_{60} (Bosi et al., 2003; Bakry et al., 2007; Partha & Conyers, 2009). However, the practical use of these types of fullerenols might be restricted because of their unfavorable contamination by Na^+ ions, which are inevitably introduced during treatment with NaOH, and the purification is rather difficult except when done by repeated gel column chromatography (Husebo et al., 2004). The relatively higher water solubility than that expected, given the number of hydroxyl groups, is reasonably explained by the corrected chemical formula $Na^+_n[C_{60}O_x(OH)_y]^{n-}$. Such a Na salt form is attributed to the weak acidity of the phenolic O–H of the low-degree fullerenols, in contrast to the alcoholic O–H of the high-degree fullerenols. Therefore, there is a demand for the development of new, facile, and scalable methods for synthesizing highly water-soluble and pure fullerenols.

$$C_{60} \xrightarrow[\text{2. } H_2O]{\text{1. } H_2SO_4 \cdot SO_3} C_{60}(OH)_{12} \qquad \text{(1-a)}$$

$$C_{60} \xrightarrow[\text{r.t. grinding}]{\text{NaOH, 30\% } H_2O_2} C_{60}(OH)_{16} \qquad \text{(1-b)}$$

$$C_{60} \xrightarrow[\text{benzene, r.t.}]{\text{NaOH, TBAH}} C_{60}(OH)_{22-26} \qquad \text{(1-c)}$$

Scheme 1. Synthesis of low- and medium-degree fullerenols

2.2 Highly polyhydroxylated fullerenols

To avoid contamination by Na^+ ions, we examined the use of hydrogen peroxide (H_2O_2) as a hydroxylation reagent instead of NaOH. Thus, we have found a new and facile approach for synthesizing high-degree fullerenols that have high water solubility without using any Na salts (Kokubo et al., 2008). The reddish brown suspension of fullerenol $C_{60}(OH)_{12}$ in 30% aqueous H_2O_2 was stirred vigorously at 60 °C under air until it turned to a transparent yellow solution, which occurred within 2–4 days (Scheme 2a). To this solution, a mixture of 2-propanol, diethyl ether, and hexane as an antisolvent was added to afford a yellowish brown to milky white precipitate of the desirable high-degree fullerenol $C_{60}(OH)_{36} \cdot 8H_2O$. A longer reaction time of up to two weeks gave the similar but more water-soluble fullerenol $C_{60}(OH)_{40} \cdot 9H_2O$.

This new approach using H_2O_2 to synthesize high-degree fullerenols was useful; however, the starting material was limited to $C_{60}(OH)_{12}$ and was not applicable to pristine C_{60}. We then improved the method in order to provide a facile, one-step method for synthesizing fullerenol from pristine C_{60}; we added an NH_3 aqueous solution to the H_2O_2 aqueous solution to give similar water-soluble fullerenols, although they contained some undesirable nitrogen-containing groups, along with hydroxyl groups (Matsubayashi et al., 2009). We further improved the method, synthesizing pure $C_{60}(OH)_{44} \cdot 8H_2O$ fullerenol with no nitrogen in one step from pristine C_{60}. To the best of our knowledge, this fullerenol has the largest number of hydroxyl groups per C_{60} among the fullerenols reported so far; this fullerenol was obtained by a two-phase synthesis in the presence of tetrabutylammonium hydroxide (TBAH) as a phase transfer catalyst (PTC) (Scheme 2b) (Kokubo et al., 2010). The fullerenol exhibits a very high water solubility of up to 64.9 mg/mL.

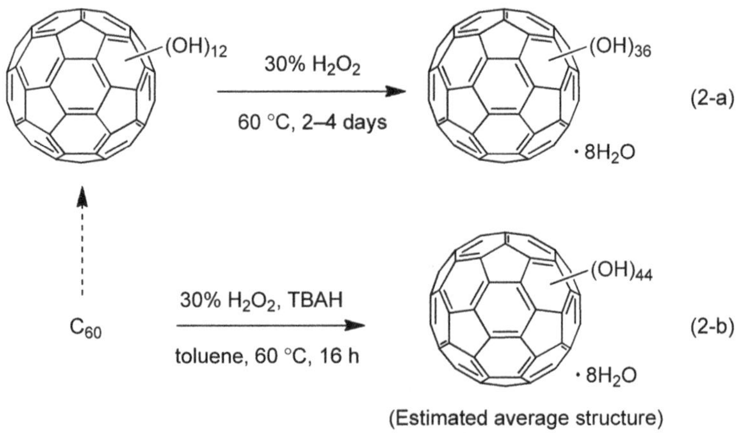

Scheme 2. Synthesis of high-degree fullerenols

2.3 Structural characterization

The structural characterization of the high-degree fullerenols was conducted by infrared spectroscopy (Fig. 1). The spectra of the fullerenols closely resembled each other, although their relative peak intensities differed somewhat, suggesting a difference in the number of introduced hydroxyl groups. Four characteristic broad bands were observed at 1080, 1370, 1620, and 3400 cm^{-1} and were assigned to $\nu C-O$, $\delta_s C-O-H$, $\nu C=C$, and $\nu O-H$, respectively. A small shoulder peak at 1720 cm^{-1} may imply the existence of a carboxylic acid group, O=C-OH, which might have formed by the further oxidation of a hydroxyl group associated with C-C bond cleavage of the fullerene nucleus. However, such partial oxidation must not be crucial because the generally strong C=O absorption is much smaller than the other generally weak or medium C=C or C-O absorptions, which is consistent with the results of the elemental analysis.

The quantitative analysis to determine the number of hydroxyl groups was conducted by elemental analysis. As shown in Table 1, the average structure of the high-degree fullerenol $C_{60}(OH)_{44} \cdot 8H_2O$ would be deduced to be $C_{60}(OH)_{52}$ if just the results of elemental analysis was used. However, the largest reported number of substituents in one C_{60} moiety is $C_{60}F_{48}$ (Tuinman et al., 1992; Troyanov et al., 2010), and thus $C_{60}(OH)_{52}$ is unlikely to be formed due to the enormous strain energy. On the other hand, it is known that the tightly entrapped water molecules, the secondary bound water, in highly hydroxylated fullerenols cannot be dissociated by the usual method of heating the fullerenols to about 120–150 °C. Therefore, water content measurements using thermogravimetric analysis was conducted (Figure 2). With a water content of 9.4 wt%, as shown in Table 1, the average structure of fullerenol was deduced to be $C_{60}(OH)_{44} \cdot 8H_2O$, using both elemental analysis and water content measurements.

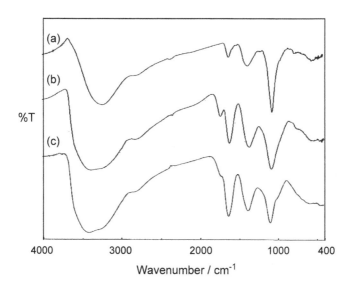

Fig. 1. IR spectra of (a) $C_{60}(OH)_{12}$, (b) $C_{60}(OH)_{36} \bullet 8H_2O$, and (c) purified $C_{60}(OH)_{44} \bullet 8H_2O$.

Average structure	Elemental analysis (%) Found (Calcd)[a]	Water content (wt%)[a,b]	Solubility (mg/mL)[c]
$C_{60}(OH)_{36} \cdot 8H_2O$	C: 48.06, H:3.61 (C: 48.79, H:3.54)	8.9 (9.7)	17.5
$(C_{60}(OH)_{44})$	(C: 49.06, H:3.02)	(0)	
$C_{60}(OH)_{44} \cdot 8H_2O$	C: 44.68, H:3.56 (C: 44.70, H:3.75)	9.4 (8.9)	64.9
$(C_{60}(OH)_{52})$	(C: 44.90, H:3.27)	(0)	

[a]Values in parentheses are calculated data. [b]Determined by TGA. [c]At 25 °C in neutral water (pH = 7).

Table 1. Average structure and water solubility of high-degree fullerenols

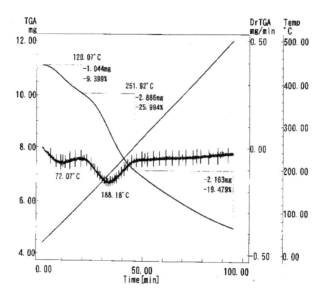

Fig. 2. Thermogravimetric analysis of fullerenol $C_{60}(OH)_{44} \cdot 8H_2O$ under N_2 flow.

2.4 Reaction mechanism

Possible reaction mechanisms for the formation of fullerenols are illustrated in Schemes 3 and 4. For the reaction using NaOH, the attack of -OH to C_{60} followed by the oxidation with molecular oxygen via one electron transfer from the C_{60} anion gives the hydroxylated C_{60}. The successive attack of -OH and repeated oxidation finally gives the medium-degree fullerenol (Scheme 3) (Husebo et al., 2004).

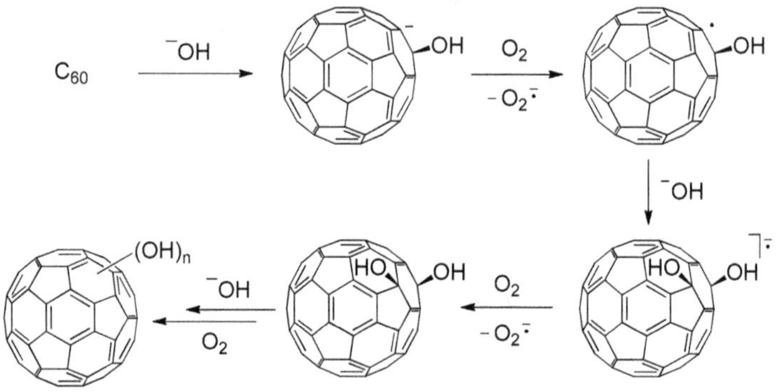

Scheme 3. A possible reaction mechanism for medium-degree fullerenols using the NaOH method

In contrast, for the reaction using H_2O_2, the basic hydroxide ion from TBAH initially induces the hydroperoxide ion -OOH because of the slightly higher acidity of H_2O_2 than that of H_2O (Scheme 4) (Wang et al., 2005; Kokubo et al., 2011). The -OOH thus formed attacks C_{60} to give fullerene oxide $C_{60}O$, followed by the attack of ^-OH and protonation. The epoxidation process may be repeated to give $C_{60}O_2$, $C_{60}O_3$, and so on (Tajima & Takeuchi, 2002), which are more susceptible than $C_{60}O$ to the subsequent nucleophilic attack of ^-OH (or ^-OOH) because of the higher strain. These fullerene oxide intermediates were detected in the reaction mixture by liquid chromatography-mass spectrometry (LC-MS) (APCI; m/z = 736, 752, and 768) and were proven to be the intermediates by their kinetic behavior. The role of the quaternary ammonium salt TBAH is that of the promotion of ^-OOH formation and its transfer from the hydrophilic aqueous phase to the hydrophobic fullerenyl sites in the organic phase as PTC.

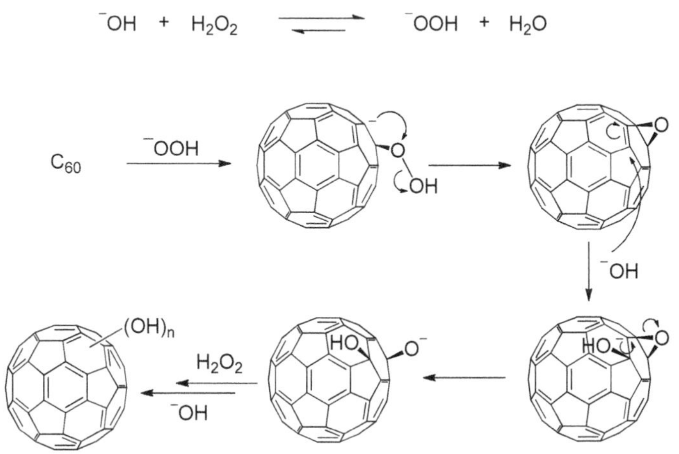

$$^-OH + H_2O_2 \rightleftharpoons {}^-OOH + H_2O$$

Scheme 4. Proposed reaction mechanism for high-degree fullerenols using the H_2O_2 method

3. Measurement of particle size distribution

3.1 Dynamic light scattering method

Although a fullerenol seems to completely dissolve in a solution, it may be aggregated in the nano-size region, as seen for many fullerene derivatives and some fullerenols (Mohan et al., 1998; Husebo et al., 2004; Brant et al., 2007; Chae et al., 2009; Su et al., 2010). The particle size of medium-degree fullerenol $C_{60}(OH)_{24}$ in aqueous solution is reported as between ca. 20 and 450 nm depending on the measurement conditions. Even such a high number of hydroxyl groups results in the formation of aggregation due to the large hydrophobic and $\pi-\pi$ interactions between fullerenyl cores.

In order to investigate the dispersant behavior, the particle size measurement of high-degree fullerenols in the 0.1 wt% aqueous solution was carried out using the common dynamic light scattering (DLS) method (Berne & Pecora, 1976). The narrow particle-size distributions around 1–2 nm in terms of the number of $C_{60}(OH)_{36}$ and $C_{60}(OH)_{44}$ molecules are essentially the same, indicating the highly dispersed nature of the fullerenols at a molecular level (Figure 3) (Kokubo et al., 2008; Kokubo et al., 2011). The average particle size of $C_{60}(OH)_{44}$ was determined to be 1.46 ± 0.38 nm (N = 8). The particle size by intensity was not applicable to the fullerenol solution because of the interference from the intensity of contamination in the range of 60–130 nm.

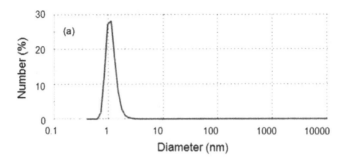

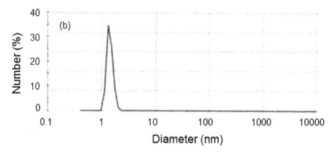

Fig. 3. DLS analysis of (a) $C_{60}(OH)_{36}$ and (b) $C_{60}(OH)_{44}$ in water (0.1 wt%) expressed by size distribution in number.

3.2 Induced grating method

Recently, the induced grating (IG) method was developed to improve the reproducibility of particle size measurements, especially in the single-nano region (Wada et al., 2006). DLS measurements provide the diffusion coefficient D, which is converted to the diameter by monitoring the fluctuations in scattering intensity due to the Brownian motion of particles. However, because the efficiency of DLS is proportional to the sixth power of the particle diameter (Kerker, 1969), the detection sensitivity strongly depends on the particle size and thus the presence of impurities. The IG method also determines the diffusion coefficient D given by the following Einstein–Stokes equation:

$$D = \frac{k_B T}{3\pi\eta d}$$

where k_B is the Boltzmann's constant, T is the temperature in Kelvin, η is the viscosity of the medium, and d is the diameter of the particles. When the radio frequency voltage is turned off, the diffraction light intensity I begins to decrease according to the following equation:

$$I = I_0 exp(-2Dq^2t)$$

where I_0 is the initial intensity and q is the value of 2π divided by the pitch of the grating. However, the measurement also includes an activation procedure induced by dielectrophoresis to form a particle grating (Pohl, 1978). Dielectrophoresis first generates a periodic density modulation of the particles, which are then relaxed to a diffuse state until they reach a steady state. Thus, the diffraction light is less affected by the presence of impurities as compared with DLS.

We therefore also conducted particle size measurements of high-degree fullerenols using the IG method (Kokubo et al., 2011). In the diffusion region, sufficient photointensity was observed for the $C_{60}(OH)_{44}$ aqueous solution (0.1 wt%) and the logarithmic value of the relative photointensity correlated well linearly with the time scale (Fig. 4).

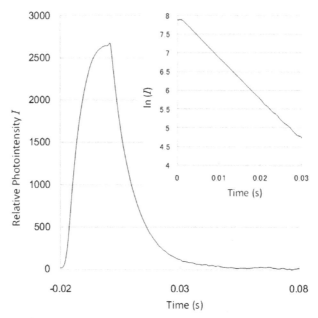

Fig. 4. Time course of relative photointensity I of $C_{60}(OH)_{44}$ aqueous solution (0.1 wt%) measured by the IG method.

The particle size distribution was narrow, in the range of 0.7–1 nm, and the average particle size was determined to be 0.806 ± 0.022 nm ($N = 8$), which was fairly consistent with the DLS results (Fig. 5). Therefore, it was confirmed that the high-degree fullerenols have high dispersion properties in water on a molecular level around their diameter of ca. 1 nm. It is remarkable that the reproducibility of the data measured over eight runs was 10 times higher for the IG method than for the DLS method.

3.3 Other methods

The particle size measurements by the DLS and IG methods were verified by a scanning probe microscope (SPM, Kokubo et al., 2011). We directly measured the particle size of the fullerenol as a function of protrusion height observed on a mica plate on which a highly diluted aqueous solution of fullerenol was applied and dried. As shown in Fig. 6, the protrusions were clearly observed as scattered spots, whereas a mica plate without the fullerenol treatment, used as a control, did not show any spots at all. From the height of eight spots, the average particle size was determined to be 1.03 ± 0.28 nm ($N = 8$).

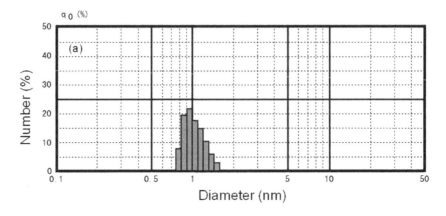

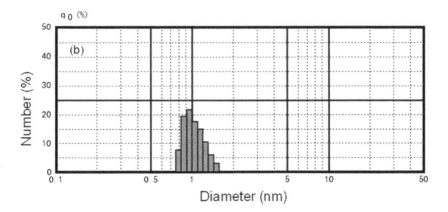

Fig. 5. IG analysis of (a) $C_{60}(OH)_{36}$ and (b) $C_{60}(OH)_{44}$ in water (0.1 wt%) expressed by size distribution in number.

The size distribution of fullerenols has also been investigated using flow field-flow fractionation (FFF) technique (Assemi et al., 2010) and transmission electron microscopy (TEM) (Wang et al., 2010). The FFF is an elution technique that analyzes ensembles of the sample that have a similar property and produces a size distribution rather than an average size. They found that the size of medium-degree fullerenol $C_{60}(OH)_{24}$ nanoparticles was ranging from about 1.8 nm (0.001 M NaCl) to 6.7 nm (0.1 M NaCl). However, this result is in contrast to some DLS data that reports sizes on the order of 100 nanometers for fullerenol nanoparticles. This is because of the fact that the impurities and the large aggregation are separated from the monodispersed fraction eluting from the FFF channel. The TEM observation revealed that the aggregation form of the low-degree fullerenol $C_{60}(OH)_{12}(ONa)_2$ in the solution with a particle size of 50–250 nm.

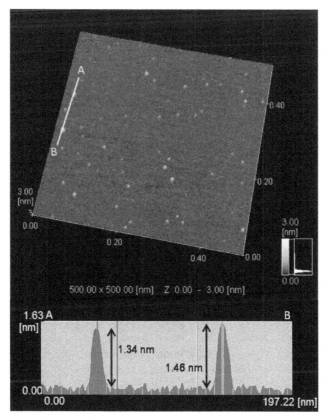

Fig. 6. SPM analysis of $C_{60}(OH)_{36}$ applied as a diluted solution and dried on a mica plate.

4. Application of fullerenols

4.1 Biochemical application

The water-soluble medium-degree fullerenols have been demonstrated to be useful as free-radical scavengers for the absorption of superoxide radicals generated by in vitro xanthine and xanthine oxidase in aqueous solution, suggesting potential use of fullerenols in biochemical or pharmaceutical applications (Chiang et al., 1995). Since then, much research has been devoted to studies on the antioxidant (Dugan et al., 1996; Djordjevic et al., 2004; Bogdanovic et al., 2008), antimicrobial (Aoshima et al., 2009), anti-cancer (Chaudhuri et al., 2009; Krishna et al., 2010), antitumor, and antimetastatic activities of medium-degree fullerenols (Jiao et al., 2010).

Recently, N. Miwa et al. reported the antioxidant activity (Kato et al., 2009) and related bioactivities (Saitoh et al., 2010; Saitoh et al., 2011) of high-degree fullerenols. Some reviews relevant to the biochemical application of fullerenes and fullerenols have also been reported (Nielsen et al., 2008; Rade et al., 2008; Partha & Conyers, 2009). Fullerenes are not considered to have highly significant acute toxicity and genotoxicity, although some toxicological results have also been reported. In contrast, fullerenols are considered to be less toxic than C_{60} due to the introduction of hydrophilic groups that reduce their cytotoxicity.

4.2 Industrial application

Because of a 1 nm grain size, high water dispersibility on a molecular level and metal-free material, the high-degree fullerenol has been proposed as a chemical mechanical polishing (CMP) slurry for use during planarization in the Cu damascene process for the fabrication of next-generation semiconductors. Y. Takaya et al. found that the Cu-surface roughness was improved from 20 to 0.6 nm root mean square (RMS) by using $C_{60}(OH)_{36}$ as functional molecular abrasive grains to achieve better polishing performance than could be achieved using conventional processes (Takaya et al., 2009). The etching ability of $C_{60}(OH)_{36}$ for a Cu surface evaluated in static etch was also found to be high in relation to the achievement of a highly planar surface by polishing experiment. Very recently, further XPS analysis and SEM observation revealed that the chemical effect of fullerenol plays a key role in high polishing performance; i.e., the fullerenol chemically reacted with the copper surface to form a complex brittle layer that was fragile enough to be removed by rubbing with a polishing pad (Takaya et al., 2011).

4.3 Other applications

Other examples of applications using low- to medium-degree fullerenols have been reported, such as polymer-based solar cells (Cao et al., 2001; Rincón et al., 2005), drug delivery and MRI contrast agents using endohedral metallofullerenol (Sitharaman et al., 2004), macromolecular materials and polymer nanocomposites (Goswami et al., 2003; Ouyang et al., 2004), proton conductors (Hinokuma et al., 2001; Maruyama et al., 2002), and electrodeposited films (Wang et al., 2010). Following these applications, water-soluble high-degree fullerenols will open the new pathways for the new fullerenol chemistry.

5. Summary

Fullerenols are one of the most important and promising fullerene derivatives that can be easily synthesized with tunable properties by varying the number of hydroxyl groups introduced. Their water solubility, high dispersing nature as single-nano carbon particles, and varied biochemical properties are extremely attractive from the viewpoint of materials chemistry as well as life science applications. The analytical methods for particle size measurement of single-nano particles will continue to be improved by the further development of these kinds of nanocarbon materials.

6. References

Aoshima, H., Kokubo, K., Shirakawa, S., Ito, M., Yamana, S. & Oshima, T. (2009). Antimicrobial activity of fullerenes and their hydroxylated derivatives, *Biocont. Sci.* 14: 69–72.

Assemi, S., Tadjiki, S., Donose, B. C., Nguyen, A. V. & Millere, J. D. (2010). Aggregation of fullerenol $C_{60}(OH)_{24}$ nanoparticles as revealed using flow field-flow fractionation and atomic force microscopy, *Langmuir* 26: 16063–16070.

Bakry, R., Vallant, R. M., Najam-ul-Haq, M., Rainer, M., Szabo, Z., Huck, C. W. & Bonn, G. K. (2007). Medicinal applications of fullerenes, *Int. J. Nanomed.* 2: 639–649.

Berne, B. J. & Pecora, R. (1976). *Dynamic light scattering with application to chemistry, biology, and physics*, General Publishing Company, Tronto.

Bogdanovic, V., Stankov, K., Icevic, I., Zikic, D., Nikolic, A., Solajic, S., Djordjevic, A. & Bogdanovic, G. (2008). Fullerenol $C_{60}(OH)_{24}$ effects on antioxidative enzymes activity in irradiated human erythroleukemia cell line, *J. Radiat. Res.* 49: 321–327.

Bosi, S., Da Ros, T., Spalluto, G. & Prato, M. (2003). Fullerene derivatives: an attractive tool for biological applications, *Eur. J. Med. Chem.* 38: 913–923.

Brant, J. A., Labille, J., Robichaud, C. O. & Wiesner, M. (2007). Fullerenol cluster formation in aqueous solutions: implications for environmental release, *J. Colloid Interf. Sci.* 314: 281–288.

Brant, J., Lecoanet, H. & Wiesner, M. R. (2005). Aggregation and deposition characteristics of fullerene nanoparticles in aqueous systems, *J. Nanopart. Res.* 7: 545–553.

Cao, T. B., Yang, S. M., Yang, Y. L., Huang, C. H. & Cao, W. X. (2001). Photoelectric conversion property of covalent-attached multilayer self-assembled films fabricated from diazoresin and fullerol, *Langmuir* 17: 6034–6036.

Chae, S.-R., Hotze, E. M. & Wiesner, M. R. (2009). Evaluation of the oxidation of organic compounds by aqueous suspensions of photosensitized hydroxylated-C_{60} fullerene aggregates, *Environ. Sci. Technol.* 43: 6208–6213.

Chaudhuri, P., Paraskar, A., Soni, S., Mashelkar, R. A. & Sengupta, S. (2009). Fullerenol-cytotoxic conjugates for cancer chemotherapy, *ACS Nano* 3: 2505–2514.

Chen, K. L. & Elimelech, M. (2007). Influence of humic acid on the aggregation kinetics of fullerene (C_{60}) nanoparticles in monovalent and divalent electrolyte solutions, *J. Colloid Interf. Sci.* 309: 126–134.

Chiang, L. Y., Wang, L.-Y., Swirczewski, J. W., Soled, S. & Cameron, S. (1994). Efficient synthesis of polyhydroxylated fullerene derivatives via hydrolysis of polycyclosulfated precursors, *J. Org. Chem.* 59: 3960–3968.

Chiang, L. Y., Lu, F.-J. & Lin, J.-T. (1995). Free radical scavenging activity of water-soluble fullerenols, *J. Chem. Soc., Chem. Commun.* 1283–1284.

Deguchi, S., Mukai, S., Tsudome, M. & Horikoshi, K. (2006). Facile generation of fullerene nanoparticles by hand-grinding, *Adv. Mater.* 18: 729–732.

Deguchi, S., Mukai, S., Yamazaki, T., Tsudome, M. & Horikoshi, K. (2010). Nanoparticles of fullerene C_{60} from engineering of antiquity, *J. Phys. Chem. C* 114: 849–856.

Djordjevic, A., Canadanovic-Brunet, J. M., Vojinovic-Miloradov, M. & Bogdanovic, G. (2004). Antioxidant properties and hypothetic radical mechanism of fullerenol $C_{60}(OH)_{24}$, *Oxid. Commun.* 27: 806–812.

Dugan, L. L., Gabrielsen, J. K., Yu, S. P., Lin, T. S. & Choi, D. W. (1996). Buckminsterfullerenol free radical scavengers reduce excitotoxic and apoptotic death of cultured cortical neurons, *Neurobiol. Dis.* 3: 129–135.

Fortner, J. D., Lyon, D. Y., Sayes, C. M., Boyd, A. M., Falkner, J. C., Hotze, E. M., Alemany, L. B., Tao, Y. J., Guo, W., Ausman, K. D., Colvin, V. L. & Hughes, J. B. (2005). C_{60} in water: nanocrystal formation and microbial response, *Environ. Sci. Technol.* 39: 4307–4316.

Goswami, T. H., Nandan, B., Alam, S. & Mathur, G. N. (2003). A selective reaction of polyhydroxy fullerene with cycloaliphatic epoxy resin in designing ether connected epoxy star utilizing fullerene as a molecular core, *Polymer* 44: 3209–3214.

Gun'kin, I. F. & Loginova, N. Y. (2006). Aggregation of fullerene C_{60} in *o*-dichlorobenzene, *Russ. J. General Chem.* 76: 1914–1915.

Hinokuma, K. & Ata, M. (2001). Fullerene proton conductors, *Chem. Phys. Lett.* 341: 442–446.

Husebo, L. O., Sitharaman, B., Furukawa, K., Kato, T. & Wilson, L. J. (2004). Fullerenols revisited as stable radical anions, *J. Am. Chem. Soc.* 126: 12055–12064.

Isaacson, C. W. & Bouchard, D. C. (2010). Effects of humic acid and sunlight on the generation and aggregation state of aqu/C_{60} nanoparticles, *Environ. Sci. Technol.* 44: 8971–8976.

Jiao, F., Liu, Y., Qu, Y., Li, W., Zhou, G., Ge, C., Li, Y., Sun, B. & Chen, C. (2010). Studies on anti-tumor and antimetastatic activities of fullerenol in a mouse breast cancer model, *Carbon* 48: 2231–2243.

Kato, S., Aoshima, H., Saitoh, Y., Miwa, N. (2009). Highly hydroxylated or γ-cyclodextrin-bicapped water-soluble derivative of fullerene: the antioxidant ability assessed by electron spin resonance method and β-carotene bleaching assay, *Bioorg. Med. Chem. Lett.* 19: 5293–5296.

Kerker, K. (1969). *The scattering of light*, Academic, New York, pp. 31–39.

Kokubo, K., Matsubayashi, K., Tategaki, H., Takada, H. & Oshima, T. (2008). Facile synthesis of highly water-soluble fullerenes more than half-covered by hydroxyl groups, *ACS Nano* 2: 327–333.

Kokubo, K., Shirakawa, S., Kobayashi, N., Aoshima, H. & Oshima, T. (2011). Facile and scalable synthesis of a highly hydroxylated water-soluble fullerenol as a single nanoparticle, *Nano Res.* 4: 204–215.

Krishna, V., Singh, A., Sharma, P., Iwakuma, N., Wang, Q., Zhang, Q., Knapik, J., Jiang, H., Grobmyer, S. R., Koopman, B. & Moudgil, B. (2010). Polyhydroxy fullerenes for non-invasive cancer imaging and therapy, *Small* 6: 2236–2241.

Kyzyma, O. A., Korobov, M. V., Avdeev, M. V., Garamus, V. M., Snegir, S. V., Petrenko, V. I., Aksenov, V. L. & Bulavin, L. A. (2010). Aggregate development in C_{60}/N-methyl-2-pyrrolidone solution and its mixture with water as revealed by extraction and mass spectroscopy, *Chem. Phys. Lett.* 493: 103–106.

Li, J., Takeuchi, A., Ozawa, M., Li, X., Saigo, K. & Kitazawa, K. (1993). Fullerol formation catalyzed by quaternary ammonium hydroxides, *J. Chem. Soc., Chem. Commun.* 1784–1785.

Maruyama, R., Shiraishi, M., Hinokuma, K., Yamada, A. & Ata, M. (2002). Electrolysis of water vapor using a fullerene-based electrolyte, *Electrochem. Solid-State Lett.* 5: A74–A76.

Matsubayashi, K., Kokubo, K., Tategaki, H., Kawahama, S. & Oshima, T. (2009). One-step synthesis of water-soluble fullerenols bearing nitrogen-containing substituents, *Fuller. Nanotub. Carbon Nanostruct.* 17: 440–456.

Mohan, H., Palit, D. K., Mittal, J. P., Chiang, L. Y., Asmus, K.-D. & Guldi, D. M. (1998). Excited states and electron transfer reactions of $C_{60}(OH)_{18}$ in aqueous solution, *J. Chem. Soc., Faraday Trans.* 94: 359–363.

Nath, S., Pal, H., Palit, D. K., Sapre, A. V. & Mittal, J. P. (1998). Aggregation of fullerene, C_{60}, in benzonitrile, *J. Phys. Chem. B* 102: 10158–10164.

Nielsen, G. D., Roursgaard, M., Jensen, K. A., Poulsen, S. S. & Larsen, S. T. (2008). In vivo biology and toxicology of fullerenes and their derivatives, *Basic Clin. Pharmacol. Toxicol.* 103: 197–208.

Oberdörster, E. (2004). Manufactured nanomaterials (fullerenes, C_{60}) induce oxidative stress in the brain of juvenile largemouth bass, *Environ. Health. Perspect.* 112: 1058–1062.

Ouyang, J., Zhou, S., Wang, F. & Goh, S. H. (2004). Structures and properties of supramolecular assembled fullerenol/poly(dimethylsiloxane) nanocomposites, *J. Phys. Chem. B* 108: 5937–5943.

Partha, R. & Conyers, J. L. (2009). Biomedical applications of functionalized fullerene-based nanomaterials, *Int. J. Nanomed.* 4: 261–275.

Pohl, H. A. (1978). *Dielectrophoresis*, Cambridge University Press.

Rade, I., Natasa, R., Biljana, G., Aleksandar, D. & Borut, S. (2008). Bioapplication and activity of fullerenol $C_{60}(OH)_{24}$, *Afr. J. Biotechnol.* 7: 4940–4950.

Rincón, M. E., Guirado-López, R. A., Rodríguez-Zavala, J. G. & Arenas-Arrocena, M. C. (2005). Molecular films based on polythiophene and fullerol: theoretical and experimental studies, *Sol. Energy Mater. Sol. Cells* 87: 33–47.

Saitoh, Y., Miyanishi, A., Mizuno, H., Kato, S., Aoshima, H., Kokubo, K. & Miwa, N. (2011). Super-highly hydroxylated fullerene derivative protects human keratinocytes from UV-induced cell injuries together with the decreases in intracellular ROS generation and DNA damages, *J. Photochem. Photobiol. B* 102: 69–76.

Saitoh, Y., Xiao, L., Mizuno, H., Kato, S., Aoshima, H., Taira, H., Kokubo, K. & Miwa, N. (2010). Novel polyhydroxylated fullerene suppresses intracellular oxidative stress together with repression of intracellular lipid accumulation during the differentiation of OP9 preadipocytes into adipocytes, *Free Radic. Res.* 44: 1072–1081.

Sitharaman, B., Bolskar, R. D., Rusakova, I. & Wilson, L. J. (2004). $Gd@C_{60}[(COOH)_2]_{10}$ and $Gd@C_{60}(OH)_x$: nanoscale aggregation studies of two metallofullerene MRI contrast agents in aqueous solution, *Nano Lett.* 4: 2373–2378.

Su, Y., Xu, J., Shen, P., Li, J., Wang, L., Li, Q., Li, W., Xu, G., Fan, C. & Huang, Q. (2010). Cellular uptake and cytotoxic evaluation of fullerenol in different cell lines, *Toxicology* 269: 155–159.

Tajima, Y. & Takeuchi, K. (2002). Discovery of $C_{60}O_3$ isomer having C_{3v} symmetry, *J. Org. Chem.* 67: 1696–1698.

Takaya, Y., Kishida, H., Hayashi, T., Michihata, M. & Kokubo, K. (2011). Chemical mechanical polishing of patterned copper wafer surface using water-soluble fullerenol slurry, *CIRP Ann. – Manuf. Techn.* 60: 567–570.

Takaya, Y., Tachika, H., Hayashi, T., Kokubo, K. & Suzuki, K. (2009). Performance of water-soluble fullerenol as novel functional molecular abrasive grain for polishing nanosurfaces, *CIRP Ann. – Manuf. Techn.* 58: 495–498.

Troyanov, S. I., Troshin, P. A., Boltalina, O. V., Ioffe, I. N., Sidorov, L. N. & Kemnitz, E. (2010). Two isomers of $C_{60}F_{48}$: an indented fullerene, *Angew. Chem. Int. Ed.* 40: 2285–2287.

Tuinman, A. A., Mukherjee, P., Adcock, J. L., Hettich, R. L. & Compton, R. N. (1992). Characterization and stability of highly fluorinated fullerene, *J. Phys. Chem.* 96: 7584–7589.

Wang, F. F., Li, N., Tian, D., Xia, G. F. & Xiao, N. (2010). Efficient synthesis of fullerenol in anion form for the preparation of electrodeposited films, *ACS Nano* 4: 5565–5572.

Wada, Y., Totoki, S., Watanabe, M., Moriya, N., Tsunazawa, Y. & Shimaoka, H. (2006). Nanoparticle size analysis with relaxation of induced grating by dielectrophoresis, *Opt. Express* 14: 5755–5764.

Wang, S., He, P., Zhang, J.-M., Jiang, H. & Zhu, S.-Z. (2005). Novel and efficient synthesis of water-soluble [60]fullerenol by solvent-free reaction, *Synth. Commun.* 35: 1803–1807.

Yevlampieva, N. P., Biryulin, Y. F., Melenevskaja, E. Y., Zgonnik, V. N. & Rjumtsev, E. I. (2002). Aggregation of fullerene C_{60} in N-methylpyrrolidone, *Colloid. Surface. A* 209: 167–171.

Ying, Q., Marecek, J. & Chu, B. (1994). Slow aggregation of buckminsterfullerene (C_{60}) in benzene solution, *Chem. Phys. Lett.* 219: 214–218.

Zhang, G., Liu, Y., Liang, D., Gan, L. & Li, Y. (2010). Facile synthesis of isomerically pure fullerenols and formation of spherical aggregates from $C_{60}(OH)_8$, *Angew. Chem. Int. Ed.* 49: 5293–5295.

5

Self-Organization and Morphological Characteristics of the Selenium Containing Nanostructures on the Base of Strong Polyacids

S.V. Valueva and L.N. Borovikova

The Institution of the Russian Academy of Science,
The Institute of High-Molecular Compounds, Saint-Petersburg,
Russia

1. Introduction

This article represents the results of the study made by methods of molecular optics of nanostructures formed in process of reduction of ionic selenium in selenite-ascorbate redox system in water solutions of high molecular polymeric stabilizers of anion type: synthetic polyacid – poly-2-acrylamide-2-methylpropansulfacid with MM $M_w = 3{\times}10^6$ and biopolyanion - deoxyribonucleic acid with $M_w = 20{\times}10^6$. It was shown that polyanion – nanoparticle complex obtained under conditions of total saturation of adsorption capacity of selenium nanoparticles (mass ratio v of the components selenium : polymer is equal to 0.1) is close to its thermodynamic stability boundary: the second virial coefficient made up $A^*_2 = -0.07{\times}10^{-4}$ cm^{3}mol/g^2 for system of deoxyribonucleic acid – nano - Se^0- H_2O and $A^*_2 = 0.2{\times}10^{-4}$ cm^{3}mol/g^2 for system of poly-2-acrylamide-2- methylpropansulfacid – nano - Se^0- H_2O. In the field of formation of stable dispersions the values of free energy ΔG^* of interaction of macromolecule – nanoparticle of selenium have been calculated for anion type nanostructures. It was found that in both cases high-molecular structures with $M^*_w = 200{\times}10^6$ (deoxyribonucleic acid as stabilizer) and with $M^*_w = 75{\times}10^6$ (poly-2-acrylamide-2-methylpropansulfacid as stabilizer) were formed with close dimensions (R^*_g – statistic dimensions of the nanoformation, R^*_h – hydrodynamic dimensions of the nanoformation) and average densities Φ^*. The values of conformational parameters ρ^* and p^* testify a form of nanostructures approximating to spherical form: $\rho^* = R^*_g / R^*_h = 1$ (for both systems), $p^* = 1.1$ for a system of deoxyribonucleic acid - nano - Se^0- H_2O and $p^* = 1.5$ for a system of poly-2-acrylamide-2- methylpropansulfacid – nano - Se^0- H_2O. On the base of the experimental data related to the values of M^*_w for spherical nanostructures assuming their mono-nuclear morphology we estimated a radius of selenium nucleus that made up $R_{nucl} = 12$ nm (deoxyribonucleic acid as stabilizer) and 9 nm (poly-2-acrylamide-2-methylpropansulfacid as stabilizer). The following conclusion has been made: under adsorption of macromolecules of strong acids on nanoparticles of selenium the spherical nanoparticles with similar types of morphology with similar dimensions, densities and width of polymer shell are formed.

At present many original articles and monographs are devoted the issues of synthesis of nanoparticles and nanosystems and studying of their properties. (see e.g. [1]). Nano-dimensional particles (NDP) in absence of stabilizer represent typical liophobic colloids characterized by very low stability. The most various substances, from which high-molecular compounds are the most important, are applied for increase of the stability of NDP.

In connection with development of biological nanotechnology the above mentioned particles represent interest as active ingredients of medicines [2, 3] as well as transport systems and adsorption matrixes for bioactive substances [4].

It should be pointed out that nanoparticles of nonmetals stabilized by polymer have not been practically studied. In this respect nanoparticles of amorphous selenium (nano- a-Se[0]) that are used as high-sensitive bio-sensors for immunoassay technology [5] and chromatography mobile affine reagents [6]. Nano-particles of amorphous selenium are characterized by the exclusive spectrum of bioactivity: even at very low concentrations in water (0.005 – 0.1%) they can adsorb antigens antibodies at their surface [7].

Previously (e.g. in [18 – 13]) it was demonstrated that under reduction of selenium ions in solutions of polymers two processes are going on: formation of selenium nanoparticles with narrow unimodal distribution by dimensions and adsorption of macromolecules on them. In the result self-organization of polymeric molecules on surface of nanoparticles and formation of nanostructures with high molecular masses of significant density and various morphologies occur.

This work is focused on studying of process of self-organization and structural-morphological characteristics of selenium containing nano-structures on the basis of strong polyacids: synthetic polyacid - poly-2-acrylamide-2- methylpropansulfacid (PAMS) and biopolyanion - deoxyribonucleic acid (DNA).

2. The objects and methods of research

Selenium represents an exclusive interest as a chemical element with unique semiconductor, photoelectric and X-ray sensitivity properties as well as bioactive substance with anti-oxidant, anti-inflammatory, anticarcinogenic, antimutagenic and detoxicant activities.

In this study nanoparticles of a-Se[0] by reduction of selenitic acid by means of ascorbic acid were selected as subject of inquiry. Compared to nanoparticles of metals a-Se[0] represents an inorganic polymer including fragments of cycles of Se_6 and Se_8 linked by covalent bonds Se-Se [14] in its backbone.

We used the following compounds as polymeric stabilizers: poly-2-acrylamide-2-methylpropansulfacid and bioplyanion - deoxyribonucleic acid (DNA).

$$-[-CH_2-CH-]_n-$$

| PAMS

$$O=C-NH-C(CH_3)_2-CH_2- SO_3Na$$

Molecular Masses (MM) of polymeric matrix made up: $M_w = 3 \times 10^6$ for PAMS (characteristic viscosity 1M $NaNO_3$ at 20[0]C made up $[\eta] = 5.2$ dl/g) and $M_w = 20 \times 10^6$ for DNA (see table).

Self-Organization and Morphological Characteristics of the Selenium Containing Nanostructures on the Base of Strong Polyacids

89

Reduction reaction of ionic selenium was carried out at concentration of polymer in water solution equal to 0.1% and mass ratio of selenium : polymer v = 0.1 i.e. under conditions of total saturation of adsorption capacity of nanoparticles as it was found for selenium – polymeric nanostructures [8].

Study of kinetics of selenium reduction (IV) was carried out at initial concentration of selenitic acid equal to 1.0 mmole/l in the regime with constant temperature (20^0 C) by means of spectrophotometer «Specord M-40» by means of registering changes of optical density of the solutions at wavelength equal to 320 nm. pH values of the solutions of the reaction mass made up 3.5. The values of the constant of the rate of reaction of formation of nanocomposites k* calculated by well known method of Guggenheim [15] are given in the table. Calculation of k* was carried out by the following formula:

$$k^* = \ln(D_k/(D_k - D_i))/t_i, \qquad (1)$$

where D_k – optical density characterizing the end of process, D_i – optical density at given moment of time t_i. Conditions of conducting of reaction of formation of nano - a-Se^0 polymer solution were kept constant: value of k* was influenced only by structure of polymeric matrix stabilizing nano - a-Se^0 being formed.

MM M_w^* and root-mean-square radiuses of inertia R_g^* of nanostructures were determined by means of elastic (static) scattering of light in solutions in water [16] and their affinity to solutions was determined also by the values of the second virial coefficient A_2^*. Quantity of N* of adsorbed macromolecules on the surface of nano-a-Se^0 was calculated by means of ratio MM for polyacids and nanostructures formed by these polyacids. The values of M_w^*, N^*, R_g^*, and A_2^* are shown in the table. Wy used photogoniodiffusiometer «Fica» for determination of the reduced intensity of scattering of solutions R_θ. Wavelength of vertical incident polarized light made up λ = 546.1nm. Measurements were carried out at scattering angles range θ = 30^0 – 150^0. Cleaning of cautions was carried out through millipore (Millex-HV) with diameter 0.45 μm. The values of increment of the refractive index dn/dc were obtained from refractometric measurements by means of instrument IRF-23. Processing of experimental data of light scattering for solutions of nanostructures was carried out by means of Zimm method (see figures 1 and 2) using double extrapolation (to c = 0 and θ = 0) of dependence of Kc/R_θ on $\sin^2(\theta/2)$ +k'c (K – calibration constant, k' – numeric constant).

Basing on data for M_w^* and root-mean-square radiuses of inertia by formula (2) we evaluated the values of average density of nanostructures

$$\Phi^* = 3 M_w^*/4\pi N_a R^3_{sph}, \qquad (2)$$

where R_{sph} = 1.29 R_g^* [17]. The values of Φ^* are given in the table.

Average hydrodynamic dimensions of nanostructures R_h^* (see table) were determined by the method of quasi-elastic (dynamic) scattering [18]. The radiuses of the equivalent hydrodynamic spheres R_h^* were calculated from the values of diffusion coefficients (D*) by Einstein-Stokes equation R_h^* = $kT/6\pi\eta_0 D^*$ (η_0 - viscosity of solvent). The value of conformation-structural parameter ρ^* (see table) was found by means of ratio of experimental values of R_g^* and R_h^* [19 – 22].

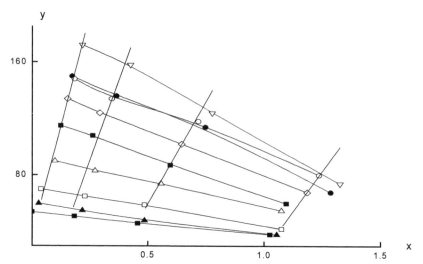

Fig. 1. Zimm Diagram for a system of DNA-Se⁰- water:
x – $\sin^2(\theta/2) + k'c$,
y- $Kc/R_\theta \times 10^8$

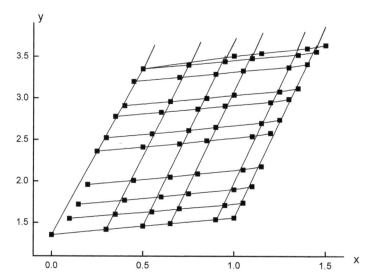

Fig. 2. Zimm diagram for a system of PAMS-Se⁰- water:
x – $\sin^2(\theta/2) + k'c$,
y- $Kc/R_\theta \times 10^8$

Optical part of the installation for measurement of quasi-elastic (dynamic) scattering was equipped with goniometer ALV-SP (Germany) with the following source of light: helium – neon (He – ne) laser Spectra-Physics with wavelength λ = 632.8 nm. Correlation function of intensity of the scattered light was obtained by means of correlator Photo Cor-FC with 288.

Analysis of correlation function was carried out by means of quasi-elastic (dynamic) scattering data processing program Dynals (Helios firm, Russia).

Molecular dispersion of the solutions of the forming nanostructures was estimated by means of birefringence (double refraction) method [23] using a character of gradient dependence of the value of double refraction Δn. The value of Δn was determined subject to rate of rotation of the rotor "g" and concentration of the solution "c" provided constant ratio selenium : polymer. Titanium dynamo-optimeter with internal rotor with 4 cm height and 0.03 cm gap between rotor and stator was used for those measurements. All measurements of double refraction Δn were carried out using thermostating at 21^0C to avoid changes of viscosity of solutions and optical distortions caused by temperature gradient. Phenethyl alcohol that has significant double refraction value ($\Delta n/g = 17 \times 10^{-12}$) and system of polystyrene – bromoform were used for calibration of the installation. Inaccuracy of determination of characteristic value of догиду refraction $[n] = \lim_{g\to 0,\, c\to 0} (\Delta n/gc\eta_0)$ did not exceed 10%. The measurements were carried out at $g < g_k$, where g_k is a gradient of velocity at which flow turbulence occurs.

In general case when $dn/dc \neq 0$ experimental value $[n]$ is formed from three effects: $[n] = [n]_e + [n]_{fs} + [n]_f$ where $[n]_e$ – intrinsic anisotropy, , $[n]_{fs}$– microform effect, $[n]_f$ – macroform effect [23]. In that a value of total segment anisotropy $[n]_{fs} + [n]_e$ is determined by equilibrium stiffness of the polymer chain and structure of elementary unit of the polymer, and value $[n]_f$ is connected with asymmetry of form of the particle by the following relation:

$$[n]_f = ((n^2_s + 2)/3)^2 \times (M_w{}^*(dn/dc)^2 f(p))/(30\pi RTn_s) =$$

$$\text{const } M_w{}^*(dn/dc)^2 f(p), \tag{3}$$

where n_s – solvent refractory index, T – absolute temperature, R – universal gas constant, $f(p)$ – tabulated function of ratio of axes of stiff ellipsoid approximating the particle [23].

Time variations of effective viscosity of PAMS solution during reduction of selenitic acid we registered by means of rheoviscosimeter «Brookfield» at rate of rotation of the rotor equal to 12 rpm.

Value of relative viscosity η_r was determined by means of Ostwald capillary viscosimeter with water outflow time equal to 120 ± 0.2 sec at 21^0C.

In this study we used for the first time a method of polarized light scattering for studying structural features of nanocomposite on the base of PAMS. The studies were carried out by means of ФПС-3М instrument with photoelectric system of registration of light intensity scattered in the range of angles $\theta = 40^0-140^0$ (wavelength of incident light = 578 nm). Calibration of the instrument was done by benzol: Rayleigh ratio for benzol at given wavelength of incident light made up 13.1×10^{-6} cm^{-1}. Angle dependencies of vertical (V_V) and horizontal (H_V) polarized components of the scattered light were measured by method described in [16]. Calculation of the parameters of scattering media was carried out within the frames of Debye-Bekey [24]. Isotropic parameters of structure such as average square of scattering micro-volume polarizability fluctuation $<\eta>^2$ and radius of polarizability fluctuation correlations α_V were determined from angle dependence V_V; the following isotropic parameters such as square of optical anisotropy average density $<\delta>^2$ and radius of correlation of optical axes of scattering elements of volume α_H. The parameters of isotropic structure are connected with dimensions α_V of macro-molecules or associates and with micro-heterogeneity $<\eta>^2$ as a function of density fluctuation [25, 26]. Nature of dependence of the parameters of anisotropic structure on concentration allowed determining of system order and statistic dimensions of oriented regions.

3. The results and discussion

When reducing selenitic acid by ascorbate in presence of polymeric stabilizers the rate constants were equal to $k^* = 0.5 \times 10^{-3}$ sec^{-1} for DNA and 0.4×10^{-3} sec^{-1} for PAMS correspondingly that significantly differs from the value of k^* in absence of stabilizers when $k^* = 1.6 \times 10^{-3}$ sec^{-1}. In addition, the values of the rate constants illustrate comparable influence of selected polymeric matrixes on process of self-organization of nanostructures.

In that during the first 10 minutes of reaction efficient viscosity of nanocomposite solutions on the base of PAMS has been changed from 165 centipoise to 55 centipoise, and after that during 2 days it reduced to 18 centipoise that verified reduction of total number of the particles in the volume of solution due to adsorption of macromolecules on the surface of Se nanoparticles being formed. Studying of characteristic viscosity of water solutions of original PAMS and formed nanostructure of PAMS –nano- Se^0 has demonstrated that estimated value [η]was reduced significantly from 70 to 12 dl/g (Figure 3). However, in that according to the data 3a static light scattering the value of molecular mass M_w^* of nanostructure made up 75×10^6 (table) i.e. it has been increased by 25 times (N* = 25) compared to free macromolecules of PAMS.

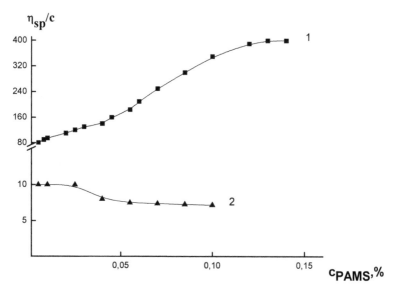

Fig. 3. Concentration dependence of reduced viscosity ηsp/c for systems PAMS-water (1) and PAMS-Se^0- water (2).

Estimation of hydrodynamic radius R_h of Se nanoparticles by the method of dynamic light scattering demonstrated that in absence of polymer big particles of selenium with radius $\sim$ 100 nm (the first day) and $\sim$ 180 nm (the second day) were formed with initial narrow unimodal distribution by dimensions. However, after $\sim$ 50 hours this distribution by dimensions became bimodal and wide. Visible aggregation of selenium occurred after two days expiration. When using PAMS as polymeric stabilizer of nanoparticles of selenium in water solutions at $v=0.1$ the nanostructures with dimensions < 100 nm (table) are formed with narrow unimodal dimension distribution without a tendency to aggregation during a week period at least.

As it follows from the data related to optical properties of PAMS polyanion in water solutions represents an non-penetrable asymmetric swelled ball with asymmetry p> 2 [27]. For system PAMS-nano- Se^0-H_2O evaluation of the value of parameter p^* in approximation $[n] \approx [n]_f$ has demonstrated that nanostructure has conformation approximating to spherical one: $p^* = 1.5$. It is compatible with the data related to parameter ρ^*: $\rho^*=1.0$ that corresponds to spherical conformation [19, 22].

Assuming spherical conformation of nanostructure on the base of PAMS we determined its packing factor k [28]:

$$k=(N^* M^*_w \Sigma_I \Delta V_I)/(v^* M_0), \qquad (4)$$

where $\Sigma_I \Delta V_I$ - intrinsic (Van der Waals) volume of the repetitive link if polymer formed from increments of Van der Waals volumes of separate atoms included into this link:

M^*_w - MM of the particle;
N^* - number of molecules of polymers in adsorbed state
M_0 – MM of the repetitive link;
v^* - volume of nanostructure equal to $(4\pi/3)\times(R_{sph})^3$ where $R_{sph}=1.29 R_g^*$.

Value of k made up 0.2 that is approximating to the value for globular protein (k = 0.6 ÷ 0.8) by the order of magnitude.

Polarized light scattering method was used for quantitative estimation of order and micro-heterogeneity of the system PAMS – nano - Se^0- H_2O.

Compared to the system PAMS - H_2O, for which parameter $<\delta>^2$ is not determined in general (water solutions do not reveal anisotropic scattering), this parameter for the system PAMS - Se^0- H_2O reaches a value $\sim 1.5\times10^{-7}$ that is more typical for solid bodies (Figure 4).

Sharp increase of parameter $<\eta^2>$ is observed for the same system in the range of concentrations of polymer c< 0.05 % (Figure 4), that verifies increase of micro-heterogeneity of the solution due to increase of scattering centers per unit volume.

In that statistic dimensions of optical dense regions α_v remain invariable in the whole range of the studied concentrations (Figure 5).

A distinctive feature of the system PAMS-Se^0-H_2O in the range of concentrations for polymers equal to 0.05 – 0.10 mass % is that inverse (abnormal) slope of angle dependence H_v scattering component is observed. It may be a result of nonrandom fluctuations of

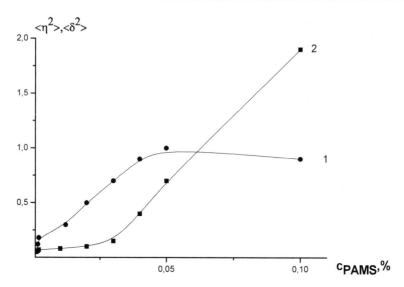

Fig. 4. Concentration dependences of mean square of polarizability fluctuations $<\eta^2>$ (1) and mean square of density of optical anisotropy $<\beta^2>$ (2) for a system PAMS-Se⁰- water.

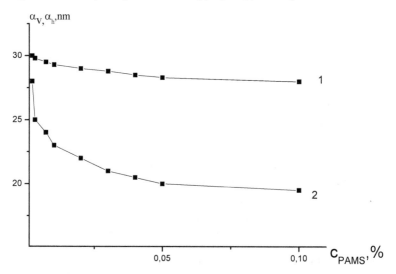

Fig. 5. Concentration dependences of correlation radius of polarizabaility fluctuations α_v (1) and correlation radius of correlation of optical axes of scattering elements α_h (2) for a system of PAMS-Se⁰- water.

anisotropy density for this system. Statistics processing of the data for anisotropic scattering within the angles 40-80⁰ has provided a possibility to determine dimensions of the oriented regions ($\alpha_H \sim 20$ nm) inside this system (Figure 5). Additional verification of high level of order of nanostructure is represented by the fact that value of local order factor F_v

Self-Organization and Morphological Characteristics of the Selenium Containing Nanostructures on the Base of Strong Polyacids

95

determined by relation (5) reaches the value equal to 0.7 that characterizes intermediate state of total disordering of macromolecules $F_v= 0.33$) and crystal structure state of the macromolecules ($F_v=1$).

$$F_v= 1/3 \ (1+ 2 \exp(- \alpha_v \ / \alpha_h \ )), \tag{5}$$

Since PAMS from structural point of view is N-substituted analog of polyacrylamide that is not adsorbed on nanoparticles of selenium [8] the main input into binding of PAMZZS with nano- a-Se^0 can be made by hydrophobic fragments as well as sulfonate groups of side links of the polymer.

According to the data of static light scattering the value of molecular mass M_w^* for nano-structure DNA-nano- Se^0 made up 200×10^6 (Table) that is it was increased by 10 times ($N^* = 10$) compared to free macromolecules of DNA. Statistics (R_g^*) and hydrodynamic (R_h^*) dimensions of the nanostructure DNA-nano- Se^0 coincide with each other ($R_g^*=R_h^* =100$ нм) and correspond to nanodimensional level (Table).

System	$k^* \times 10^{-3} s^{-1}$	$M_w \times 10^{-6}$	$M^*_w \times 10^{-6}$	N^*	$A_2^* \times 10^4$, cm^{3}mole/ g^2	R_g^*, nm	R_h^*, nm	$R_я$, nm	ΔR, nm	Φ^* g/cm^3	ρ^*	$-G^* \times 10^5$, J/m^2
DNA	0.50	20.00	200.00	10.00	-0.07	100.00	100.00	12.00	88.00	0.04	1.10	0.02
PAMS	0.4	3	75	25	0.2	90	90	9	81	0.02	1,5	0.01

Table 1. The constants of the rate of reaction of formation of nanococmposites and structural-conformation parameters of the corresponding nanostructures.

It is known that in the field of acid pH (pH<4.5) oxygen destabilization/denaturation of the DNA macromolecules takes place. Within the range of pH from 3.0 to 4.5 partial destabilization of twin helix is observed, further decrease of pH results in denaturation of DNA already that is verified by studies of UV absorption of DNA, circular dichroism spectra of DNA, viscosimetry data, and double refraction (Δn) data in flow [29 – 32].

For the system DNA-nano- Se^0–H$_2$O at pH = 3.5 destabilization of twin helix in process of self-organization of the structure is observed: the given value of double refraction (Δn) has a positive sign compared to corresponding value for negative DNA (Figure 6). If for isolated DNA macromolecule in native condition [n] ≈ [n]$_e$ [23] then for the system DNA-nano-Se^0–H$_2$O approximation [n] ≈ [n]$_f$ is fulfilled. Estimation of the asymmetry level of the form of the nanostructure resulted in the value p*= 1.1. It agrees to the light scattering data ($\rho^* =R_g^*/R_h^* =1.0$).

It seems that formation of adsorbates of polymers on nanoparticles of nonmetals with the form close to spherical one is a universal phenomenon. For nanostructures, which are different from morphology point of view in the range of saturation of adsorption capacity v = 0.1, the particles with a form close to spherical one were found experimentally [8-13, 33, 34].

Calculation of average density of nanostructures at v = 0.1 by formula (2) demonstrated that though the nanostructures with density Φ^*=0.02 and 0.04 g/cm^3 are formed (Table) that exceed a density of polymer ball [23] but less than for selenium containing spherical nanostructures on the base of nonionic rigid chain molecules (Φ^*=(0.12 -0.14) g/cm^3) [12].

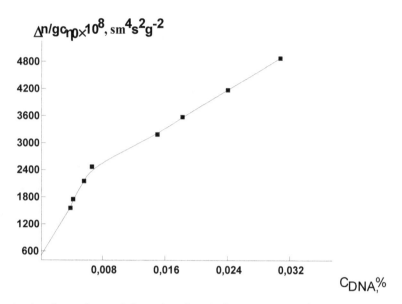

Fig. 6. Concentration dependence of the reduced optical anisotropy $\Delta n / gc\eta_0 \times 10^8$, $sm^4s^2g^{-2}$ for a system of DNA-Se^0- water.

Thus it was revealed that in both cases super high molecular nanostructures are formed ($M^*_w = 200 \times 10^6$ /DNA as stabilizer/ and $M^*_w = 75 \times 10^6$ /PAMS as stabilizer/) with close dimensions (R^*_g – statistic dimensions of nanostructure, R^*_h – hydrodynamic dimensions of nanostructure) and average densities Φ^* (Table). The values of conformation parameters ρ^* and p^* testify a form of nanostructures approaching to spherical one: $\rho^* = R^*_g / R^*_h = 1$ (for both systems); $p^* = 1.1$ for system DNA-nano- Se^0- H_2O and $p^* = 1.5$ for system PAMS-nano-Se^0- H_2O (see Table).

Basing on experimental data for values of M^*_w for spherical nanostructures and assuming their mononuclear morphology we evaluated a radius of selenium nucleus that made up $R_{nucl} = 12$ nm (DNA as stabilizer) and 9 nm (PAMS as stabilizer). Thus in water solution a width of polymer shell ΔR on the particles of nano- Se^0 stabilized by strong polyacids has small differences and makes up ~ 80 – 90 nm (see Table).

With the purpose of experimental determination of the dimensions of nuclei in nanostructures PAMS-Se^0 and DNA-Se^0 their water solutions were passed through cool dehumidification, and the samples obtained in the form of pellets were studied by X-ray diffraction method. It was found that amorphous particles of nano- Se^0 of spherical form for both nanocomposites had unimodal distribution by dimensions and average radius of selenium nucleus made up 10 – 15 nm. These values have good correlation with design quantity of R_{nucl} for nanocomposites.

It was shown that a complex polyanion – nanoparticle obtained under conditions of total saturation of adsorption capacity of selenium nanoparticles (mass ratio v of the components of the selenium: polymer complex was equal to 0.1) was close to its thermodynamic limit:

Self-Organization and Morphological Characteristics of the Selenium Containing Nanostructures on the Base of Strong Polyacids

97

the second virial coefficient made up $A^*_2 = -0.07 \times 10^{-4}$ cm^{3}mole/g^2 for DNA-nano- Se^0- H$_2$O system and $A^*_2 = 0.2 \times 10^{-4}$ cm^{3}mole/g^2 for PAMS-nano- Se^0- H$_2$O system. It agrees with the data obtained at $v = 0.1$ for selenium containing nanostructures on the base of the system of nonionic polymers and polybases [8, 12, 35].

Basing in relation [36] we have:

$$\Delta G^* = kTlnC_e / \pi D_{c\varphi}{}^2, \qquad (6)$$

that is applicable in the field of stable dispersions for spherical nanostructures of arbitrary morphologies, we obtained the values of free energy ΔG^* of interaction of macromolecule – nanoparticle per unit area of the surface of the particle (see Table). In equation (6) C_e – equilibrium concentration of polymer in molar fractions of monomeric units, $D_{sph} = 2R_{sph}$, ($R_{sph} = 1.29$ R_g^*). You may see from the Table that quantity of free energy at $v = 0.1$does not depend practically on structure of monomeric unit and nature of polyacids. It seems that in both cases the same mechanism of adsorption of macromolecules on selenium nanoparticles is realized. So, for ionogenic polymeric matrix (at pH = 3.5 strong polyacids are in ionized state) it is observed electrosteric stabilization stipulated for by electrostatic effects from one side and entropic and osmotic effects from another side [37] appearing due to hydrophobic interactions in aqueous medium between hydrophobic fragments of organic macromolecules and energy saturated surface if selenium nanoparticles.

From totality of the presented data we can make conclusion that at adsorption of macromolecules of strong acids on selenium nanoparticles spherical nanostructures of similar morphology types with close dimensions, densities and width of polymer shell are formed.

Thus, selenium nanoparticles represent adsorption matrixes where high local concentrations of polyanions containing hydrophobic fragment can be reached.

4. Conclusions

1. It has been determined by means of molecular optics that macromolecules of PAMS and DNA adsorbed on nano-Se^0 form high molecular nanostructures (with MM M$_w^*$=75$\times$10^6 и 200$\times$10^6) with dimensions equal to 90 – 100 nm including 10/25 macromolecules densely packed into its composition.
2. It was shown that polyanion – nanoparticle complex obtained under conditions of total saturation of adsorption capacity of selenium nanoparticles (mass ratio v of the components of selenium: polymer complex was equal to 0.1) is close to the limit of its thermodynamic stability. It agrees with the data obtained at $v = 0.1$ for selenium containing nanostructures on the base of the systems of non-iongenic polymers and polybases.
3. In saturation range of adsorption capacity $v = 0.1$ universality of formation selenium – polymeric nanostructures of spherical form was experimentally determined.

5. References

[1] Pomogailo A.D., Rosenberg A.S., Uflyand I.E. (2000). *Metal nanoparticles in polymers.* M. Chemistry.

[2] Kopeikin V.V., Panarin E.F. (2001). Water soluble nanocomposites of zero-valent silver with increased antiinfection activity, *J.Reports of the Academy of Science*. Vol. 380, No. 4, 497 -500.

[3] Balogh L., Swanson D. R., Tomalia D. A., Hagnauer G. L., McManus A. T. (2001). Antimicrobial Dendrimer Nanocomposites and a Method of Treating Wounds, US 6 224 898 B1.

[4] Connelly S., Fitzmaurice D. (1999). Programmed Assembly of Gold Nanocrystals in Aqueous Solution, *Adv.Mater.*, Vol. 11, No. 14, 1202-1205.

[5] Yost D. A., Russell J.C., Yang H. (1990). Non-Metal Colloidal Particle Immunoassay, US Patent 4 954 452.

[6] Ching S., Gordon J., Billing P.A. (1988). Process for Immunochromatography with Colloidal Particles, EP App. 0 299428 A2.

[7] Zhang J.S., Gao X.Y., Zhang L.D., Bao Y.P. (2001). Biological effects of a nano red elemental selenium, *Biofactors*. Vol. 15, No. 1, 27-38.

[8] Kopeikin V.V., Valueva S.V., Kipper A.I., Borovikova L.N., Filippov A.P. (2003). Synthesis of selenium nanoparticles in water solutions of polyvinylpyrrolidone and morphologic characateristcs of the nanocomposites being formed, *High-molecular compounds*, Vol. 45 A, No. 4, 615 - 619.

[9] Kopeikin V.V., Valueva S.V., Kipper A.I., Kalinina N.A., Silinskaya I.G., Khlebosolova E.N., Shishkina G.V., Borovikova L.N. (2003) Study of processes of self-organization of macromolecules of poly-2-acrylamido-2-methylpropan of sulfacids and sodium dodecyl sulfate on nanoparticles of zero valent selenium, *High molecular compounds*, Vol. 45 A, No. 6, 963-967.

[10] Valueva S.V., Kopeikin V.V., Kipper A.I., Filippov A.P., Shishkina G.V., Khlebosolova E.N., Rumyantseva N.V., Nazarkina Y.I., Borovikova L.N. (2005). Formation of nanoparticles of zero valent selenium in water solutions of polyampholit in presence of different redox systems, *High-molecular compounds*, Vol. 47Б, No. 5, 857-860.

[11] Valueva S.V., Kipper A.I., Kopeikin V.V., Borovikova L.N., Ivanov D.A., Filippov A.P. (2005). Influence of molecular mass of polymeric matrix on morphologic characteristics of the selenium containing nanostructures and their resistance to impact of hydrodynamic field, *High-molecular compounds*, Vol. 47A, No. 3, 438-443.

[12] Valueva S.V., Kipper A.I., Kopeiki V.V., Borovikova L.N., Lavrentyev V.K., Ivanov D.A., Filippov A.P. (2006). Studying of process of formation of selenium containing nanostructures and their characteristics on the base of the rigid molecules of cellulose derivatives, *High molecular compounds*, Vol. 48A, No. 8, 1403 - 1409

[13] Valueva S.V., Borovikova L.N., Koreneva V.V, Nazarkina Y.I., Kipper A.I., Kopeikin V.V. (2007). Structural-morphologic and biologic properties of nanoparticles of selenium stabilized by bovine serum albumin, *Journal of physical chemistry*, Vol.81, No. 7, 1329 - 1333.

[14] Handbook of Chemistry and Physics (2001). Ed. Lide R.D., 81th Edition, Chapman & Hill CRC.

[15] Berezin I.V., Klesov A.A. (1976). *Practical course of chemical and enzyme kinetics*, M.: Moscow State University.

[16] Eskin V.E. (1986). *Light scattering by polymer solutions and properties of macromolecules*, L.: Nauka, Science.

Self-Organization and Morphological Characteristics of the Selenium Containing Nanostructures on the Base of Strong Polyacids

99

[17] Pogodina N.V., Tsvetkov N.V. (1997). Structure and Dynamics of the Polyelectrolyte Complex Formation, *Macromolecules*. Vol. 30, No. 17, 4897-4904.

[18] Brown W. (1993). *Dynamic Light Scattering: the Method and Some Application*. Oxford: Clarondon Press.

[19] Meewes M., Ricka J., De Silva M., Nuffengger R., Binkert Th. (1991). Coll-globule transition of poly (N-isopropylacrylamide). A study of surfactant effects by light scattering, *Macromolecules*. Vol. 24, No. 21, 5811-5816.

[20] Nishio I., Shao Thang Sun, Swislow G., Tanaka T. (1979). First observation of the coll-globule transition in a single polymer chain, *Nature*. Vol. 281, No. 5728, 208-209.

[21] Konishi T., Yoshizaki T., Yamakawa H. (1991). On the "universal constants" p and Φ of flexible polymers, *Macromolecules*. Vol. 24, No. 20, 5614-5622.

[22] Burchard B.W. (1992). *Static and dynamic light scattering approaches to structure determination of biopolymers// Laser Light Scattering in Biochemistry* Eds. by Harding S.E., Satelle D.B., Bloomfild V.A. Cambridge : Royal Soc. Chem. Information Services.

[23] Tsvetkov V.N., Eskin V.E., Frenkel S.Y. (1964). *Structure of macromolecules in solutions*. M.: Nauka, Science.

[24] Stein P. (1978). *Polymer Blends*. Ed. by Poul D.P., Newman S. New York; San Francisko; London: Acad. Press.

[25] Kalinina N.A., Kallistov O.V., Kuznetsov N.P., Batrakova T.V., Romashkova K.A., Gusinskaya V.A., Sidorovich A.V. (1990). Light scattering and structure of moderate concentrated solutions and films of polyamides, *High molecular compounds*, Vol. 32 A, No. 4, 695 -700.

[26] Kallistov O.V., Krivibokov V.V., Kalinina N.A., Silinskaya I.G., Kutuzov Y.I., Sidorovich A.V. (1985). Structural peculiarities of the moderate concentrated solutions of polymers with different rigidity of molecular chain, *High molecular compounds*, Vol. 27 A, No. 5, 968 - 971.

[27] Valuev S.V., Kipper S.V. (2001). Influence of ionic force on asymmetry of p form of macromolecule of poly-2-acrylamido-2-methylpropan sulfacids in diluted water-salt solutions, *Journal of applied chemistry*, Vol. 74, No. 9, 1513 - 1517.

[28] Askadsk A.A., Matveyev Y.I. (1983). *Chemical structure and physical properties of polymers*. M.: Chemistry.

[29] Dove W.F., Wallace T., Davidson N. (1959). Spectrofotometric stady of the protonation of undenatured DNA, *Biochem. Biophys. Res. Com.*, Vol. 1, No. 6. 312-317.

[30] Luck G., Zimmer Ch., Snatzke G. (1968). Circular dichroism of protonated DNA, *Biophys. Biochem. Acta*, Vol. 169, 548-549.

[31] Struts A.V., Slonitsky S.V. (1983). Study of conformation of DNA molecule in water solutions of acrylamide and semicarbazide, *Bulletin of the Leningrad State University, Physics, Chemistry*, Vol. 4, No. 22, 33 - 39.

[32] Kasyanenko N.A., Bartoshevich S.F., Frisman E.V. (1985). Study of influence of pH media on conformation of DNA molecule, *Molecular biology*, Vol. 19, No. 5, 1386 - 1393.

[33] Storhoff J.J., Lazaorides A.A., Mucic R.C., Mirkin C.A., Letsinger R.L., Schatz G.C. (2000). DNA-linked gold nanoparticle assemblies, *J.Amer.Chem.Soc.* Vol. 122, 4640 - 4651.

[34] Bronstein L.M., Valetsky P.M., Antonietti M. (1997). Formation of nanoparticles of metals in organized polymeric stsructures, *High molecular compounds*, Vol. 39 A, No. 11, 1847 - 1855.

[35] Kopeikin V.V., Valueva S.V., Kipper A.I., Filippov A.P., Khlebosolova E.N., Borovikova L.N., Lavrentyev V.K. (2003). Study of formation of the particles of nano- Se^0 in water solution of cation polyelectrolyte, *Journal of applied chemistry*, Vol. 76, No. 5, 847 – 851.

[36] Litmanovich O.E., Papisov I.M. (1999). Influence of length of macromolecule on dimension of the particle of metal reduced in polymeric solution, *High molecular compounds*, Vol. 41 A, No. 11, 1824 – 1830.

[37] Mayer A.B.R. (2001). Colloidal Metall Nanoparticles Dispersed in Amphiphilic Polymers, *Polym. Adv. Technol.* Vol. 12, 96 - 102.

6

The Reactivity of Colloidal Inorganic Nanoparticles

Neus G. Bastús[1], Eudald Casals[1], Isaac Ojea[1],
Miriam Varon[1] and Victor Puntes[1,2]
[1]Institut Català de Nanotecnologia, Barcelona,
[2]Institut Català de Recerca i Estudis Avançats (ICREA) Barcelona,
Spain

1. Introduction

The development of functional colloidal inorganic nanoparticles (INPs) has increased exponentially over the past decades offering a "toolbox" ready to be used in a wide range of applications such as materials science, catalysis, biology and medicine (Freitas, 1999; Alivisatos, 2001). This applicability relies on their unique size, morphology and structure, which determines not only their properties but also its reactivity. At the nanometer scale, the confinement of electrons, phonons and photons leads to a new generation of materials which have improved or new physico-chemical properties in comparison with bulk materials. Well-known examples are the size and shape dependence absorption and scattering in noble metal INPs (e.g. Au or Ag), the enhanced fluorescence in semiconductor quantum dots (QDs) (e.g. CdSe or PbS) and the superparamagnetic moment in magnetic NPs (e.g. iron oxide or cobalt) (Burda et al., 2005). Additionally, as the size of the material is reduced and percentage of atoms at the surface becomes significant, the entire particles become more reactive (Bastús et al., 2008). Thus, in very small crystals, both the thermodynamics and kinetics of reactions can change with size. For instance, a large surface-to-volume ratio can be accompanied by a lowering of phase transition temperatures (Goldstein et al., 1992). Additionally, this high fraction of unsaturated atoms at INP's surface may lead to some instabilities which can further lead to degradation and corrosion processes. Although this secondary processes are often uncontrollable, the feature is extremely useful for catalysis applications (e.g. Pt NPs) allowing reactions at the active sites of their surfaces (Li & Somorjai, 2010).

All in all, the reactivity of an INP is not only determined by the size, composition and structure of its core. In fact, an INP is the combination of an inorganic core, which determines its physico-chemical properties, and an organic/inorganic shell, which dictates the interfacial interactions by the chemical nature of the organic layer. Therefore, the surface coating play a dominant role when controlling and tuning the reactivity of the particle, determining its solubility and selectivity against a desired target.

In this context, we identify some critical factors that one should take into account when studying the reactivity of INPs **(Figure 1)**. Thus, by reducing the size (or tuning the shape) of a nanostructured material, the atomic surface-to-volume ratio increases. In this process, most of the atoms are preferentially located at the edges and kinks of the particle, which dramatically increases its reactivity. Another consequence of the high surface-to-volume ratio of INPs is their high ability to suffer chemical transformations, degradation and corrosion, either caused by the removal of ions which they are in equilibrium with, or by the addition of other ions, leading to its growth. In all cases, these chemical reactions lead to accidental or intended morphological transformations of INPs. Independent of the particular reactivity of the inorganic core, INP's surface coating, its substitution and degradation drastically determine the final interaction between the particle and the surrounding medium. Thus, same core material can interact distinctly depending on the surface nature of the coating layer or its time-dependent evolution. Finally, it is important to consider the interactions between INPs in a colloidal solution. INPs are not isolated entities. On the contrary, they are in a constant Brownian movement which forces their collisions towards an aggregated state. Although this reactivity is often undesirable (physico-chemical properties of INPs are restricted to individual particles in non-physical contact) it can be used as driving force for the formation of more complex structures either by an oriented attached mechanism or by their use as catalysts.

An important prerequisite for boosting the applicability of INPs is developing an in-depth study of INP's reactivity. This condition is especially important in biological and medical systems where the interactions and interferences of INPs with cells and tissues determine not only the potential toxicity of engineered materials but also its biodistribution, degradation and biocompatibility (Casals et al., 2008). The response of nanostructured materials in biological fluids is extremely complex and diverse and depends on a variety of

Reactivity of Colloidal Inorganic Nanoparticles

1 Reactivity and Size/Shape of the Inorganic Core

2 Surface Chemistry and Ligand Exchange Reactions

3 Reactivity and (Intended or Accidental) Functionalization: From Targeting to Protein Corona

4 Interactions with Ions: Chemical Transformation, Degradation and Corrosion

5 Cooperative Effects with other Particles in Solution: From Shape Control to Aggregation

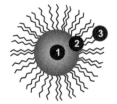

Fig. 1. Reactivity of Colloidal Inorganic Nanoparticles. The critical factors to consider on the study of INP's reactivity are i) Effects of inorganic core's size and shape, ii) the surface molecule substitution and ligand exchange, iii) the particular reactivity of the coating molecule, iv) the interactions with ions present in the colloidal solution and v) cooperative effects with other NPs present in solution.

involved parameters. INPs can aggregate into microscopic structures or may be encapsulated into surrounding material. NP's surface, which determine its final activity, may experience constant modifications such as corrosion or degradation.

In this context, this chapter explores the reactivity of colloidal INPs. It focuses on the particular physical-chemical properties of INPs and in the understanding of the factors that determine its chemical reactivity and therefore its applicability, in nanobiomedicine and catalysis among others. We summarize the potential reactivity of INPs attributed either to their size, shape, chemical composition and structure, focussing of the special features that make them unique and potentially more toxic and risky. Thus, interactions of INPs with surrounding environments will determine their stability while (intended or accidental) surface coating defines much of their bioactivity. Consequences of this are the different final bioactivity of the same element depending on their surface charge (INPs positively charged predominantly attach negative biological surfaces as cell membranes, leading to cell death) and the nature of the non-specific coating that evolves as time progresses leading to different responses at different exposure times. Additionally, toxicological aspects associated to the reactivity of INPs (size, shape, chemical composition and structure) will be widely discussed. The aim of this chapter is not to provide in-depth insights into INP's features, but to tackle the study from a holistic perspective, in the framework of INP's full life cycle.

2. Reactivity and morphology

2.1 Reactivity and size

The size of nanostructured materials determines, among others, its reactivity. The reasons are clear and easily defined. Whereas in a bulk material, surface atoms contribute only a relatively small fraction (only a few per cent) of the total number of atoms, in INPs the curvature radii is so high that all the atoms of the particle lay close or at the surface of the kinks, steps and edges. Therefore, this small particles contain all (or almost all) surface atoms, and those atoms have lower coordination numbers than in the bulk material. Since the reactivity of a single atom tend to increase as its coordination decreases, it follows that the reactivity of nanostructured materials increases as particle size decreases. Thus, in the case of a nanocrystal (NC) with a face-centered cubic (f.c.c.) structure, when the size of the crystal is small enough (1 nm) all the atoms of the particle are located at the surface (and edges) while it decreases to 50% when the size is 2 nm. For larger diameters, the number of surface atoms is extremely low (< 5% for 20 nm) and the reactivity of the crystal is determined by the quality of the surface, i.e., the number steps, kinks and terraces of its surface (**Figure 2-A**). Thus, there is a combined effect in INPs. First, greater accessibility to all constituent (surface) atoms and secondly, enhanced reactivity because of the low coordination number they exhibit.

Among other features, the size of INPs has been extensively investigated for their implications in catalysis applications since it deeply impacts on both surface structure and electronic properties. Rather than by the exact number of atoms in a cluster (particle size and size distribution), the catalytic reactivity is determined by the arrangement of the surface atoms (number of edges and corners) and the exposure of certain crystallographic planes. Especially interesting is the case of Au NPs which, although considered chemically inert, has been found to be a very effective catalyst for sizes below 5 nm (Hvolbaek et al., 2007). Catalysts based on Au NPs allow significant lower reaction temperatures than used in existing cases which is promising for the development of energy efficient processes. Thus, Xu et al. (Xu et al., 2006) showed that unsupported nanoporous Au, with pores sizes less than 6 nm, made by selective dissolution (de-alloying) of Ag from Ag/Au alloy, are active catalysts for CO oxidation. Further evidences for the scaling of the catalytic activity with the number of corner atoms were provided by Overbury et al. (Overbury et al., 2006) by correlating the CO oxidation activity of a Au/TiO$_2$ catalyst with Au particle size.

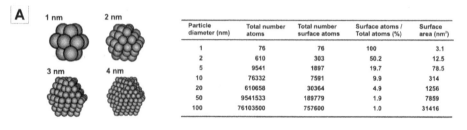

Particle diameter (nm)	Total number atoms	Total number surface atoms	Surface atoms / Total atoms (%)	Surface area (nm^2)
1	76	76	100	3.1
2	610	303	50.2	12.5
5	9541	1897	19.7	78.5
10	76332	7591	9.9	314
20	610658	30364	4.9	1256
50	9541533	189779	1.9	7859
100	76103500	757600	1.0	31416

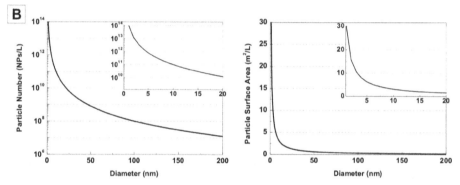

Fig. 2. Core, surface and edge distribution of atoms in inorganic nanoparticles. (A) Atomic distribution in icosahedral nanoparticles from 1 to 4 nm in diameter (left) and total number of atoms, surface atoms, surface-to-volume ratio and total surface area for Au nanoparticles of different sizes. (B) Particle number (left) and total particle surface area (right) for a colloidal solution which contains 1g/1 L of Au atoms.

Another effect of reducing the size of the nanocrystal is the increase of the particle number for a given mass. Thus, considering a colloidal solution of Au NPs with a fixed mass (1 g/1 L) and density (19g/cm^3), as the particle size decreases 10 times, the concentration increases by a factor 1000 (**Figure 2-B**). This phenomenon has been a source of controversy when studying the toxicity of INPs as a function of their size, being often shown that smaller particles exhibit more acute toxicity and had higher inflammatory effects that small ones. Toxicity and INP's dose are directly interrelated. Almost any type of nanomaterial has been found to be toxic at high concentrations, and *vice versa*, nothing is toxic at enough small doses. It is often observed that doses used in *in vitro* studies are very high while there is little consideration or discussion about the realistic *in vivo* exposures (Oberdorster et al., 2005). For instance, 100 µg of INPs/ml in cell culture media, labelled as a low dose, is high and unlikely to be encountered *in vivo*. Considering the concentration of as-synthesized engineered INPs, where colloidal synthesis represents the major technique to get individual size and shape controlled NPs, the standard concentrations are ranging from 10^{12} to 10^{16} INPs/ml, depending on the material, that corresponds to an upper limit for very dense materials of few mg/ml. Alternatively, total surface area has been found to be a better dose-parameter than particle mass or particle number in toxicity studies (Oberdörster, 2000; Oberdörster et al., 2000), especially when comparing particles of different sizes. Furthermore, it has been hypothesized that phenomena that operate at low realistic doses are likely to be different from those operating at very high doses when organism or cell defences are overwhelmed (Oberdorster et al., 2005). In addition to that, it is important pointing out that INP's concentration may change in different parts of the body, either by aggregation or accumulation (trapping) in special organs as the liver or kidneys (Bastús et al., 2008; Casals et al., 2008).

The increase of INP's concentration as decreasing INP's size withstands with an increase of the total surface area, and hence the reactivity of the system. This enhancement on reactivity could be either beneficial (antioxidant, carrier capacity for drugs, increase uptake and interaction with biological tissues) or disadvantageous (toxicity, instability, induction of oxidative stress) depending on the intended use and the material used (Bastús et al., 2008). For instance, small CeO_2 INPs are known to exhibit superior antioxidant properties due to the presence of oxygen vacancies on its surface and a cycle of dual oxidation states (Ce^{3+} and Ce^{4+}), which makes them extremely useful as a quenchers of Reactive Oxygen Species (ROS) in biological systems (Karakoti et al., 2008; Garcia et al., 2010). On the other hand, INP's degradation increases as INP's size decreases.

2.2 Reactivity and shape

Similarly, the reactivity of a particle also depends on its shape. As explained above, INP's reactivity is directly associated to surface atoms, whose number is determined not only by particle's size but also by the distribution of atoms within the particle. Therefore, particle's shape is as important as size when studying INP's reactivity. This dependence can be easily evaluated by calculating the number of surface atoms for different particle's geometry while

keeping constant the total number of atoms in the particle (**Figure 3**). Thus, in the case of 10 nm spherical Au NPs, only ~ 10% of atoms are located at the surface of the particle while this number increases up to ~ 29% when same number of atoms are distributed forming a disk. Although obtaining much more smaller values, same tendency was observed in larger particles. Thus, ~ 0.47% are surface atoms in a 200 nm spherical Au NP whereas this value increases up to ~ 0.53% or to ~ 1.46% when moving to anisotropic shapes such as cubes and disks. Beyond the number of surface atoms, INP's reactivity is also directly determined by the quality of the surface. Therefore, atoms in flat surfaces, which have a higher degree of coordination than spheres, have a lower reactivity compared to the atoms on vertices and edges present in the anisotropic shapes. This allows a greater portion of atoms or molecules to be oriented on the surface rather than within the interior of the material, hence allowing adjacent atoms and molecules to interact more readily. As further discussed, this fact has important consequences on the functionalization of INPs, in particular when studying how its surface curvature affects the organization of molecules (as alkanethiols) attached at their surfaces. Thus, the degree of packaging of molecules attached to relatively small Au NPs (less than 20 nm) is significantly lower than their large particle counterparts, allowing for less density of molecules to be attached to the highly curved surfaces than to the larger flatter surfaces, when equal areas are compared.

Moreover, anisotropic shapes expose different crystal faces. In fact, shape control is usually achieved by the selective interaction of surfactant molecules with a particular crystallographic facet, which leads to a preferential growth along a certain direction. Well known examples are the synthesis of anisotropic semiconductor NCs (Kumar & Nann, 2006). For semiconductor systems, where different crystal planes are chemically distinct, this has been demonstrated by using mixtures of surfactants. Thus, Alivisatos' group induced anisotropy in CdSe spherical NCs by using a tri-n-octylphosphine oxide (TOPO) and phosphonic acids, instead of TOPO alone (Peng et al., 2000). Resultant CdSe nanorods displayed many important property modifications compared to spherical NCs, such as polarized light emission (Hu et al., 2001). In continuation of this success, other complex morphologies such as arrows and tetrapods, were successfully synthesized (Manna et al., 2000), all of them explained on the basis of the selective adsorption of surfactants on different crystallographic faces. For the fcc noble metals, the low-energy {111} and {100} facets possess similar surface energies and, therefore, chemical reactivities, which limit the number of candidate molecules for directing shape control by selective binding preference for different crystal planes. Specially interesting is the case of the polymer poly(vinyl pyrrolidone) (PVP) since it has been demonstrated to stabilize the lowest-energy crystal facets of these fcc crystal systems to give colloidal structures bounded by {111}, {100}, and {110} planes (Tao et al., 2008).

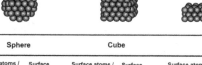

	Sphere		Cube		Rod		Disk	
Particle diameter (nm)	Total number atoms	Surface atoms / Total atoms (%)	Surface area (nm²)	Surface atoms / Total atoms (%)	Surface area (nm²)	Surface atoms / Total atoms (%)	Surface area (nm²)	Surface atoms / Total atoms (%)	Surface area (nm²)
10	76080	9.4	314	10.6	354	13.5	451	29.2	971
20	608642	4.7	1256	5.3	1416	6.8	1804	14.5	3886
30	2054169	3.1	2826	3.5	3187	4.5	4059	9.7	8743
50	9512937	1.9	7850	2.1	8854	2.7	11274	5.8	24287
70	26095556	1.35	15394	1.5	17364	1.9	22109	4.2	47629
100	76080340	0.94	31400	1.06	35419	1.35	45097	2.9	97151
200	608828006	0.47	125600	0.53	141676	0.67	180386	1.46	388686

Number of total atoms for each size was kept constant while ratio and total surface area was calculated for each geometry

Fig. 3. Shape dependent surface atomic distribution of nanoparticles . Total number of atoms was kept constant (considering Au spheres of different diameters) while the number of surface atoms and the surface area was calculated by different geometries. Rod particles were considered to have a cubic section with a long length 10 times larger than the short length and disk particles have a diameter 10 times larger than their height.

The possibility of controlling the exposition of different crystal faces by tuning NC's morphology has been widely exploited for its applicability in catalysis, where the yield and selectivity of the catalyzed reaction depends on the nature of the catalyst. This dependence has been observed in the Pt-catalyzed hydrogenation of benzene where the {100} surface yields only cyclohexane while the {111} surface gives both cyclohexene and cyclohexane (Bratlie et al., 2007). Similarly, electron-transfer reaction between hexacyannoferrate and thiosulfate has been catalyzed by Pt nanocrystals with different shapes obtaining that tetrahedral NCs, completely bound by {111} crystal facets, exhibit the highest catalytic activity whereas cubic nanocrystals exhibit the lowest activity (Tao et al., 2008). Although promising results have been obtained, the largest obstacles facing shape-controlled nanocatalysis are compatible surface chemistry and shape retention. Catalytic studies need naked surfaces, which is never the case of colloidal NCs, protected by a layer of organic material. Although removing these stabilizing agents may be required to create accessible active sites, it is difficult to do it without inducing significant morphological changes via surface reconstruction, particle ripening, melting, or oxidation (Barghorn et al., 2005).

The correlation between reactivity and INP's shape has also relevant consequences on the study of the potential toxicity of nanomaterials. A typical example is the common synthesis of Au NPs in the presence of hexadecyltrimethylammonium bromide (CTAB). CTAB is a toxic surfactant that strongly binds to the surface of Au NPs confining its growth and conferring stability. Whereas at relative low CTAB concentrations spherical Au NPs are obtained, at high concentrations (beyond the second critical micellar concentration threshold), CTAB forces crystal growth to anisotropic shapes leading to rod-like structures (Nikoobakht & El-Sayed, 2003). This correlation between CTAB concentration and morphology can be understood by a stronger interaction between the atoms at the surface of the rod particle and CTAB molecules. In fact, these hypothesis correlates with shape-dependent toxicity results. Thus, after observation that CTAB-stabilized Au NPs were toxic (rods and spheres), similar cleaning processes lead to non-toxic spheres but still toxic rods (Connor et al., 2005).

3. Corrosion and interactions with ionic species

3.1 Degradation and corrosion

Another consequence of the high surface-to-volume ratio and low atomic surface coordination of nanostructured materials is its degradation, much more relevant at the nanoscale than in bulk materials. INPs are constantly exposed to degradation processes, which reduce them to their constituent atoms either by chemical reactions with the surrounding media or easily because the degradated state is thermodynamically more favorable.

All chemical reactions follow the laws of the chemical equilibrium between reactants and products even though this equilibrium may be displaced to one of the components. This simple idea can be applied to an INP's colloidal solution. A colloidal INP's solution is a chemical reaction where the reagent is the precursor compound and the product is the synthesized colloidal solution. In this regard, it has been observed when synthesizing INPs that only a percentage of initial precursor used is finally converted to INPs. There is therefore an equilibrium between INPs and precursor ion **(Figure 4)**. This fact has important implications not only on the final INP's morphology but also when studying its toxicity. First, because it can lead to the morphological evolution of colloidal INPs (such as a broadening of the size distribution), following an Ostwald ripening process. Additionally, it is especially important when calculating INP's concentration from the initial amount of precursor weighted. As previously mentioned, toxicity and INP's concentration are directly interrelated and hence concentration value must be always corrected by the precursor to INPs conversion rate. Moreover, this chemical equilibrium is also important when purifying INPs (by means of precipitation or other washing methods) since this process displace the equilibrium towards colloidal species, which leads to the new release of ions at the expense of INPs' dissolution and further modifications on INPs' morphology. This decrease of ion concentration in solution is, in fact, a general phenomenon when studying INP's full life cycle since scavengers (as serum proteins), or the simple sample dilution can deplete ionic concentration, forcing the partial or complete INP's degradation. Although degradation process is difficult to prevent, it is possible to protect INPs by designing engineered coatings. Thus, several studies show how the encapsulation of INPs in micelles or their functionalization with self-assembled monolayers (SAM) of molecules could protect them from corrosion.

3.2 Toxicity and ion release

A critical parameter when studying the potential toxicity of INPs is the composition and reactivity of the materials used in their synthesis. Due to the abundance and high reactivity of metals in the presence of oxygen, metal oxide INPs are especially interesting. Metal oxide surfaces are known to be very biocompatible due to their charge organization at the surface (Weissleder et al., 1989). However, not all the oxide INPs display the same toxicity (Gojova et al., 2007). Different stochiometries give different crystal structures with different properties and different surfaces, leading to different potential toxicities. This is the case of magnetic conductors (magnetite), magnetic insulators (maghemite) or non-magnetic insulators (hematite) where the reactivity of conducting and non-conducting materials and the degree of aggregation of magnetic and non-magnetic particles can be significantly different.

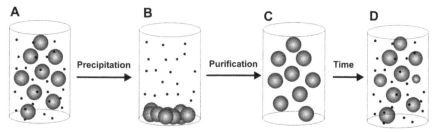

Fig. 4. **Degradation of inorganic nanoparticles.** As-synthesized INPs in equilibrium with precursor species (A). Purification and washing processes displace the equilibrium towards colloidal species (B, C), which leads to the new release of ions at the expense of NP's dissolution. On this stage, INP's morphology is modified due to an Ostwald ripening process (D).

Toxicity is frequently related to the ability of nanostructured materials to release ions. In nanostructured metallic implants, were wear-corrosion greatly contributes to ion release processes, it often correlated with health problems (Ito et al., 2001). Even fairly stable oxides, as in the case of magnetite (Fe_3O_4), may continue oxidizing when exposed to biological environments (Lazaro et al., 2005). In these cases, released ions may undergo chemical transformations which may lead to different chemical species, with altered isotopic composition, electronic or oxidation state and molecular structure (Auffan et al., 2009), which may modify its biological impact (Gupta & Gupta, 2005). In other cases, the corrosion process can be toxic by itself, modifying the redox biochemistry activity of the surrounding biological media and leading to an oxidative stress both *in vivo* and *in vitro* (Oberdorster et al., 2005). For example, Derfus et al. (Derfus et al., 2004) reported the intracellular oxidation and toxicity of CdSe QDs associated to the release of Cd^{2+} cations (Cd^{2+} binds to mitochondria proteins leading to hepatic injuries) while Franklin et al. (Franklin et al., 2007) revealed comparable toxicity of ZnO INPs (30 nm) and $ZnCl_2$ salts, therefore presuming that effects may be attributed to Zn ions released from ZnO INPs. All in all, the environmental and health effects of INPs have contributed to study corrosion processes on nanomaterials.

3.3 Reactivity and ionic/cationic Interactions

Relevant consequences of INP's reactivity are their chemical transformations *via* insertion or exchange of atoms. In extended solids, reactions involving chemical transformations are in general very slow because of high activation energies for the diffusion of atoms and ions in the solid. However, it has been proved that cation exchange reaction can occur completely and reversibly in ionic NCs at room temperature with unusually fast reactions rates. A clear example is the interaction of semiconductor nanocrystals with metal cations, CdSe reacts with Ag^+ ions to yield Ag_2Se NCs by the forward cation exchange reaction, and *vice versa*, Ag_2Se reacts with Cd^{2+} ions to yield CdSe, for the reverse cation exchange reaction (**Figure 5-A**) (Son et al., 2004). While this reaction is extremely fast (< 1s) for NCs, similar experiments carried out with micrometer-sized powders of CdSe found the cation exchange to be virtually prohibited. The speed of the reaction can be explained on the basis of simple scaling of the size in diffusion-controlled reaction schemes, when the reaction time is roughly proportional to the square of the size. This cation exchange reaction, initially investigated with Ag^+ ions, was easily extended not only to other cations but also to the formation of complex hybrid structures with precise structural and compositional tailoring (Zhang et al., 2010). Thus, CdSe NCs can be transformed into CuSe and PbSe though the cations exchange reaction with Cu^{2+} and Pb^{2+} ions while Au-CdS-CdSe core-shell-shell hybrid structures can be obtained from Au-Ag core-shell INPs.

Another chemical transformation extremely interesting is the galvanic corrosion between two different metals and/or alloys having electrical contact with each other. This effect is a coupled reaction where the more active metal corrodes at an accelerated rate and the more noble metal corrodes at a retarded rate. Galvanic replacement has been widely exploited for the obtention of hollow nanostructures of metals by reacting metals solutions (such as Au^{3+}, Pt^{2+}, and Pd^{2+} salts) with pre-formed NCs of a more reactive metal (Sun et al., 2002). An example is the synthesis of Au nanoboxes from Ag nanocubes (**Figure 5-B**). Because the standard reduction potential of $AuCl_4^-/Au$ pair is higher than the one of Ag^+/Ag pair, Ag nanocubes suspended in solution can be oxidized by $HAuCl_4$. The elemental Au produced in this reaction is confined to the vicinity of the template surface, nucleating into very small particles and growing, forming a thin shell around the Ag template. This process continues until the Ag template has been completely consumed. At the last stages of the reaction, Au shells reconstruct their walls into highly crystalline structures via processes such as Ostwald ripening. These Au shells have a morphology similar to that of the Ag templates, with their void sizes mainly determined by the dimensions of the templates. These chemically transformed particles exhibit plasmonic properties completely different from those solid (even made of the same metal).

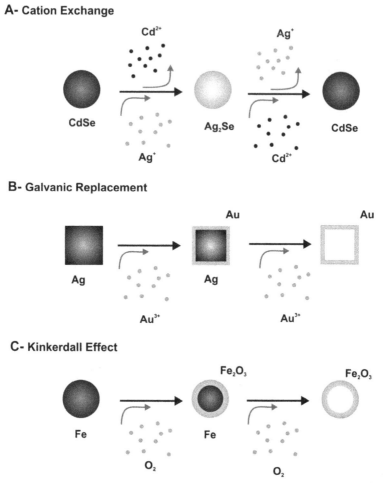

Fig. 5. **Chemical transformations of inorganic nanoparticles via insertion or exchange of atoms.** (A) Cation exchange reactions in semiconductor nanocrystals. (B) Galvanic replacement in metal nanoparticles, (C) Kinkerdall effect in magnetic nanoparticles.

Solid diffusion processes are also a synthetic strategy rather used to prepare hollow solid materials. This phenomenon, called Kirkendall effect, results from the difference of the solid-state diffusion rates of the reactants in an alloying or oxidation reaction (**Figure 5-C**). A significant difference in the diffusion coefficients produce an accumulation of vacancies at the interface of the two components that can lead to the formation of cavities. Thus, the atomic diffusion occurs through vacancy exchange and not by interchange of atoms. The first experimental proof of this phenomenon, originally identified in metal alloys, was reported by Smigelkas and Kirkendall in 1947 (Smigelskas & Kirkendall, 1947). Subsequently, this process was extended into other systems such as oxides, sulfides and nitrides, in which the metallic atoms diffuse faster. To mention some examples at the nanoscale, Alivisatos and co-workers described the formation of hollow oxides (CoO) and chalcogenides (Co_3S_4) nanostructures through the Kirkendall effect employing Co NCs as starting material and reacting them with oxygen and sulfur, respectively, at 182 °C (Yin et al., 2004). As well, Cabot et al. studied the vacancy coalescence during the oxidation process of iron INPs in solution with a controlled oxygen flow and temperature, which leads to the formation of hollow iron oxide INPs. (Cabot et al., 2007). Also has to be considered that the properties of the INPs exposed to transformation processes can be affected. Thus, INPs can lose the applicability in where they are interesting for. Worthy to mention is the work reported by Cabot et al. who studied the magnetic properties and surface effects in hollow maghemite INPs where the single dipole cannot stay at the core of the INPs (Cabot et al., 2009).

3.4 Morphology control and seeded growth methods

The reactivity of colloidal INPs against ionic precursor molecules can be used to precisely control INP's morphology following seeded-growth methods in which pre-synthesized INP's are used as catalysts for the growth of larger particles. The catalyst is usually a colloidal INP that is either injected into a solution together with the precursor molecules needed to growth the desired crystal, or is nucleated *in situ*. Well known examples are the synthesis of nanowires of CdSe, InP, Si and Ge (Trentler et al., 1995; Ahrenkiel et al., 2003; Grebinski et al., 2004) from semiconductor seeds and the size and shape control of Au NPs (Bastús et al.; Pérez-Juste et al., 2005). In the later case, the reduction of the metal salt, such as $HAuCl_4$, was typically carried out in an aqueous system in the presence of ligands such as cetyltrimethylammonium bromide (CTAB) or sodium citrate. **(Figure 6)**. Even more complex structures can be synthesized by taking advantage of NP's reactivity. This is the case of hybrid NCs (HNCs), composed by the combination of different materials assembled together in a single nanostructure. A clear example is the combination of fluorescence and magnetism into a single nanostructure which represents the possibility of bio-detect and bio-separate biological entities. Specially interesting are HNCs based on semiconductor materials, mainly for their utility on optoelectronic, photovoltaic and catalysis applications, where the combination of metal and semiconductor domains allows photo-generated charge carriers performing redox reactions with high efficiency (Kamat, 2007). In the case of semiconductor materials with anisotropic shapes (such as rods, tetrapods or even pyramids), the higher reactivity of the polar facets opens up the possibility of nucleating a second material exclusively at these locations. This idea has been clearly demonstrated for the growth of Au domains onto CdSe NCs (Mokari et al., 2004; Meyns et al., 2010).

3.5 Adsorption, alloying and environmental remediation

Taking advantage from their reduced size, large surface area and highly reactive surface sites, nanomaterials have become particularly attractive as separation media for environmental remediation (Diallo & Savage, 2005; Savage & Diallo, 2005; Karn et al., 2009). For certain reduced diameters, INPs are sustained by Brownian dispersion avoiding sedimentation, which means that they can be maintained exploring the volume without additional agitation. This capacity to be dispersed in groundwater allows them to travel farther than larger macro-sized particles, thus achieving a wider distribution and permitting the whole volume to be quickly scanned with a relatively low number of particles. For example, in a rough estimation, a 10 nm metallic INP in water at room temperature (RT) will experience Brownian relaxation in the order of the nanosecond, and after each Brownian step in solution it will move about 10 to 20 nm. Therefore, a typical INP concentration of few nM will explore the total volume in the time order of the cent-second (assuming a 10 % efficiency, i.e. the INP would repeat a previous position up to 10 times before visiting a new one).

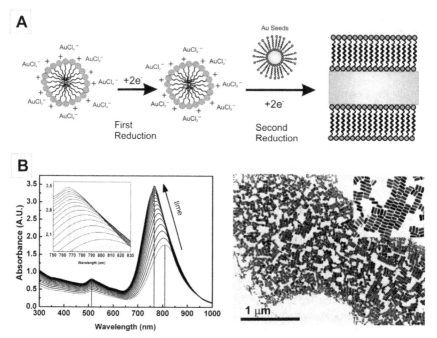

Fig. 6. Reactivity and shape control. Growth of Au nanorods in the presence of a seed NCs used as catalyst. **(A)** The transport of Au ions bound to the CTAB micelles to the growing seed particles is controlled by the double layer interaction. The first reduction is confined in metallomicelles while the second reduction only begins after the seed solution is added. **(B)** Vis-NIR absorption spectra of a representative Au NRs solution growing from Au seeds as a function of time from 20 to 60 min after the addition of seed to the growth solution and morphological characterization of resultant particles by transmission electron microscopy.

The high adsorption capacity of nanomaterials for certain pollutants has been demonstrated in many cases. For example, heavy metals have been effectively removed from contaminated water, as is the case of Cr^{+6} and Cr^{+3} employing maghemite (Hu et al., 2005) and silica NPs (López & Castaño, 2008) respectively, or Hg^{+2} with alumina INPs (Pacheco et al., 2006). In addition, the magnetic properties of some nanomaterials have been employed for magnetic separation after the absorption process, which allowed mixtures of different-sized INPs to be separated by the application of different magnetic fields (Yavuz et al.,

2006). With this methodology As^{+3} and As^{+5} could be eliminated from drinking water supplies employing iron oxide INPs, which involved the formation of weak arsenic-iron oxide complexes at INP's surface. An overview on the use of nano-size iron particles for environmental remediation has been reported by Zhang et al. (Zhang, 2003) Furthermore, the incorporation of some other metal ions, such as TiO_2, into the lattice structure of iron oxide enhances the material properties namely surface adsorption and photo-induced catalysis (Gupta & Ghosh, 2009). The surface functionalization is also a very important factor in environmental remediation since it can enhance the affinity towards specific target molecules. A well known example is the case of magnetite INPs functionalized with di-mercaptosuccinic acid, where the thiolated ligand acts as an effective sorbent for toxic soft metals such as Hg, Ag, Pb, Cd and Ti while As binds to the iron oxide lattices (Yantasee et al., 2007). Alternatively, specific functionalization can also be used on organic–inorganic nanocomposite semiconductor systems to trap organic molecules and then promote their complete degradation making use of the photocatalytic properties (Beydoun et al., 1999). A comprehensive overview of the different manufactured nanomaterials along with the pollutants they could potentially remediate has been reviewed elsewhere (Karn et al., 2009).

Apart from the adsorption capacity, the chemistry of metal alloying has also been used for sequestration of mercury. Thus, in the presence of Au (Henglein & Giersig, 2000; Lisha et al., 2009) or Ag NPs (Katsikas et al., 1996), the Hg^{2+} atoms present in solution can easily be reduced followed by adsorption on the INP's surface and further amalgam formation. As a consequence of the mercury uptake, remarkable morphological transformations can be observed leading to non-spherical shapes and coalescence of particles to an increasing extent. The optical absorption spectra of the mercury containing Au particles is shifted and damped. Similar chemical and optical transformations of spherical CdTe and CdSe NCs into anisotropic alloyed CdHgTe and CdHgSe particles respectively have been observed in the presence of mercury, which have been attributed to the positive redox potential of mercury species (Taniguchi et al., 2011).

4. Reactivity and surface chemistry

Independent of particle size and core composition and structure, the surface coating of the nanostructured material play a dominant role when controlling and tuning the reactivity of the particle and defining its bioactivity. Although as-synthesized INPs present excellent physical and chemical properties, they do not always possess suitable surface properties for specific applications. Then, it is necessary to modify their surface for several reasons such as solubility and stabilization against aggregation, selectivity against a desired target or for surfactant-mediated self-organization. INPs are necessary coated by an intended or accidental coating which contributes to the final reactivity. Thus, the same core material could exhibit different behavior depending on the nature of its surface coating and its aggregation state.

4.1 Surface chemistry and phase transfer

Although there have been many significant developments in the synthesis of INPs, maintaining their stability for long times avoiding its aggregation it is an important issue yet to be deeply explored. Pure metals, such as Fe, Co, Ni and their metal alloys as well as quantum dots (QDs) and Au NPs are usually synthesized in organic solvents and then stabilized using hydrophobic surfactants which provide steric stabilization of the particles (Peng et al., 2000; Puntes et al., 2001; Talapin et al., 2001). The general structure of the surfactant consists of a head group that "sticks" (adsorbed or chemically linked) to the INP's surface while the other end points towards the solution and provides colloidal stability. For biological applications, surface coating should be polar to provide high aqueous solubility and prevent INPs' aggregation and hence the hydrophobic surfactant has to be either replaced by others hydrophilic surfactant molecules in a ligand exchange reaction (Aldana et al., 2001; Gerion et al., 2001) or, alternatively, encapsulated within a polymeric sphere (Pöselt et al., 2009). In this way, by choosing the surfactant molecules, it is possible not only to adjust the surface properties of the particles but also tune its final reactivity. All of these surface modifications alter the chemical composition of INP's surface and therefore, their integration in different environments. An interesting example was the possibility to tune the toxicity of C_{60} by modifying the chemical nature of their surface (Sayes et al., 2004). C_{60} is an hydrophobic organic nanomaterial that forms aggregates when, pristine, gets in contact with water. These aggregates are cytotoxic at low concentration (20 parts per billion) causing cell death via disruption of cell membranes and oxidation of lipids. However, the increase of the solubility of fullerene derivatives decreases the toxic pronounced effects, obtaining a toxicity seven orders of magnitude lower in the case of $C_{60}(OH)_{24}$. Other similar examples are the coating of carbon nanotubes (CNTs) with heparin (Murugesan et al., 2006) or poly(L-lactide) (Zhang et al., 2006) to yield biocompatible structures. Although this increase of solubility may lower the toxicity of nanomaterials, the biopersistence of these coatings is as essential as the knowledge about the bioavailability of the core material, which could exhibit intrinsic toxic properties.

4.2 Surface chemistry and functionalization

One of the most challenging ligand exchange reactions is the INPs functionalization for biomedical applications. In biomedicine, INPs have to be functionalized in order to selectively interact with biological targets. For this purpose, they are usually coated with biological molecules such as antibodies (Wang et al., 2002; El-Sayed et al., 2005), aptamers (Farokhzad et al., 2006), peptides (Kogan et al., 2006; Bastús et al., 2009), and oligonucleotides (Rosi et al., 2006). Some relevant examples are the functionalization of INPs with uptake ligands that specifically bind to receptors of the cell membrane (ex. transferrin (Wagner et al., 1994)), vector ligands that specifically interact with receptors which are over-expressed on the cell surface (Dixit et al., 2006) and peptides that facilitates crossing the cell membrane, either using cell penetrating peptides (Tkachenko et al., 2003; Pujals et al., 2009) or achieving the release of the INP to the cytosol with membrane-disruptive peptides (Han et al., 2006).

However, coatings used for INPs may not be persistent and could be metabolized exposing INP's core. Vector and targeting agents bind usually to INPs' surface by reactive (terminal) moieties. Thus, thiols, amines and carboxilic acids strongly bind to metal, semiconductor and magnetic cores with different and specific affinities. As a consequence of these different activities it is necessary to be aware of the surfactant exchange possibilities: If INP's media is rich in new species that bind stronger than those already present at the INP's surface, a ligand exchange process would take place. This is the case of Au NPs coated with amine or carboxilic groups and further exposed to thiol molecules (Woehrle et al., 2005). In fact, it has even been observed that Au NPs coated with thiol molecules may undergo a ligand exchange process if other thiols are present in solution at similar or higher concentrations. Additionally, this process could lead to a decrease of INP's solubility and a consequent non-reversible aggregation. In more sophisticated situations, as in the case of alloys presenting different types of surface atoms, they may spontaneously segregate forming a core-shell structure (atoms with lower surface tension migrates outwards at room temperature conditions (Margeat et al., 2007)), specially for smaller INPs. This phenomenon directly represents important modification of the INP's surface coating. Other studies have been shown that surface coating can be weathered either by its exposition to oxigen-rich environments or by ultraviolet (UV) irradiation (Rancan et al., 2002; Derfus et al., 2004).

4.3 Surface chemistry and protein corona

As previously mentioned, surface coating modifications are not always intended. They can eventually be accidental. One interesting case is the modification of INP's surface when exposed to biological fluids and their non-specific binding of media proteins (Bastús et al., 2009). This is the case of loosely coated INPs, such as citrate-stabilized Au NPs, where citrate molecules are immediately replaced by media proteins. This process, known as INP's protein corona (INP-PC), is one of the most significant alterations and may, in turn, strongly influence the biocompatibility and biodistribution of these INPs (Casals et al., 2010). INP-PC is a process governed by the competitive time-dependent adsorption of complex mixture of proteins (as serum) by a finite number of surface sites (Vroman, 1962). Thus, highest mobility proteins arrive first and are later (may take several hours) replaced by less motile proteins that have a higher affinity for the surface (Slack & Horbett, 1995). Therefore, after exposition to cell culture media, INPs coating evolves in a time-dependent manner from a soft PC, a transient agglomeration of proteins onto INP's surface, to a hard PC, a dense albumin-rich proteic coating (Casals et al., 2010). This modifications of INP's coating alter INP's reactivity and strongly determine INP-biology interactions, since they are expected to happen at different time-scales. Whereas the removal of particles from the bloodstream takes place in a few minutes, interactions with cells of distant organs may be days after the exposure. Although this PC coating may increase the biocompatibility of the material, it may also alter their toxicity profile, something that has to be carefully considered before using INPs in biomedical applications. All in all, the increasing number of recently publications (Dobrovolskaia & McNeil, 2007), which include the study of different materials such as Au (Casals et al., 2010), Ag, Fe_2O_3, CoO, CeO_2 (Casals et al., 2011) FePt and CdSe/ZnS (Rocker et al., 2009) and the biological implications of the process, indicates the increasing importance of the topic.

5. Reactivity between Inorganic nanoparticles in a colloidal solution

INPs are in a constant Brownian movement which force their collisions towards an aggregated state. Although this reactivity is often undesirable (some physico-chemical properties of INPs are restricted to individual particles in non-physical contact) it can be used as driving force for the formation of more complex structures by an oriented attached mechanism.

5.1 Reactivity and agglomeration

The revolutionary control of physico-chemical properties of nanostructured materials is restricted to individual particles in non-physical contact. Therefore, INP's surface needs to be stabilized in order to avoid irreversible agglomeration and, consequently, there is a significant effort to obtain isolated INPs during long periods of time, which is usually achieved by electrostatic, steric or electro-steric means (Roucoux et al., 2002). In fact, this may be one of the critical differences with the previously existing INPs, since natural and un-intentional occurring INPs tend to agglomerate readily. The degree of agglomeration has relevant consequences on the final behavior of the nanometric material. For instance, INPs used for sunscreens are usually composed of large agglomerates of nanometric domains (Bastús et al., 2008). Although these agglomerates are still efficient sun blockers, their optical properties are less intense than the individual INPs. Additionally, their large size avoids the penetration of such particles beneath the external skin.

INPs' aggregation can be exploited to be used as test to probe either the success of its functionalization or the presence of biological molecules (**Figure 7**). Metal particles

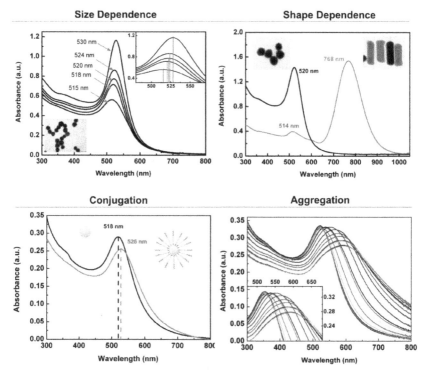

Fig. 7. **Optical properties of gold nanoparticles.** The position of the surface plasmon resonant band depends on the size (A) and shape (B) of Au NPs. This peak position depends on the chemical composition of particle surfactant shell (by changing the refractive index of the surrounding environment), being therefore sensible to functionalization processes (C). The aggregation of Au NPs correlates with a red-shift of the SPR band position along with an increase in the absorbance from 600 to 800 nm (D).

(especially Au and Ag) interact resonantly with visible light absorbing a certain fraction of the visible wavelengths while another fraction is reflected. Small particles absorbs resonantly light in the blue region (~ 500 nm) while as Au NP's size increases, surface plasmon band (SPR) shifts to longer wavelengths, yielding Au NPs a pale-blue colour. SPR band is not only influenced by INP's size, it depends as well on its shape, the chemical composition of the INP's surfactant shell and its agglomeration state (due to dipole coupling between the plasmons of neighbouring particles). This property allows the use of Au NPs as sensors. Thus, when particles aggregate absorption wavelengths red-shifts and the colloidal solution changes its color from red to purple. One of the most typical examples that takes advantage of SPR changes is the colorimetric test to detect single-stranded oligonucleotide targets by using DNA-modified Au NP probes. Thus, two species are present in solution such that each is functionalized with a DNA–oligonucleotide complementary to one half of a given target oligonucleotide. Mixing the two probes with the target results in the formation of a polymeric network of DNA–Au-NPs with a concomitant red-to-purple colour change (Mirkin et al., 1996). Similarly, INP's aggregation can be use as a test to evaluate the INP's stabilization mechanism which allow, under certain conditions, determine the success of INP's functionalization process. Thus, electrostatically-stabilized citrate-coated Au NPs easily aggregate when changing the ionic strength and/or the pH of the medium in which they are solved, while peptidic conjugates with significant steric repulsion (stabilized by the interactions of molecules side chains or domains) withstand much higher changes without compromising stability.

5.2 Reactivity and shape control

The reactivity of colloidal INPs against other INPs is the driving force for the formation of complex structures by an oriented attachment mechanism. Thus, nearly isotropic NCs previously formed in solution coalesce and fuse along some preferential crystallographic directions leading to wires (Tang et al., 2002), rings (Cho et al., 2005) or even sheets (Schliehe et al., 2010). Although this mechanism is not completely understood, one possible explanation is the NC's coalescence along certain directions, which might reduce the overall surface energy by eliminating some high energy facets. Additionally, this process may be facilitated when the organic shell is partially removed from the surface of the initial seed particle or when weakly coordinating molecules are used as stabilizers, so that dipole-dipole inter-particle interactions are enhanced and one-directional NC attachment is spontaneously promoted (Cozzoli et al., 2006).

6. Conclusion

This chapter overviews the particular physical-chemical properties of INPs in order to understand what factors determine its chemical reactivity and thus its applicability in nanobiomedicine and catalysis among others. According to this, the critical factors that determine the interactions of INPs with surrounding environment are: i) Effects of inorganic core's size and shape, ii) the particular reactivity of the coating molecule, iii) the surface molecule substitution and ligand exchange, iv) the interactions with ions present in the colloidal solution and v) the cooperative effects with other NPs present in solution. Along with this study, toxicological aspects associated to the reactivity of INPs (size, shape, chemical composition and structure) have been widely discussed.

7. Acknowledgment

The authors acknowledge financial support from the grants "Plan Nacional" from the Spanish Ministry of Science and Innovation (MAT2009-14734-C02-01) and "Nanotecnologías en biomedicina (NANOBIOMED)" from the Consolider-Ingenio 2010 Program (CSD2006-00012).

8. References

Ahrenkiel, S. P., Micic, O. I., Miedaner, A., Curtis, C. J., Nedeljkovic, J. M. & Nozik, A. J. (2003). Synthesis and Characterization of Colloidal InP Quantum Rods. *Nano Letters*, Vol.3, No. 6, pp. 833-837, ISNN 1530-6984

Aldana, J., Wang, Y. A. & Peng, X. G. (2001). Photochemical instability of CdSe nanocrystals coated by hydrophilic thiols. *Journal of the American Chemical Society*, Vol.123, No. 36, pp. 8844-8850, ISNN 0002-7863

Alivisatos, P. (2001). Less is more in medicine. *Scientific American*, Vol.285, No. 3, pp. 66-73, ISNN 0036-8733

Auffan, M., Rose, J., Bottero, J.-Y., Lowry, G. V., Jolivet, J.-P. & Wiesner, M. R. (2009). Towards a definition of inorganic nanoparticles from an environmental, health and safety perspective. *Nat Nano*, Vol.4, No. 10, pp. 634-641, ISNN 1748-3387

Barghorn, S., Nimmrich, V., Striebinger, A., Krantz, C., Keller, P., Janson, B., Bahr, M., Schmidt, M., Bitner, R. S., Harlan, J., Barlow, E., Ebert, U. & Hillen, H. (2005). Globular amyloid-peptide 1-42 oligome ;a homogenous and stable neuropathological protein in Alzheimer's disease. *Journal of Neurochemistry*, Vol.95, No. 3, pp. 834-847, ISNN

Bastús, N. G., Casals, E., Vázquez-Campos, S. & Puntes, V. (2008). Reactivity of engineered inorganic nanoparticles and carbon nanostructures in biological media. *Nanotoxicology*, Vol.2, No. 3, pp. 99 - 112, ISNN 1743-5390

Bastús, N. G., Comenge, J. & Puntes, V. Kinetically Controlled Seeded Growth Synthesis of Citrate-Stabilized Gold Nanoparticles of up to 200 nm: Size Focusing versus Ostwald Ripening. *Langmuir*, Vol.27, No. 17, pp. 11098-11105, ISNN 0743-7463

Bastús, N. G., Sanchez-Tillo, E., Pujals, S., Farrera, C., Lopez, C., Giralt, E., Celada, A., Lloberas, J. & Puntes, V. (2009). Homogeneous Conjugation of Peptides onto Gold Nanoparticles Enhances Macrophage Response. *ACS Nano*, Vol.3, No. 6, pp. 1335-1344, ISNN 1936-0851

Bratlie, K. M., Lee, H., Komvopoulos, K., Yang, P. & Somorjai, G. A. (2007). Platinum Nanoparticle Shape Effects on Benzene Hydrogenation Selectivity. *Nano Letters*, Vol.7, No. 10, pp. 3097-3101, ISNN 1530-6984

Burda, C., Chen, X., Narayanan, R. & El-Sayed, M. A. (2005). Chemistry and Properties of Nanocrystals of Different Shapes. *Chemical Reviews*, Vol.105, No. 4, pp. 1025-1102, ISNN 0009-2665

Cabot, A., Alivisatos, A. P., Puntes, V. F., Balcells, L., Iglesias, O. & Labarta, A. (2009). Magnetic domains and surface effects in hollow maghemite nanoparticles. *Physical Review B*, Vol.79, No. 9, pp. 094419

Cabot, A., Puntes, V. F., Shevchenko, E., Yin, Y., Balcells, L. s., Marcus, M. A., Hughes, S. M. & Alivisatos, A. P. (2007). Vacancy Coalescence during Oxidation of Iron

Nanoparticles. *Journal of the American Chemical Society*, Vol.129, No. 34, pp. 10358-10360, ISNN 0002-7863

Casals, E., Pfaller, T., Duschl, A., Oostingh, G. J. & Puntes, V. (2010). Time Evolution of the Nanoparticle Protein Corona. *ACS Nano*, Vol.4, No. 7, pp. 3623-3632, ISNN 1936-0851

Casals, E., Pfaller, T., Duschl, A., Oostingh, G. J. & Puntes, V. (2011). Hardening of the Nanoparticle Protein Corona in Metal (Au, Ag) and Oxide (Fe_3O_4, CoO and CeO_2) Nanoparticles. *Small*, Vol.7, No. 24, pp. 3479-3486

Casals, E., Vázquez-Campos, S., Bastús, N. G. & Puntes, V. (2008). Distribution and potential toxicity of engineered inorganic nanoparticles and carbon nanostructures in biological systems. *TrAC Trends in Analytical Chemistry*, Vol.27, No. 8, pp. 672-683, ISNN 0165-9936

Cho, K.-S., Talapin, D. V., Gaschler, W. & Murray, C. B. (2005). Designing PbSe Nanowires and Nanorings through Oriented Attachment of Nanoparticles. *Journal of the American Chemical Society*, Vol.127, No. 19, pp. 7140-7147, ISNN 0002-7863

Connor, E. E., Mwamuka, J., Gole, A., Murphy, C. J. & Wyatt, M. D. (2005). Gold nanoparticles are taken up by human cells but do not cause acute cytotoxicity. *Small*, Vol.1, No. 3, pp. 325-327, ISNN 1613-6810

Cozzoli, P. D., Pellegrino, T. & Manna, L. (2006). Synthesis, properties and perspectives of hybrid nanocrystal structures. *Chemical Society Reviews*, Vol.35, No. 11, pp. 1195-1208, ISNN 0306-0012

Derfus, A. M., Chan, W. C. W. & Bhatia, S. N. (2004). Probing the cytotoxicity of semiconductor quantum dots. *Nano Letters*, Vol.4, No. 1, pp. 11-18, ISNN 1530-6984

Diallo, M. S. & Savage, N. (2005). Nanoparticles and Water Quality. *Journal of Nanoparticle Research*, Vol.7, No. 4, pp. 325-330, ISNN 1388-0764

Dixit, V., VandenBossche, J., Sherman, D. M., Thompson, D. H. & Andres, R. P. (2006). Synthesis and Grafting of Thioctic Acid-PEG-Folate Conjugates onto Au Nanoparticles for Selective Targeting of Folate Receptor-Positive Tumor Cells. *Bioconjugate Chem.*, Vol.17, No. 3, pp. 603-609, ISNN 1043-1802

Dobrovolskaia, M. A. & McNeil, S. E. (2007). Immunological properties of engineered nanomaterials. *Nat Nano*, Vol.2, No. 8, pp. 469-478, ISNN 1748-3387

El-Sayed, I. H., Huang, X. & El-Sayed, M. A. (2005). Surface Plasmon Resonance Scattering and Absorption of anti-EGFR Antibody Conjugated Gold Nanoparticles in Cancer Diagnostics: Applications in Oral Cancer. *Nano Lett.*, Vol.5, No. 5, pp. 829-834, ISNN 1530-6984

Farokhzad, O. C., Cheng, J. J., Teply, B. A., Sherifi, I., Jon, S., Kantoff, P. W., Richie, J. P. & Langer, R. (2006). Targeted nanoparticle-aptamer bioconjugates for cancer chemotherapy in vivo. *Proceedings of the National Academy of Sciences of the United States of America*, Vol.103, No. 16, pp. 6315-6320, ISNN 0027-8424

Franklin, N. M., Rogers, N. J., Apte, S. C., Batley, G. E., Gadd, G. E. & Casey, P. S. (2007). Comparative Toxicity of Nanoparticulate ZnO, Bulk ZnO, and ZnCl2 to a Freshwater Microalga (Pseudokirchneriella subcapitata): The Importance of Particle Solubility. *Environmental Science & Technology*, Vol.41, No. 24, pp. 8484-8490, ISNN 0013-936X

Freitas, R. A. J. (1999). Nanomedicine, Volume I: Basic Capabilities. Georgetown, Landes Bioscience., ISNN 978-1-57059-680-3

Garcia, A., Espinosa, R., Delgado, L., Casals, E., Gonzalez, E., Puntes, V., Barata, C., Font, X. & Sanchez, A. (2010). Acute toxicity of cerium oxide, titanium oxide and iron oxide nanoparticles using standardized tests. *Desalination*, Vol.269, No. 1-3, pp. 136-141, ISSN 0011-9164

Gerion, D., Pinaud, F., Williams, S. C., Parak, W. J., Zanchet, D., Weiss, S. & Alivisatos, A. P. (2001). Synthesis and Properties of Biocompatible Water-Soluble Silica-Coated CdSe/ZnS Semiconductor Quantum Dots. *J. Phys. Chem. B*, Vol.105, No. 37, pp. 8861-8871, ISSN 1520-6106

Gojova, A., Guo, B., Kota, R. S., Rutledge, J. C., Kennedy, I. M. & Barakat, A. I. (2007). Induction of inflammation in vascular endothelial cells by metal oxide nanoparticles: Effect of particle composition. *Environmental Health Perspectives*, Vol.115, No. 3, pp. 403-409, ISSN 0091-6765

Goldstein, A. N., Echer, C. M. & Alivisatos, A. P. (1992). Melting in Semiconductor Nanocrystals. *Science*, Vol.256, No. 5062, pp. 1425-1427,

Grebinski, J. W., Hull, K. L., Zhang, J., Kosel, T. H. & Kuno, M. (2004). Solution-Based Straight and Branched CdSe Nanowires. *Chemistry of Materials*, Vol.16, No. 25, pp. 5260-5272, ISSN 0897-4756

Gupta, A. K. & Gupta, M. (2005). Cytotoxicity suppression and cellular uptake enhancement of surface modified magnetic nanoparticles. *Biomaterials*, Vol.26, No. 13, pp. 1565-1573, ISSN 0142-9612

Gupta, K. & Ghosh, U. C. (2009). Arsenic removal using hydrous nanostructure iron(III)-titanium(IV) binary mixed oxide from aqueous solution. *Journal of Hazardous Materials*, Vol.161, No. 2-3, pp. 884-892, ISSN 0304-3894

Han, G., You, C.-C., Kim, B.-j., Turingan, R. S., Forbes, N. S., Martin, C. T. & Rotello, V. M. (2006). Light-Regulated Release of DNA and Its Delivery to Nuclei by Means of Photolabile Gold Nanoparticles. *Angewandte Chemie International Edition*, Vol.45, No. 19, pp. 3165-3169, ISSN 1521-3773

Henglein, A. & Giersig, M. (2000). Optical and Chemical Observations on Gold-Mercury Nanoparticles in Aqueous Solution. *The Journal of Physical Chemistry B*, Vol.104, No. 21, pp. 5056-5060, ISSN 1520-6106

Hu, J., Chen, G. & Lo, I. M. C. (2005). Removal and recovery of Cr(VI) from waste water by maghemite nanoparticles. *Water Research*, Vol.39, No. 18, pp. 4528-4536, ISSN 0043-1354

Hu, J., Li, L.-s., Yang, W., Manna, L., Wang, L.-w. & Alivisatos, A. P. (2001). Linearly Polarized Emission from Colloidal Semiconductor Quantum Rods. *Science*, Vol.292, No. 5524, pp. 2060-2063

Hvolbaek, B., Janssens, T. V. W., Clausen, B. S., Falsig, H., Christensen, C. H. & Norskov, J. K. (2007). Catalytic activity of Au nanoparticles. *Nano Today*, Vol.2, No. 4, pp. 14-18, ISSN 1748-0132

Ito, A., Sun, X. & Tateishi, T. (2001). In-vitro analysis of metallic particles, colloidal nanoparticles and ions in wear-corrosion products of SUS317L stainless steel. *Materials Science and Engineering: C*, Vol.17, No. 1-2, pp. 161-166, ISSN 0928-4931

Kamat, P. V. (2007). Meeting the Clean Energy Demand: A Nanostructure Architectures for Solar Energy Conversion. *The Journal of Physical Chemistry C*, Vol.111, No. 7, pp. 2834-2860, ISSN 1932-7447

Karakoti, A., Monteiro-Riviere, N., Aggarwal, R., Davis, J., Narayan, R., Self, W., McGinnis, J. & Seal, S. (2008). Nanoceria as antioxidant: Synthesis and biomedical applications. *JOM Journal of the Minerals, Metals and Materials Society*, Vol.60, No. 3, pp. 33-37, ISSN 1047-4838

Karn, B., Kuiken, T. & Otto, M. (2009). Nanotechnology and *in Situ* Remediation: A Review of the Benefits and Potential Risks. *Environ Health Perspect*, Vol.117, No. 12, pp. 1813-1831

Katsikas, L., Gutierrez, M. & Henglein, A. (1996). Bimetallic Colloids: Silver and Mercury. *The Journal of Physical Chemistry*, Vol.100, No. 27, pp. 11203-11206, ISSN 0022-3654

Kogan, M. J., Bastus, N. G., Amigo, R., Grillo-Bosch, D., Araya, E., Turiel, A., Labarta, A., Giralt, E. & Puntes, V. F. (2006). Nanoparticle-mediated local and remote manipulation of protein aggregation. *Nano Letters*, Vol.6, No. 1, pp. 110-115, ISSN 1530-6984

Kumar, S. & Nann, T. (2006). Shape Control of II–VI Semiconductor Nanomaterials. *Small*, Vol.2, No. 3, pp. 316-329, ISSN 1613-6829

Lazaro, F. J., Abadia, A. R., Romero, M. S., Gutierrez, L., Lazaro, J. & Morales, M. P. (2005). Magnetic characterisation of rat muscle tissues after subcutaneous iron dextran injection. *Biochimica Et Biophysica Acta-Molecular Basis of Disease*, Vol.1740, No. 3, pp. 434-445, ISSN 0925-4439

Li, Y. & Somorjai, G. A. (2010). Nanoscale Advances in Catalysis and Energy Applications. *Nano Letters*, Vol.10, No. 7, pp. 2289-2295, ISSN 1530-6984

Lisha, K., Anshup, A. & Pradeep, T. (2009). Towards a practical solution for removing inorganic mercury from drinking water using gold nanoparticles. *Gold Bulletin*, Vol.42, No. 2, pp. 144-152, ISSN 0017-1557

López, X. & Castaño, V. M. (2008). Chromium Removal from Industrial Water Through Functionallized Nanoparticles *J. Nanosci. Nanotechnol*, Vol.8, No., pp. 5733–5738 ISNN

Manna, L., Scher, E. C. & Alivisatos, A. P. (2000). Synthesis of Soluble and Processable Rod-, Arrow-, Teardrop-, and Tetrapod-Shaped CdSe Nanocrystals. *Journal of the American Chemical Society*, Vol.122, No. 51, pp. 12700-12706, ISSN 0002-7863

Margeat, O., Ciuculescu, D., Lecante, P., Respaud, M., Amiens, C. & Chaudret, B. (2007). NiFe nanoparticles: A soft magnetic material? *Small*, Vol.3, No. 3, pp. 451-458, ISSN 1613-6810

Meyns, M., Bastus, N. G., Cai, Y., Kornowski, A., Juarez, B. H., Weller, H. & Klinke, C. (2010). Growth and reductive transformation of a gold shell around pyramidal cadmium selenide nanocrystals. *Journal of Materials Chemistry*, Vol. 20, pp. 10602-10605, ISSN 0959-9428

Mirkin, C. A., Letsinger, R. L., Mucic, R. C. & Storhoff, J. J. (1996). A DNA-based method for rationally assembling nanoparticles into macroscopic materials. *Nature*, Vol.382, No. 6592, pp. 607-609, ISSN 0028-0836

Mokari, T., Rothenberg, E., Popov, I., Costi, R. & Banin, U. (2004). Selective growth of metal tips onto semiconductor quantum rods and tetrapods. *Science*, Vol.304, No. 5678, pp. 1787-1790, ISSN 0036-8075

Murugesan, S., Park, T. J., Yang, H. C., Mousa, S. & Linhardt, R. J. (2006). Blood compatible carbon nanotubes - Nano-based neoproteoglycans. *Langmuir*, Vol.22, No. 8, pp. 3461-3463, ISSN 0743-7463

Nikoobakht, B. & El-Sayed, M. A. (2003). Preparation and growth mechanism of gold nanorods using seed-mediated growth method. *Chemistry of Materials*, Vol.15, No. 10, pp. 1957-1962, ISNN 0897-4756

Oberdörster, G. (2000). Toxicology of ultrafine particles: in vivo studies. *Philosophical Transactions of the Royal Society of London Series a-Mathematical Physical and Engineering Sciences*, Vol.358, No. 1775, pp. 2719-2739, ISNN 1364-503X

Oberdörster, G., Finkelstein, J., Johnston, C., Gelein, R., Cox, C., Baggs, R. & Elder, A. (2000). Acute Pulmonary Effects of Ultrafine Particles in Rats and Mice. *Res Rep Health Eff Inst*, Vol.96, No., pp. 5-74

Oberdorster, G., Oberdorster, E. & Oberdorster, J. (2005). Nanotoxicology: an emerging discipline evolving from studies of ultrafine particles. *Environ Health Perspect*, Vol.113, No. 7, pp. 823-39, ISNN 0091-6765 (Print)

Overbury, S. H., Schwartz, V., Mullins, D. R., Yan, W. & Dai, S. (2006). Evaluation of the Au size effect: CO oxidation catalyzed by Au/TiO_2. *Journal of Catalysis*, Vol.241, No. 1, pp. 56-65, ISNN 0021-9517

Pacheco, S., Medina, M., Valencia, F. & Tapia, J. (2006). Removal of Inorganic Mercury from Polluted Water Using Structured Nanoparticles, ASCE.

Peng, X., Manna, L., Yang, W., Wickham, J., Scher, E., Kadavanich, A. & Alivisatos, A. P. (2000). Shape control of CdSe nanocrystals. *Nature*, Vol.404, No. 6773, pp. 59-61, ISNN 0028-0836

Peng, X. G., Manna, L., Yang, W. D., Wickham, J., Scher, E., Kadavanich, A. & Alivisatos, A. P. (2000). Shape control of CdSe nanocrystals. *Nature*, Vol.404, No. 6773, pp. 59-61, ISNN 0028-0836

Pérez-Juste, J., Pastoriza-Santos, I., Liz-Marzán, L. M. & Mulvaney, P. (2005). Gold nanorods: Synthesis, characterization and applications. *Coordination Chemistry Reviews*, Vol.249, No. 17-18, pp. 1870-1901, ISNN 0010-8545

Pöselt, E., Fischer, S., Foerster, S. & Weller, H. (2009). Highly Stable Biocompatible Inorganic Nanoparticles by Self-Assembly of Triblock-Copolymer Ligands *Langmuir*, Vol.25, No. 24, pp. 13906-13913, ISNN 0743-7463

Pujals, S., Bastus, N. G., Pereiro, E., Lopez-Iglesias, C., Punte, V. F., Kogan, M. J. & Giralt, E. (2009). Shuttling Gold Nanoparticles into Tumoral Cells with an Amphipathic Proline-Rich Peptide. *Chembiochem*, Vol.10, No. 6, pp. 1025-1031, ISNN 1439-4227

Puntes, V. F., Krishnan, K. M. & Alivisatos, A. P. (2001). Colloidal nanocrystal shape and size control: The case of cobalt. *Science*, Vol.291, No. 5511, pp. 2115-2117, ISNN 0036-8075

Rancan, F., Rosan, S., Boehm, F., Cantrell, A., Brellreich, M., Schoenberger, H., Hirsch, A. & Moussa, F. (2002). Cytotoxicity and photocytotoxicity of a dendritic C-60 mono-adduct and a malonic acid C-60 tris-adduct on Jurkat cells. *Journal of Photochemistry and Photobiology B-Biology*, Vol.67, No. 3, pp. 157-162, ISNN 1011-1344

Rocker, C., Potzl, M., Zhang, F., Parak, W. J. & Nienhaus, G. U. (2009). A quantitative fluorescence study of protein monolayer formation on colloidal nanoparticles. *Nat Nano*, Vol.4, No. 9, pp. 577-580, ISNN 1748-3387

Rosi, N. L., Giljohann, D. A., Thaxton, C. S., Lytton-Jean, A. K. R., Han, M. S. & Mirkin, C. A. (2006). Oligonucleotide-modified gold nanoparticles for intracellular gene regulation. *Science*, Vol.312, No. 5776, pp. 1027-1030, ISNN 0036-8075

Roucoux, A., Schulz, J. r. & Patin, H. (2002). Reduced Transition Metal Colloids: Novel Family of Reusable Catalysts? *Chemical Reviews*, Vol.102, No. 10, pp. 3757-3778, ISNN 0009-2665

Savage, N. & Diallo, M. S. (2005). Nanomaterials and Water Purification: Opportunities and Challenges. *Journal of Nanoparticle Research*, Vol.7, No. 4, pp. 331-342, ISNN 1388-0764

Sayes, C. M., Fortner, J. D., Guo, W., Lyon, D., Boyd, A. M., Ausman, K. D., Tao, Y. J., Sitharaman, B., Wilson, L. J., Hughes, J. B., West, J. L. & Colvin, V. L. (2004). The differential cytotoxicity of water-soluble fullerenes. *Nano Letters*, Vol.4, No. 10, pp. 1881-1887, ISNN 1530-6984

Schliehe, C., Juarez, B. H., Pelletier, M., Jander, S., Greshnykh, D., Nagel, M., Meyer, A., Foerster, S., Kornowski, A., Klinke, C. & Weller, H. (2010). Ultrathin PbS Sheets by Two-Dimensional Oriented Attachment. *Science*, Vol.329, No. 5991, pp. 550-553

Slack, S. M. & Horbett, T. A. (1995). The Vroman Effect. Proteins at Interfaces II, American Chemical Society. 602: 112-128.

Smigelskas, A. D. & Kirkendall, E. O. (1947). Zinc Diffusion in Alpha Brass. *Transactions of the American Institute of Mining and Metallurgical Engineers*, Vol.171, No. 130-142

Son, D. H., Hughes, S. M., Yin, Y. & Paul Alivisatos, A. (2004). Cation Exchange Reactions in Ionic Nanocrystals. *Science*, Vol.306, No. 5698, pp. 1009-1012

Sun, Y., Mayers, B. T. & Xia, Y. (2002). Template-Engaged Replacement Reaction: A One-Step Approach to the Large-Scale Synthesis of Metal Nanostructures with Hollow Interiors. *Nano Letters*, Vol.2, No. 5, pp. 481-485, ISNN 1530-6984

Talapin, D. V., Rogach, A. L., Kornowski, A., Haase, M. & Weller, H. (2001). Highly luminescent monodisperse CdSe and CdSe/ZnS nanocrystals synthesized in a hexadecylamine-trioctylphosphine oxide-trioctylphospine mixture. *Nano Letters*, Vol.1, No. 4, pp. 207-211, ISNN 1530-6984

Tang, Z., Kotov, N. A. & Giersig, M. (2002). Spontaneous Organization of Single CdTe Nanoparticles into Luminescent Nanowires. *Science*, Vol.297, No. 5579, pp. 237-240

Taniguchi, S., Green, M. & Lim, T. (2011). The Room-Temperature Synthesis of Anisotropic CdHgTe Quantum Dot Alloys: A Molecular Welding Effect. *Journal of the American Chemical Society*, Vol.133, No. 10, pp. 3328-3331, ISNN 0002-7863

Tao, A. R., Habas, S. & Yang, P. (2008). Shape Control of Colloidal Metal Nanocrystals. *Small*, Vol.4, No. 3, pp. 310-325, ISNN 1613-6829

Tkachenko, A. G., Xie, H., Coleman, D., Glomm, W., Ryan, J., Anderson, M. F., Franzen, S. & Feldheim, D. L. (2003). Multifunctional Gold Nanoparticle-Peptide Complexes for Nuclear Targeting. *J. Am. Chem. Soc.*, Vol.125, No. 16, pp. 4700-4701, ISNN 0002-7863

Trentler, T. J., Hickman, K. M., Goel, S. C., Viano, A. M., Gibbons, P. C. & Buhro, W. E. (1995). Solution-Liquid-Solid Growth of Crystalline III-V Semiconductors: An Analogy to Vapor-Liquid-Solid Growth. *Science*, Vol.270, No. 5243, pp. 1791-1794

Vroman, L. (1962). Effect of Adsorbed Proteins on the Wettability of Hydrophilic and Hydrophobic Solids. *Nature*, Vol.196, No. 4853, pp. 476-477

Wagner, E., Curiel, D. & Cotten, M. (1994). Delivery of drugs, proteins and genes into cells using transferrin as a ligand for receptor-mediated endocytosis. *Advanced Drug Delivery Reviews*, Vol.14, No. 1, pp. 113-135

Wang, S., Mamedova, N., Kotov, N. A., Chen, W. & Studer, J. (2002). Antigen/Antibody Immunocomplex from CdTe Nanoparticle Bioconjugates. *Nano Lett.*, Vol.2, No. 8, pp. 817-822, ISNN 1530-6984

Weissleder, R., Stark, D. D., Engelstad, B. L., Bacon, B. R., Compton, C. C., White, D. L., Jacobs, P. & Lewis, J. (1989). Superparamagnetic Iron-Oxide - Pharmacokinetics and Toxicity. *American Journal of Roentgenology*, Vol.152, No. 1, pp. 167-173, ISNN 0361-803X

Woehrle, G. H., Brown, L. O. & Hutchison, J. E. (2005). Thiol-functionalized, 1.5-nm gold nanoparticles through ligand exchange reactions: Scope and mechanism of ligand exchange. *Journal of the American Chemical Society*, Vol.127, No. 7, pp. 2172-2183, ISNN 0002-7863

Xu, C., Su, J., Xu, X., Liu, P., Zhao, H., Tian, F. & Ding, Y. (2006). Low Temperature CO Oxidation over Unsupported Nanoporous Gold. *Journal of the American Chemical Society*, Vol.129, No. 1, pp. 42-43, ISNN 0002-7863

Yavuz, C. T., Mayo, J. T., Yu, W. W., Prakash, A., Falkner, J. C., Yean, S., Cong, L., Shipley, H. J., Kan, A., Tomson, M., Natelson, D. & Colvin, V. L. (2006). Low-Field Magnetic Separation of Monodisperse Fe_3O_4 Nanocrystals. *Science*, Vol.314, No. 5801, pp. 964-967

Yin, Y., Rioux, R. M., Erdonmez, C. K., Hughes, S., Somorjai, G. A. & Alivisatos, A. P. (2004). Formation of Hollow Nanocrystals Through the Nanoscale Kirkendall Effect. *Science*, Vol.304, No. 5671, pp. 711-714

Zhang, D. H., Kandadai, M. A., Cech, J., Roth, S. & Curran, S. A. (2006). Poly(L-lactide) (PLLA)/multiwalled carbon nanotube (MWCNT) composite: Characterization and biocompatibility evaluation. *Journal of Physical Chemistry B*, Vol.110, No. 26, pp. 12910-12915, ISNN 1520-6106

Zhang, J., Tang, Y., Lee, K. & Ouyang, M. (2010). Nonepitaxial Growth of Hybrid Core-Shell Nanostructures with Large Lattice Mismatches. *Science*, Vol.327, No. 5973, pp. 1634-1638

Zhang, W. X. (2003). Nanoscale Iron Particles for Environmental Remediation: An Overview. *Journal of Nanoparticle Research*, Vol.5, No. 3, pp. 323-332, ISNN 1388-0764

Section 2

Nano-Technology

7

Characteristics of the Laser-Induced Breakdown Detection of Colloidal Nanoparticles for Determining Particle Size

E.C. Jung and H.R. Cho
Nuclear Chemistry Research Division, Korea Atomic Energy Research Institute
Republic of Korea

1. Introduction

Laser-induced breakdown detection (LIBD) of colloidal nanoparticles has been investigated in various fields, such as the measurement of natural colloids in drinking water (Bundschuh et al., 2005; Kaegi et al., 2008; Wagner et al., 2005; Walther et al., 2006), the in-situ observation of colloid mediated pollutant transport (Hauser et al., 2002; Möri et al., 2003) and the real-time measurement of the solubility of radioactive elements (Bundschuh et al., 2000; Cho et al., 2008; Knopp et al., 1999; J.I. Kim, 2006; Neck et al., 2001, 2003; Opel et al., 2007; Walther et al., 2007). LIBD is an established technique that measures the size and concentration of colloidal particles in aqueous media using plasma formation, which is induced by focusing a short pulse laser beam into the solution (Kim & Walther, 2007). LIBD is especially efficient for detecting small particles less than 100 nm in diameter, which are not easily detectable using commercially available devices that adopt the measurement of a scattered light intensity, such as photon correlation spectroscopy (PCS) (Bundschuh et al., 2001b). Although several different LIBD systems have been developed over the last decades, a commercial instrument that adopts the LIBD method is currently unavailable.

This chapter is composed of a brief review of LIBD systems that adopt different detection schemes (Section 2), followed by the particle size determination methods (Section 3). Although these methods in principle allow the determination of particle size, until now, particle sizing capabilities in most LIBD experiments have been tested only for polystyrene reference particles with a well-defined size. Thus it should be verified that these LIBD methods are suitable for determining the particle size of different materials (Jung et al., 2011). Newly investigated characteristics of LIBD of colloidal uranium and alumina particles were reported in the last section (Section 4) followed by the summary (Section 5).

2. LIBD systems

In a recent publication, two different LIBD systems and their applications were reviewed in detail (Kim & Walther, 2007). One was based on the optical detection of plasma flash using either a photomultiplier tube (Ajiro et al., 1992; Fujimori et al., 1992) or a charge-coupled device (CCD) camera (Bundschuh et al., 2001a, 2005; Hauser et al., 2002; Jung et al., 2006, 2007; J.W. Kim et al., 2008; Walther et al., 2002). A second system was based on the acoustic detection of a shock wave using a piezoelectric transducer (PZT) (Bundschuh et al., 2001b; Izumida et al., 1998; Kitamori et al., 1988, 1989; Satio et al., 1999; Scherbaum et al., 1996; Walther et al., 2004). Recently, we developed a modified LIBD system based on the optical detection of a shock wave; this system measures the deflection of a probe laser beam due to a shock wave-induced change in the refractive index of the liquid medium (Cho et al., 2008; Jung et al., 2009). In the next subsections, these different LIBD systems are described briefly based on the experimental apparatuses installed in our laboratory (Jung et al., 2006, 2007, 2009, 2011; Cho et al., 2008).

2.1 Optical detection of plasma flash using a CCD camera

A schematic diagram of the experimental arrangement of the LIBD system using a CCD camera is shown in Fig. 1 (Jung et al., 2006, 2007, 2011). The second harmonic (wavelength=532 nm, pulse width=6 ns, and pulse repetition rate=20 Hz) of the pulsed Nd:YAG laser (Continuum, Surelite II) was used for laser-induced breakdown. In order to adjust the laser beam diameter to approximately 4 mm, an iris diaphragm was installed on the travel path of the pulsed laser beam. Two Glan-Thompson polarizers, required to adjust the laser pulse energy, were arranged downstream of the iris diaphragm. The first polarizer was installed in such a fashion as to allow it to rotate; the second polarizer was installed to fix the direction so that only a polarized laser beam perpendicular to a bottom plane could pass through it. Accordingly, the laser pulse energy incident on the sample could be adjusted by rotating the first polarizer. The laser beam profile measured by a laser beam profiler (Newport, LBP-1) showed an approximately 80% Gaussian profile and the pulse-to-pulse fluctuation of the laser pulse energy was within 2.5% for 3,000 laser pulses. The laser beam was focused on the interior of a sample cell (Hellma, 111-QS) using a plano-convex

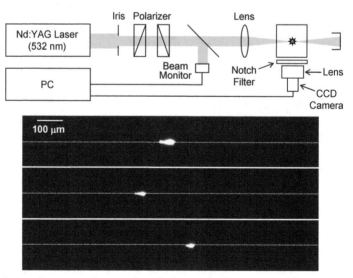

Fig. 1. Experimental setup and examples of plasma flashes acquired for each laser shot.

lens with a focal length of 40 mm. Laser-induced breakdowns occurred at the location on which the pulsed laser beam travelling in the Z-axis direction was focused.

The plasma flashes of the breakdown events were detected by a CCD camera (Hitachi Kokusai Electric, KP-F100BCL) with a variable macro-microscope. To prevent scattered laser light from reaching the CCD camera, a notch filter of ~1% transmission at 532 nm was inserted between the macro-microscope and the sample cell. Data from the CCD camera were recorded by a frame grabber card and processed by a home-made Labview software program. A typical example of the plasma flash is demonstrated in the lower part of Fig. 1, in which the optical image of a breakdown event is shown. The breakdown events occurred here and there along the laser beam propagation axis because of the Brownian motion of colloidal particles. In this study, the spatial distributions of 3,000 breakdown events were imaged on a pulse-to-pulse basis. A computer recorded the coordinates of plasma emission in the X-Z plane, the location and time of each breakdown event and the breakdown probability.

2.2 Acoustic detection of a shock wave using a PZT

A schematic diagram of the experimental arrangement of the LIBD system using a PZT is shown in Fig. 2. The role of each optical component was described in detail in previous subsection 2.1. When a laser-induced plasma is generated, a strong shock wave is propagated inside the liquid medium due to explosive expansion. To detect this shock wave, a PZT encapsulated in a cylindrical metal tube was attached to the wall of the sample cell. The coupling of metal tube housing to quartz sample cell was achieved by spiral springs. The waveform of the PZT signal was monitored using a digital oscilloscope (Tektronix TDS460A). The magnitude of the PZT signal was measured using a gated integrator (SRS SR250) and a boxcar averaging system (SRS SR280). The output was collected with a computer.

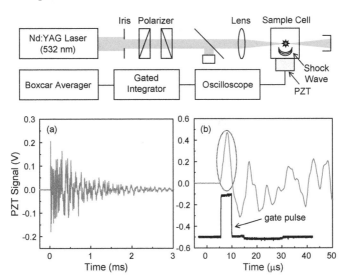

Fig. 2. Experimental setup and (a) PZT signal waveform acquired for a single laser shot, (b) early part of waveform (upper trace) and boxcar gate for signal detection (lower trace).

Fig. 2(a) and (b) shows the time trace as recorded by the PZT for a single laser shot and the early part of the waveform, respectively. The signal begins at a well-defined delay time t after the occurrence of the breakdown at $t = 0$, as shown in Fig. 2(b). This delay time was found to be linearly dependent on the distance between the focal point of the laser beam and the PZT surface. From these results, we are able to obtain a shock wave velocity of $\sim 1.48 \times 10^5$ cm/s. This is in good agreement with the established ultrasonic velocity at room temperature in water, which is $\sim 1.486 \times 10^5$ cm/s (Zapka et al., 1982).

In this experiment, the pulse height of the first signal in the waveform of the PZT signal, designated as a dotted circle in Fig. 2(b), showed a tendency to be proportional to the particle size (Kitamori et al., 1988, 1989). Thus the magnitude of the PZT signal was measured using a gated integrator with appropriate time delay and gate width as shown in the lower trace in Fig. 2(b). Because this signal magnitude was much higher than the magnitude of the electronic noise, each PZT signal magnitude greater than a certain threshold level was used to measure the breakdown event.

2.3 Optical detection of a shock wave using a probe beam

A schematic diagram of the experimental arrangement of the LIBD system using a probe beam is shown in Fig. 3 (Cho et al., 2008, Jung et al., 2009). The role of each optical component was described in detail in previous subsection 2.1. To detect the shock wave, a CW He-Ne laser (Uniphase, 20 mW) was used as a probe beam. When a laser-induced shock wave propagates through the medium, the refractive index of that medium changes. This change results in the deflection of the probe beam passing through the medium, and therefore the probe beam intensity reached at a high speed silicon photodiode (Thorlabs Det110) changes. A notch filter at 532 nm was used in order to prevent the photodiode from receiving the scattered light of the Nd:YAG laser. A pinhole (diameter of 0.5 mm) was used to detect the change in probe beam intensity for better sensitivity. The waveform of the probe beam signal was monitored using a digital oscilloscope (Tektronix TDS460A). The signal magnitude of the photodiode was measured using a gated integrator (SRS SR250) and a boxcar averaging system (SRS SR280). The output was collected with a computer.

Fig. 3(a) and (b) shows a typical waveform of a probe beam signal generated by a laser-induced shock wave and the early part of the waveform, respectively (Cho et al., 2008, Jung et al., 2009). The probe beam passed the 9.1 mm position below the focal point of the Nd:YAG laser beam, at which laser-induced breakdown occurred. The first peak at ~ 6.2 µs, which appears after the occurrence of the breakdown at $t = 0$, represents the deflected signal of the probe beam due to the shock wave. The second peak at ~ 9 µs results from the shock wave being reflected at the wall of the sample cell. The magnitude of the probe beam signal was measured using a gated integrator of appropriate time delay and gate width, as can be seen in the lower trace in Fig. 3(b). Comparing this technology with previously developed acoustic detection technology that used a PZT enables us to obtain a remote measurement in a non-contact manner. The probe beam signal begins at a well-defined delay time t after the occurrence of the breakdown at $t = 0$, as can be seen in Fig. 3(a) and (b). The delay time increases as the distance between the focal point of the laser beam and the probe beam position increases.

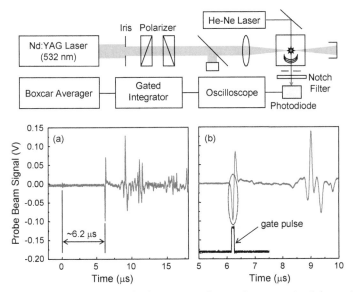

Fig. 3. Experimental setup and (a) probe beam signal waveform acquired for a single laser shot, (b) early part of waveform (upper trace) and boxcar gate for signal detection (lower trace).

Both a plasma emission and a shock wave can be measured simultaneously by constructing the detection components as shown in Fig. 4 (Jung et al., 2009). The laser beam propagated along the Z-axis direction was focused on an internal location of a sample cell. The probe beam propagated along the Y-axis direction passed by that location, which was lower than the focus of the pulsed laser beam by a distance "d" in the X-axis direction. By adjusting the distance downward from the focal point of the pulsed laser beam by approximately 12 mm, we were able to install the CCD camera and the PZT as can be seen in Fig. 4, to allow for simultaneous measurements.

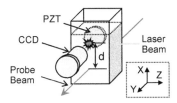

Fig. 4. Simultaneous detection using CCD camera, PZT and probe beam.

3. Particle size determination methods

The information on particle size is obtained by measuring the breakdown threshold energy which is defined as the minimum laser pulse energy required to generate a laser-induced plasma (Walther et al., 2002; Bundschuh et al., 2001a; Yun, 2007; Jung et al., 2011). The threshold energy can be determined from the breakdown probability measured as a function of the laser pulse energy (denoted as the "s-curve" in this chapter) (Walther et al.,

2002). Here, the breakdown probability is defined as the number of measured breakdown events divided by the total number of incident laser pulses.

Fig. 5 shows the breakdown probabilities as a function of the laser pulse energy for the 21 nm polystyrene particles at a laser wavelength of 532 nm. The empty and filled symbols represent the data derived from two different particle number densities, $5\times10^8/cm^3$ and $5\times10^7/cm^3$, for the same size particles. At a fixed particle number density, higher laser pulse energy resulted in higher breakdown probability. The threshold energy, corresponding to the laser pulse energy at which the breakdown probability is 0.01 in the s-curve, was determined in order to obtain the calibration curve for determining the particle size (Walther et al., 2002; Jung et al., 2011). At a fixed laser pulse energy, higher particle number density led to higher breakdown probability. As is apparent from the data shown in Fig. 5, the threshold energy is dependent on the particle concentration. In this study, the threshold energy was determined under experimental conditions in order to show that the increase in particle number density did not result in any further change of the threshold energy.

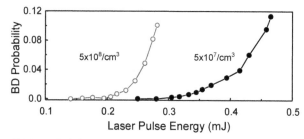

Fig. 5. Breakdown (BD) probability as a function of laser pulse energy for two different particle number densities. Data were obtained from polystyrene particles (size of 21 nm) at a laser wavelength of 532 nm.

Fig. 6 shows the dependence of the threshold energy on polystyrene particle size measured at a laser wavelength of 532 nm (Jung et al., 2011). At the fixed particle size, the threshold energy decreases with increasing particle number density, as depicted by the empty symbols in Fig. 6. The filled symbols indicate the minimum value of the threshold energy for each particle size. It is apparent that the threshold energy decreases with increasing particle size in this calibration curve.

The aforementioned particle sizing method, implemented by measuring the breakdown threshold energy, should be performed with a variation of the laser pulse energy. Under experimental conditions in which the laser pulse energy is fixed, the information on particle size can be obtained by measuring the spatial distribution of breakdown events (Bundschuh et al., 2001a; Hauser et al., 2002; Jung et al., 2007) and the frequency distribution curve of the magnitude of a laser-induced shock wave (Kitamori et al., 1989; Jung et al., 2009). In the next subsections, these methods are reported based on the experimental results obtained in our laboratory. In order to count only breakdown events induced by colloidal particles, the laser pulse energy was set to such a value that no breakdown event is observed in pure water.

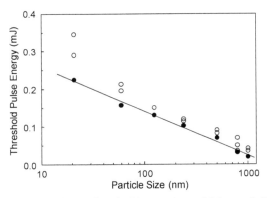

Fig. 6. Dependence of the breakdown threshold energies, which are defined as the energies at which the breakdown probability amounts to 0.01, on the particle size. Empty symbols represent the threshold energies determined from samples with different concentrations.

3.1 Particle sizing by measuring spatial distribution of breakdown events

The hyperbolic lines in Fig. 7 denote a focused laser beam propagated along the Z-axis; the spatial mode of the laser beam has a Gaussian profile with a beam waist of 5 µm at Z=0 (Jung et al., 2007). The exact focal point of the laser beam is designated as the origin of the coordinate axes. The contour lines in Fig. 7 represent the irradiance (equivalent to the power density) distribution of the laser beam. The regions indicated as "A" and "C" correspond to the upstream and downstream of the irradiance. The small ellipse indicated as "B" corresponds to the central region of the highest irradiance. As the laser beam waist increases along the Z-axis, the irradiance decreases, as depicted by the outer peanut-like shape.

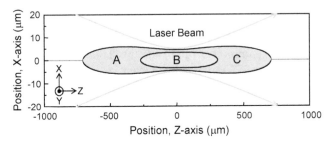

Fig. 7. Hyperbolic lines indicate the focused laser beam and contour lines indicate the laser irradiance distribution (A: upstream, B: central region of the highest irradiance, C: downstream).

When the CCD camera detects the plasma flash in the X-Z plane, as can be seen in Fig. 1, the spatial distribution of the breakdown events for small particles resembles the small ellipse shown in Fig. 7 because a breakdown is only induced at a region of a high irradiance, whereas the spatial distribution for large particles resembles the outer peanut-like shape due to their relatively low threshold energy for a breakdown (Bundschuh et al., 2001a).

Fig. 8 shows spatial distributions of 3,000 breakdown events for two different polystyrene particles (diameters of 21 nm and 60 nm) measured with a laser wavelength of 532 nm. Each data point represents the exact position on the X-Z plane at which a breakdown occurred. The length of the spatial distribution determined for 95% of the breakdown events (denoted the "effective focal length" in this manuscript) was measured in order to obtain the calibration curve for determining the particle size. At an incident laser pulse energy of 0.5 mJ, the breakdown probabilities were 0.2 for the 21 nm particles and 0.35 for the 60 nm particles, of which the concentrations were 1 parts per billion (ppb) and 7 ppb, respectively. The effective focal lengths were approximately 200 μm for the 21 nm particles and approximately 290 μm for the 60 nm particles. The spatial distributions of all data reported in this study exhibited good rotational symmetries centered on the origin of the Z-axis. Because the spatial distribution of the breakdown events along the Z-axis expands in proportion to a particle's diameter, a calibration for the particle sizing can be established by using these effective focal lengths (Bundschuh et al., 2001a).

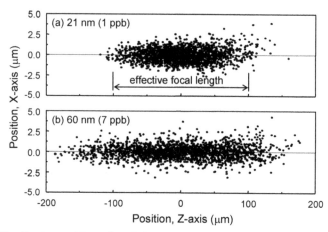

Fig. 8. Spatial distributions of 3,000 breakdown events for two different polystyrene particles (a) diameters of 21 nm (concentration of 1 ppb) and (b) 60 nm (concentration of 7 ppb). The length determined for the 95% of breakdown events was defined as the effective focal length.

3.2 Particle sizing by measuring frequency distribution of shock wave magnitude

Since each data point in Fig. 8 has its own acoustic signal magnitude, the main issue of this subsection is to investigate the size determination method by analyzing the acoustic signal magnitudes. Fig. 9 shows the distribution of PZT signal magnitude measured by the gated integrator and the boxcar averaging system, as explained in subsection 2.2. When a total of 3,000 laser pulses were incident to the sample cell containing the 21 nm polystyrene particles, 684 breakdown events, corresponding to a breakdown probability of 0.23, were measured. As is apparent from the data in Fig. 9, the signal magnitudes for the breakdown events were much higher (> 0.5 V) than the magnitude of the electrical noise of the PZT signal (< 0.2 V).

The frequency distribution curve of the PZT signal magnitudes, which were distributed in a wide voltage range, showed a Gaussian distribution, as plotted by solid line in Fig. 9. As an example of the frequency distribution curve, the data obtained from two different particles with diameters of 21 nm and 60 nm are shown in Fig. 10. Solid lines illustrate the Gaussian curves fitted to the experimental data points. The dependence of the peak and width of the distribution curve on the particle size is distinctive. The peak and width of the distribution curve increase as the particle size increases (Kitamori et al., 1989).

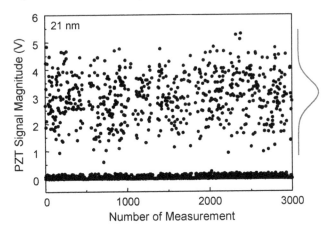

Fig. 9. Distribution of PZT signal magnitude for a total of 3,000 incident laser pulses.

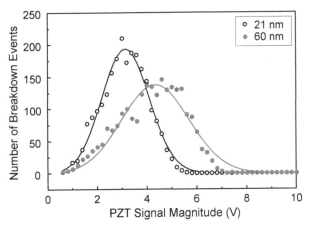

Fig. 10. The number of breakdown events as a function of the PZT signal magnitude for two different polystyrene particles (21 and 60 nm). Solid lines illustrate the Gaussian curves fitted to the experimental data points.

Although the detection method using PZT is very simple, this method has a drawback. It is difficult to obtain reproducible data due to the physical characteristics that are present when a PZT is directly attached to a sample cell to measure shock waves. Because several samples having particles of different sizes are contained in different sample cells, it is not easy to firmly attach the PZT to each sample cell in exactly the same manner. Therefore, difficulty arises when the magnitudes of the PZT signals are compared with each other for a plurality of different sample cells (Jung et al., 2009).

Comparing the optical detection method using a probe beam with the acoustic detection method using a PZT, the former method is advantageous for detecting harmful colloidal particles (such as radioactive substances), which must be located in a special environment isolated from their surroundings (in a glove box or in a cleaning booth), because of its capability for remote measurement of a shock wave in a non-contact manner (Cho et al., 2008). At first, the effect of the particle size on the waveform of the probe beam signal was investigated. The pulse height and width of the first signal in the waveform, designated as a dotted circle in Fig. 3(b), were different for particles with different sizes (data not shown). It was observed that the pulse height and width increased as the particle size increased (Jung et al., 2009).

Fig. 11 shows the results obtained by measuring the frequency distribution of the probe beam signal magnitude with respect to standard polystyrene nanoparticles having different sizes under conditions in which the delay and width of the gate pulse were fixed, as can be seen in Fig. 3(b). The concentrations of the particles having sizes of 21, 33, and 60 nm were 1, 2 and 7 ppb, respectively. In Fig. 11, the X-axis denotes the magnitude of the probe beam signal; the Y-axis denotes the number of breakdown events of the respective polystyrene nanoparticles having different sizes. The interval of the X-axis during which the data was processed to form a frequency distribution curve is 0.1 V. As the size of the polystyrene nanoparticles increases, the peak of the frequency distribution curve increases. Therefore, the calibration curve required for the determination of the particle size can be obtained by using the peaks of different frequency distribution curves (Jung et al., 2009).

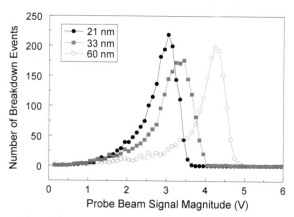

Fig. 11. The number of breakdown events as a function of the probe beam signal magnitude for three different polystyrene particles (21, 33 and 60 nm).

4. Material dependent characteristics of LIBD

Although the aforementioned LIBD methods in principle allow the determination of particle size, until now, particle sizing capabilities in most LIBD experiments have been tested only for polystyrene reference particles with a well-defined size. Because physical processes such as multiphoton ionization (MPI) and electron cascade growth affect the generation of laser-induced plasma, characteristics of LIBD (the breakdown threshold energy in the s-curve, the spatial distribution of breakdown events and the frequency distribution curve of the shock wave magnitude) may appear in different ways for different materials having the same particle size. Thus, it should be verified that these LIBD methods are suitable for determining the particle size of different materials (Jung et al., 2011).

In this section, LIBD experiments were performed for several different materials: polystyrene, uranium, silica and alumina nanoparticles. Colloidal uranium particles are generated from a spent nuclear fuel, thus, the measurement of their sizes is of interest to understand their behavior in groundwater, to allow for the safety assessment of nuclear waste disposal. We were also motivated to examine the size of silica and alumina nanoparticles, which can form natural colloids in groundwater, because these colloidal particles serve as crucial carriers for the migration of radionuclides in groundwater.

4.1 Sample preparation

Monodisperse polystyrene (21, 33, 60, 125, 240, 500, 800 and 1000 nm, Duke Scientific), silica (60, 238, 490, 730 and 990 nm, Corpuscular and Duke Scientific) and alumina (50, 240, 500, 800 and 1080 nm, Corpuscular) particles of a well-defined size were used in this study. Before sample preparation, the company-certified particle sizes of these reference particles were confirmed once again in our laboratory by TEM (Transmission Electron Microscopy, Jeol Ltd., JEM-2000FXII) and PCS (Malvern, Zetasizer, Nano ZS90). These particles were suspended in ultra-pure water from a water purification apparatus (Millipore, Milli-Q-element).

For the uranium sample, a pure U(VI) solution in $HClO_4$ (Merck, analysis grade) was prepared from natural UO_2 powder dissolved in $HClO_4$. A U(VI) stock solution of 1.0 mM at pH 3.8 in 0.1 M $NaClO_4$ was prepared by addition of a 1.0 M NaOH solution (Sigma-Aldrich, 99.99% decarbonated NaOH) at room temperature (Cho et al., 2008; Jung et al., 2009). The stock solution was equilibrated with air for one year. The uranium concentration of the stock solution was determined by using a kinetic phosphorescence analyzer (Chemcheck, KPA-11). U(VI) samples were prepared in a glove box purged with Ar gas. An aliquot of the stock solution was slowly titrated with four different 0.1 M $NaClO_4$ solutions at pH levels of 3.8, 9, 10, and 11 to maintain the ionic strength of the samples at 0.1 M $NaClO_4$. The uranium concentration of each sample was calculated from the measured concentration of the stock solution and the dilution factor during titration. The basic solutions (pH 9, 10, 11) of 0.1 M $NaClO_4$ were prepared with 99.99% decarbonated NaOH (Sigma-Aldrich) in a glove box. For purification, $NaClO_4:H_2O$ (Merck, analysis grade) was carefully recrystallized, diluted in pure water and filtered through a membrane filter (Advantec, mixed cellulose esters) with the pore size of 100 nm. By using LIBD system before the titration of the samples, it was proved that all the chemicals, including the uranium stock solution, contained no detectable colloidal particles. To minimize the adsorption of the formed uranium colloids on the surface of the test tubes, U(VI) samples were prepared in Teflon FEP tubes (Oak Ridge). U(VI) samples were equilibrated at 298±2 K during over one year in a glove box. The pH measurement was carried out with a glass combination pH electrode (Orion, Ross type, 6 mm diameter) calibrated with four pH buffer solutions (Mettler Toledo, pH 2.00-9.21). For the size measurement of uranium colloidal particles, a portion of the uranium sample was delivered in a sealable quartz cell out of a glove box.

4.2 Particle sizing of uranium colloids with a laser wavelength of 532 nm

Fig. 12 shows the breakdown probability as a function of the pH of the uranium sample at the concentration of approximately 1 parts per million (ppm). The breakdown probabilities were almost zero below a pH value of 5.5, at which the given uranium concentration exceeded the solubility limit of U(VI) hydrolysis compounds. At a pH of 5.6 there was an observable increase in breakdown probability and uranium colloids were formed. After the onset of breakdown, the breakdown probability increased with the increase of the pH value. Using LIBD, we determined the mean particle size to be approximately 600 nm in the sample at a pH of 7.0 just after preparing the sample. In this experiment, the particle size of colloidal uranium was determined by using the calibration curve made from the effective focal length of the spatial distribution of breakdown events of polystyrene nanoparticles.

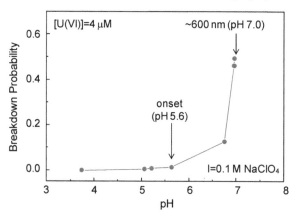

Fig. 12. Breakdown probability as a function of pH shows the formation of uranium colloidal particles.

One year after the sample preparation, the LIBD experiment was performed again to measure the size of the colloidal particles for the same sample. The solid symbol in Fig. 13 indicates the effective focal length obtained from the colloidal uranium particles, which were acidified to a pH of 5.8 after one year. When the effective focal length determined from the uranium sample was compared directly with the effective focal length values of the reference polystyrene particles, the mean particle size was found to be approximately 43 nm. From the change of the pH value from 7.0 to 5.8 and the change of the particle size from approximately 600 to 50 nm, we speculate that only small particles remained as suspended colloidal particles in the solution (Jung et al., 2009).

Because the characteristics of LIBD are dependent on not only the particle size but also the material properties, it is essential to compare the particle size determined using LIBD with the size determined using other methods, such as PCS and TEM. We tried to measure the particle size of uranium colloids, designated by the solid symbol in Fig. 13, using PCS, but we could not measure the particle size of uranium colloids because of their extremely low concentration. For the purpose of comparison, we prepared another sample that showed a strong enough light scattering intensity of uranium particles for PCS (Jung et al., 2009). The supernatant of the uranium sample (1.0 mM of U(VI) at pH 6.12) was carefully concentrated by ultrafiltration using a cellulose filter (10 kD, Centrion YM-10, Millipore). The concentrated supernatant was collected in a sealable quartz cell.

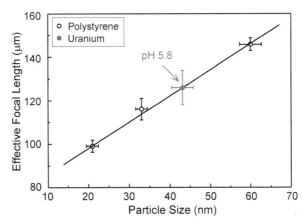

Fig. 13. Size measurement of colloidal uranium particles (solid symbol). The calibration curve was obtained from the effective focal lengths of the polystyrene reference particles with different sizes of 21, 33 and 60 nm in diameter (open symbols).

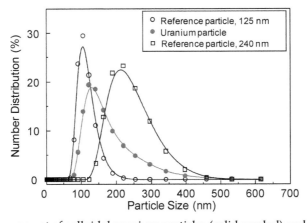

Fig. 14. Size measurement of colloidal uranium particles (solid symbol) and polystyrene reference particles (open symbols) using PCS.

The experimental number distributions for three different particles (two reference polystyrene samples and concentrated uranium sample) measured by PCS are presented in Fig. 14. The average sizes of the polystyrene particles were 143 and 257 nm with polydispersity indexes ranging from 0.05 to 0.08, respectively. Due to the hydrodynamic effect, these sizes were slightly larger than the sizes of 125 and 240 nm in diameter determined by the manufacturer using TEM. The average size of uranium colloids in Fig. 14 was 210 nm with a polydispersity index of 0.15. The polydispersity index of uranium colloids, higher than those of reference particles, indicates the heterogeneity of uranium colloids in size and morphology. Nevertheless, the cumulated mean sizes obtained by repeating the experiments for one week were in the range of 205-215 nm (Jung et al., 2009).

A TEM image of colloidal uranium particles and the number-weighted particle size distribution from the TEM image analysis are shown in Fig. 15. The average size of the uranium colloids shown in Fig. 15 is approximately 200 nm, with a high polydispersity. Even though the shapes of the uranium particles are not ideal spheres, the average size of the uranium particles obtained from TEM is found to reasonably agree with the size determined by PCS.

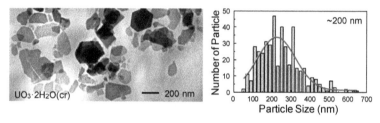

Fig. 15. TEM image of colloidal uranium particles and number-weighted particle size distribution obtained from TEM analysis.

Fig. 16 shows the particle sizes of concentrated uranium colloids determined by using the calibration curves made from the spatial distribution of 3,000 breakdown events for the reference polystyrene particles (21, 33, 60, 125 and 240 nm in diameter). The open and filled symbols in Fig. 16 represent the mean values obtained by repeating the experiments on the polystyrene and uranium particles, respectively. When the effective focal lengths determined in the uranium sample were compared directly with those of the reference particles, the estimated mean particle size was approximately 158 nm. Taking into account the polydispersity of the uranium colloids, the mean particle size determined using LIBD was in reasonable agreement with the sizes determined using PCS and TEM. Thus, we can obtain information on the relative differences or changes in the particle size of uranium colloids by using the present LIBD method (Jung et al., 2009).

To minimize the polydispersity of uranium particles, the large particles in the supernatant of the uranium sample were eliminated by filtration using a membrane filter (mixed cellulose ester, Advantec) having a pore size of 100 nm. The solid symbol designated by arrow in Fig. 16 represents the datum obtained from the filtrated sample. When the uranium datum was compared directly with the polystyrene data, the mean particle size of the filtrated uranium colloids was found to be approximately 31 nm. We tried to measure the particle size of the filtrated uranium colloids using PCS and TEM, but we could not measure the particle size of uranium colloids because of their extremely low concentration. Fig. 17 illustrates the spatial distributions of breakdown events for the same uranium particles as can be seen in Fig. 16. As is apparent from Fig. 17, the uranium colloids in the filtered sample show shorter effective focal length in the spatial distribution of breakdown events than that measured from the unfiltered, supernatant sample.

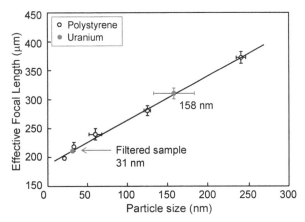

Fig. 16. Size measurements of colloidal uranium particles using effective focal lengths. The calibration curve was obtained from the size of the polystyrene reference particles having diameters of 21, 33, 60, 125 and 240 nm.

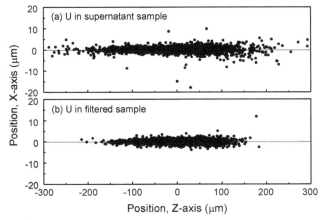

Fig. 17. Spatial distributions of 3,000 breakdown events for two different uranium particles in supernatant sample (a) and filtered sample (b).

4.3 Particle sizing of alumina colloids with a laser wavelength of 441 nm

Fig. 18 shows the effective focal lengths measured at a laser wavelength of 532 nm as a function of particle size for three different materials: polystyrene, silica and alumina colloidal particles. As already reported in the previous section, the effective focal length for polystyrene particles, designated as empty circles, is observed to be directly correlated to the particle size. In contrast, the effective focal lengths, designated as solid circles for alumina and square symbols for silica particles, are not correlated to the particle size. Although the solid lines fitted to the data points show slightly increasing behavior with poor correlation coefficients, it seems that the effective focal lengths for silica (Jung et al., 2011) and alumina are almost unchanged as the particle size increases.

The reason for these phenomena can be understood in terms of the difference in the ionization potentials (IPs) of these materials: ~7.8 eV for polystyrene, ~11.7 eV for silica and ~9.1 eV for alumina (Yun, 2007; Porter et al., 1955; B.H. Stephan, 1991). It is generally accepted that higher irradiance is required for the breakdown of a material with a higher IP. When a laser pulse at a wavelength of 532 nm (photon energy of ~2.33 eV) is used for the breakdown, simultaneous 4-photon absorption is required to induce MPI of polystyrene, while at least simultaneous 5-photon absorption is required for the MPI of silica and alumina. Thus, higher laser pulse energy is a prerequisite for the breakdown of silica and alumina, compared with that for polystyrene particles. It was reported that the breakdown threshold energy of silica was approximately 1.8-2.2 times higher than that of polystyrene particles (Bundschuh et al., 2001; Yun, 2007). The results shown in Fig. 18 imply that a size determination of silica and alumina particles using the spatial distribution of breakdown events measured at a laser wavelength of 532 nm is invalid under the present experimental conditions (Jung et al., 2011).

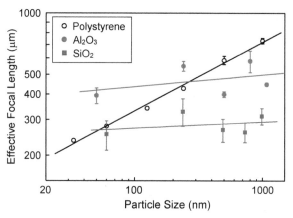

Fig. 18. Effective focal length as a function of particle size for polystyrene, silica and alumina particles at the laser wavelength of 532 nm.

As illustrated in Fig. 7, the irradiance distribution along the laser beam propagation axis can be divided into three distinct regions: the central region of highest irradiance near the focal point (region A), and the outside upstream and downstream regions of weak irradiance (regions B and C). The correlation between the effective focal length and the particle size for polystyrene reflects the fact that the breakdown of large particles is induced at the outside upstream and downstream regions, as well as at the central region, because the threshold energies of large particles are lower than those of small particles. The unchanged effective focal lengths for several silica and alumina particles indicate that breakdowns are only induced at the central region of high irradiance. It is speculated that these phenomena occur because the irradiance of the outside upstream and downstream regions is too low for breakdown, which requires simultaneous 5-photon absorption for the MPI (Jung et al., 2011).

The results shown in Fig. 19 illustrate the dependence of the effective focal length on the laser pulse energy, corroborating this speculation. Fig. 19 shows the effective focal length as a function of laser pulse energy for polystyrene, silica and filtrated uranium particles. For polystyrene particles with different diameters (21, 33 and 60 nm), the effective focal length increases almost linearly with the increase of the laser pulse energy, ranging from 0.4 to 0.6 mJ. Similar to that of polystyrene particles, the effective focal length of filtrated uranium shows increasing behavior with increasing laser pulse energy. Our recent measurement of solubility product of U(VI) hydrolysis compounds suggests that the colloidal particles are $UO_3 \cdot 2H_2O(cr)$ (Cho et al., 2008). The information on the IP of $UO_3 \cdot 2H_2O(cr)$ particles is unknown at this time. However, under the assumption that the IP of $UO_3 \cdot 2H_2O(cr)$ is not much different from the IP of UO_2 (IP=~6.17 eV) (Han et al., 2003), the MPI of the uranium compound can easily occur with 3- or 4-photon absorption at the photon energy level of ~2.33 eV. For silica (60 nm in diameter), however, the effective focal length does not change as the laser pulse energy increases, ranging from 0.5 to 0.6 mJ (Jung et al., 2011). With laser pulse energy above 0.6 mJ, breakdown occurs even for ultra-pure water.

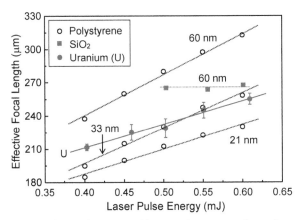

Fig. 19. Effective focal length as a function of laser pulse energy for polystyrene, uranium and silica particles at the laser wavelength of 532 nm.

When the laser photon energy is high enough, the MPI processes of silica and alumina may occur via simultaneous 4-photon absorption. As a result, breakdown events may occur in the upstream and downstream regions, as well as in the central region, because 4-photon absorption occurs at relatively low irradiance compared with 5-photon absorption. In this case, the correlation between the effective focal length in the spatial distribution of breakdown events and the particle size is expected even for silica and alumina particles.

To observe this effect, the dependence of the effective focal length on the particle size was investigated with a laser wavelength of 441 nm (photon energy of ~2.81 eV), and the results are shown in Fig. 20. It should be noted that the measurement of the incident wavelength-dependent characteristics of LIBD was performed by using a laser pulse generated from a wavelength tunable OPO (Optical Parametric Oscillator, OPOTEK, Vibrant) system. In this experiment, the OPO laser beam was focused on the interior of a sample cell using a bi-convex lens with a focal length of 15 mm. To prevent scattered laser light from reaching the CCD camera, a notch filter of ~1% transmission at 441 nm was inserted between the macro-microscope and the sample cell. When breakdown was induced by 441 nm laser radiation, the 4-photon absorption energy of approximately 11.2 eV exceeded the IP of ~9.1 eV for alumina. Therefore, the dependence of the effective focal length on the particle size was obviously observed for alumina, which is designated as filled symbols in Fig. 20, as expected. More recently, the results on the calibration curve obtained with 355 nm laser radiation (photon energy of ~3.49 eV) was reported for silica particles and the similar dependence of the effective focal length on the particle size was observed (Jung et al., 2011).

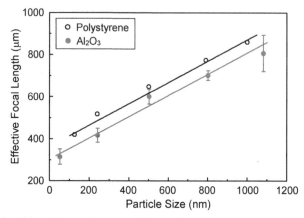

Fig. 20. Effective focal length as a function of particle size for polystyrene and alumina particles at the laser wavelength of 441 nm.

5. Summary

Several different LIBD systems using different detection devices, such as a PZT, a CCD camera and an optical probe beam, were described with their own particle size determination methods. In this chapter, material-dependent characteristics of LIBD were discussed for four different colloidal particles: polystyrene, $UO_3 \cdot 2H_2O(cr)$, silica, and alumina. When a focused laser pulse of 532 nm wavelength is used to generate laser-induced plasma, simultaneous 4-photon absorption is required to induce MPI of polystyrene and $UO_3 \cdot 2H_2O(cr)$; however, at least 5-photon absorption is required to induce MPI of silica and alumina. Thus, it was observed that the spatial distribution of the breakdown events measured with 532 nm laser radiation increased with increasing particle size for polystyrene and uranium compound, while spatial distribution of breakdown events was nearly unchanged for silica and alumina. The unchanged effective focal length with increase of particle size for silica and alumina resulted from the laser irradiance in the outside upstream and downstream regions, which was insufficient to induce MPI of these particles via simultaneous 5-photon absorption. When breakdown was induced by a laser wavelength of 441 nm, a correlation between the effective focal length and the particle size was observed for alumina because the simultaneous 4-photon absorption energy exceeds the IP of this material. Thus, the calibration curve obtained with an appropriate laser wavelength can be used to determine unknown particle sizes, even for materials with a high IPs.

6. Acknowledgment

This work was supported by the nuclear research and development program (2007-2011) through the National Research Foundation (NRF) of Korea funded by the Ministry of Education, Science and Technology (MEST). Part of this work was supported by the research and education program (2007) through the NRF of Korea funded by the MEST. The authors thank Mr. Hyungsoo Jung, Mr. Kilho Lee, and Mr. Sanghyun Park for obtaining data in section 3.1 and 3.2. The authors thank Ms. M.R. Park for obtaining data in section 4.3. The authors thank Dr. H.M. Kim for analyzing TEM results in section 4.2.

7. References

Ajiro, T.; Fujimori, H.; Matsui, T. & Izumi, S. (1992). Particle Size Dependence of Correlation Between Plasma Emission Delay Time and Plasma Emission Intensity of Laser Breakdown Induced by a Particle. *Japanese Journal of Applied Physics*, Vol. 31, No. 9, pp. 2760-2761, ISSN 0021-4922

Bundschuh, T.; Knopp, R.; Müller, R.; Kim, J.I.; Neck, V. & Fanghänel, Th. (2000). Application of LIBD to the Determination of the Solubility Product of Thorium(IV)-colloids. *Radiochimica Acta*, Vol. 88, No. 9-11, pp. 625-629, ISSN 0033-8230

Bundschuh, T.; Hauser, W.; Kim, J.I.; Knopp, R. & Scherbaum, F.J. (2001a). Determination of Colloid Size by 2-D Optical Detection of Laser Induced Plasma. *Colloids and Surfaces A: Physicochemical and Engineering Aspects*, Vol. 180, pp. 285-293, ISSN 0927-7757

Bundschuh, T.; Knopp, R. & Kim, J.I. (2001b). Laser-induced Breakdown Detection (LIBD) of Aquatic Colloids with Different Laser Systems. *Colloids and Surfaces A: Physicochemical and Engineering Aspects*, Vol. 177, pp. 47-55, ISSN 0927-7757

Bundschuh, T.; Wagner, T. U. & Koster, R. (2005). Laser-induced Breakdown Detection (LIBD) for the Highly Sensitive Quantification of Aquatic Colloids. Part I: Principle of LIBD and Mathematical Model. *Particle and Particle Systems Characterization*, Vol. 22, No. 3, pp. 172-180, ISSN 1521-4117

Cho, H.-R.; Jung, E.C. & Jee, K.Y. (2008). Probe Beam Detection of Laser-induced Breakdown for Measuring Solubility of Actinide Compounds. *Japanese Journal of Applied Physics*, Vol. 47, No. 5, pp. 3530-3532, ISSN 0021-4922

Fujimori, H.; Matsui, T.; Ajiro, T.; Yokose, K.; Hsueh, Y-M. & Izumi, S. (1992). Detection of Fine Particles in Liquids by Laser Breakdown Method. *Japanese Journal of Applied Physics*, Vol. 31, No. 5, pp. 1514-1518, ISSN 0021-4922

Han, J.; Kaledin, L.A.; Goncharov, V.; Komissarov, A.V.; Heaven, M.C. (2003). Accurate Ionization Potentials for UO and UO_2: A Rigorous Test of Relativistic Quantum Chemistry Calculations. *Journal of American Chemical Society*, Vol. 125, No. 24, pp. 7176-7177, ISSN 0002-7863

Hauser, W.; Geckeis, H.; Kim, J.I. & Fierz, Th. (2002). A Mobile Laser-induced Breakdown Detection System and Its Application for the in situ-monitoring of Colloid Migration. *Colloids and Surfaces A: Physicochemical and Engineering Aspects*, Vol. 203, 37-45, ISSN 0927-7757

Izumida, S.; Onishi, K. & Saito, M. (1998). Estimation of Laser-induced Breakdown Threshold of Microparticles in Water. *Japanese Journal of Applied Physics*, Vol. 37, No. 4, pp. 2039-2042, ISSN 0021-4922

Jung, E.C.; Yun, J.-I.; Kim, J.I.; Park, Y.J.; Park, K.K.; Fanghänel, Th. & Kim, W.H. (2006). Size Measurement of Nanoparticles Using the Emission Intensity Distribution of Laser-induced Plasma. *Applied Physics B*, Vol. 85, pp. 625-629, ISSN 0946-2171

Jung, E.C.; Yun, J.-I.; Kim, J.I.; Bouby, M.; Geckeis, H.; Park, Y.J.; Park, K.K.; Fanghänel, Th. & Kim, W.H. (2007). Measurement of Bimodal Size Distribution of Nanoparticles by Using the Spatial Distribution of Laser-induced Plasma. *Applied Physics B*, Vol. 87, pp. 497-502, ISSN 0946-2171

Jung, E.C.; Cho, H.-R.; Park, K.K.; Yeon, J.-W. & Song, K. (2009). Nanoparticle Sizing by a Laser-induced Breakdown Detection Using an Optical Probe Beam Deflection. *Applied Physics B*, Vol. 97, pp. 867-875, ISSN 0946-2171

Jung, E.C.; Cho, H.-R.; Park, M.R. (2011). Laser-Induced Breakdown Detection of Colloidal Silica Nanoparticles. Sumitted to *Applied Physics B*, ISSN 0946-2171

Kaegi, R.; Wagner, T.; Hetzer, B.; Sinnet, B.; Tzvekov, G. & Boller, M. (2008). Size, Number and Chemical Composition of Nanosized Particles in Drinking Water Determined by Analytical Microscopy and LIBD. *Water Research*, Vol. 42, pp. 2778-2786, ISSN 0043-1354

Kim, J.I. (2006). Significance of Actinide Chemistry for the Long-term Safety of Waste Disposal. *Nuclear Engineering and Technology*, Vol. 38, No. 6, pp. 459-482, ISSN 1738-5733

Kim, J.I. & Walther C. (2007). Laser-induced Breakdown Detection, In *Environmental Colloids and Particles: Behaviour, Separation and Characterisation*, Wilkinson, K.J. & Lead, J.R. (Ed.), pp. 555-612, Wiley, ISBN 978-0-470-02432-4, Chichester, U.K.

Kim, J.W.; Son, J.A.; Yun, J.-I.; Jung, E.C.; Park, S.H. & Choi, J.G. (2008). Analysis of Laser-induced Breakdown Images Measuring the Sizes of Mixed Aquatic Nanoparticles. *Chemical Physics Letters*, Vol. 462, No. 1-3, pp. 75-77, ISSN 0009-2614

Kitamori, T.; Yokose, K.; Suzuki, K.; Sawada, T. & Gohshi, Y. (1988). Laser Breakdown Acoustic Effect of Ultrafine Particle in Liquids and Its Application to Particle Counting. *Japanese Journal of Applied Physics*, Vol. 27, No. 6, pp. L983-L985, ISSN 0021-4922

Kitamori, T.; Yokose, K.; Sakagami, M. & Sawada, T. (1989). Detection and Counting of Ultrafine Particles in Ultrapure Water Using Laser Breakdown Acoustic Method. *Japanese Journal of Applied Physics*, Vol. 28, No. 7, pp. 1195-1198, ISSN 0021-4922

Knopp, R.; Neck, V. & Kim, J.I. (1999). Solubility, Hydrolysis and Colloid Formation of Plutonium(IV). *Radiochimica Acta*, Vol. 86, pp. 101-108, ISSN 0033-8230

Möri, A.; Alexander, W.R.; Geckeis, H.; Hauser, W.; Schäfer, T.; Eikenberg, J.; Fierz, T.; Degueldre, C. & Missana, T. (2003). The Colloid and Radionuclide Retardation Experiment at the Grimsel Test Site: Influence of Bentonite Colloids on Radionuclide Migration in a Fractured Rock. *Colloids and Surfaces A: Physicochemical and Engineering Aspects*, Vol. 217, No. 1-3, pp. 33-47, ISSN 0927-7757

Neck, V.; Kim, J.I.; Seidel, B.S.; Marquardt, C.M.; Dardenne, K.; Jensen, M.P. & Hauser, W. (2001). A Spectroscopic Study of the Hydrolysis, Colloid Formation and Solubility of Np(IV). *Radiochimica Acta*, Vol. 89, No. 7, pp. 439-446, ISSN 0033-8230

Neck, V.; Altmaier, M.; Müller, R.; Bauer, A. & Fanghänel, Th. (2003). Solubility of Crystalline Thorium Dioxide. *Radiochimica Acta*, Vol. 91, No. 5, pp. 253-262, ISSN 0033-8230

Opel, K.; Weiss, S.; Hübener, S.; Zänker, H. & Bernhard, G. (2007). Study of the Solubility of Amorphous and Crystalline Uranium Dioxide by Combined Spectroscopic Methods. *Radiochimica Acta*, Vol. 95, No. 3, 143-149, ISSN 0033-8230

Porter, R.F.; Chupka, W.A. & Inghram, M.G. (1955). Mass Spectrometric Study of Gaseous Species in the Si-SiO$_2$ System. *The Journal of Chemical Physics*, Vol. 23, pp. 216-217, ISSN 0032-9606

Saito, M.; Izumida, S.; Onishi, K. & Akazawa, J. (1999). Detection Efficiency of Microparticles in Laser Breakdown Water Analysis, *Journal of Applied Physics*, Vol. 85, No. 9, pp. 6353-6357, ISSN 0021-8979

Scherbaum, F.J.; Knopp, R. & Kim, J.I. (1996). Counting of Particle in Aqueous Solutions by Laser-induced Photoacoustic Breakdown Detection. *Applied Physics B*, Vol. 63, pp. 299-306, ISSN 0946-2171

Stephan, B.H.; Stephen, W.M. (1991). Determination of the Ionization Potentials of Aluminum Oxides via Charge Transfer. *The Journal of Physical Chemistry*, Vol. 95, No. 23, pp. 9091-9094, ISSN 0022-3654

Wagner, T. U.; Bundschuh, T. & Koster, R. (2005). Laser-induced Breakdown Detection (LIBD) for the Highly Sensitive Quantification of Aquatic Colloids. Part II: Experimental Setup of LIBD and Applications. *Particle and Particle Systems Characterization*, Vol. 22, No. 3, pp. 181-191, ISSN 1521-4117

Walther, C.; Bitea, C.; Hauser, W.; Kim, J.I. & Scherbaum, F.J. (2002). Laser Induced Breakdown Detection for the Assessment of Colloid Mediated Radionuclide Migration. *Nuclear Instruments and Methods in Physics Research B*, Vol. 195, pp. 374-388, ISSN 0168-583X

Walther, C.; Cho, H.-R. & Fanghänel, Th. (2004). Measuring Multimodal Size Distributions of Aquatic Colloids at Trace Concentrations. *Applied Physics Letters*, Vol. 85, No. 26, pp. 6329-6331, ISSN 0003-6951

Walther, C.; Büchner, S.; Filella, M. & Chanudet, V. (2006). Probing particle Size Distributions in Natural Surface waters from 15 nm to 2 mm by a Combination of LIBD and Single-particle Counting. *Journal of Colloid and Interface Science*, Vol. 301, No. 2, pp. 532-537, ISSN 0021-9797

Walther, C.; Cho, H.-R.; Marquardt, C.M.; Neck, V.; Seibert, A.; Yun, J.-I. & Fanghänel, Th. (2007). Hydrolysis of Plutonium(IV) in Acidic Solutions: No Effect of Hydrolysis on Absorption-spectra of Mononuclear Hydroxide Complexes. *Radiochimica Acta*, Vol. 95, No. 1, pp. 7-16, ISSN 0033-8230

Yun, J.-I. (2007). Material Dependence of Laser-induced Breakdown of Colloidal Particles in Water. *Journal of the Optical Society of Korea*, Vol. 11, No. 1, pp. 34-39, ISSN 1226-4776

Zapka, W.; Tam, A.C. (1982). Photoacoustic pulse generation and probe-beam deflection for ultrasonic velocity measurements in liquids. *Applied Physics Letters*, Vol. 40, No. 4, pp. 310-312, ISSN 0003-6951

8

Platinum Fuel Cell Nanoparticle Syntheses: Effect on Morphology, Structure and Electrocatalytic Behavior

C. Coutanceau, S. Baranton and T.W. Napporn

e-Lyse, Laboratoire de Catalyse en Chimie Organique, UMR CNRS-Université de Poitiers,
France

1. Introduction

Because of high cost and low availability of platinum, which yet remains unavoidable as catalyst in proton exchange membrane fuel cells for achieving acceptable electric performances, numerous synthesis methods of Pt nanoparticles were developed. The decrease of the particle size in a certain extent leads to a decrease of the noble metal content in the fuel cell electrodes and also to an increase of their real surface area. Physical methods such as plasma sputtering of metals [Brault et al., 2004; Caillard et al., 2007, Cho et al., 2008], laser ablation [Perrière et al., 2001; Boulmer et al., 2006], metal organic chemical vapor deposition [Billy et al., 2010], etc., have been and are still developed for preparing low loaded platinum based electrodes for Proton Exchange Membrane Fuel Cell (PEMFC). Although such methods present important interests due to the high control of metal deposition parameters, they involve also important material losses in the deposition chamber and on the electrode mask. Metal electrodeposition can also be classified as a physical method of electrode preparation [Chen et al., 2003; Ayyadurai et al., 2007]. This method consists in the electrochemical reduction of a metallic salt on an electron conductive support. It is then possible by varying the current or potential sequence applied to the electrode to control or at least to change the structure of the catalyst. However, the faradic yield is often very low because hydrogen evolution can occur as soon as platinum is deposited on the substrate [Coutanceau et al., 2004].

Chemical methods for platinum nanoparticle synthesis allow obtaining platinum nanoparticles deposited and well disseminated on a carbon powder in a simplest and more versatile manner. Amongst the chemical methods for nanostructured catalyst preparation, the impregnation-reduction method is often used in the field of heterogeneous catalysis. The principle of such method consists in the reduction of metallic salts impregnating a carbon powder with either a reducing gas or by thermal treatment under inert atmosphere [Vigier et al., 2004; Coutanceau et al., 2008]. Different possibilities exist. For example, the cationic exchange method [Richard & Gallezot, 1987] consists in activating a carbon support with sodium hypochlorite to form surface carboxylic acid groups, which are transformed into ammonium salts after treatment with ammonia. The ammonium groups are exchanged by contact with a $Pt(NH_3)_4OH_2$ alkaline complex, and the catalyst precursor is reduced at 300°C under pure hydrogen flow to form metallic particles. Pt/C catalyst prepared by this method Were well disseminated on the support, and displayed a mean particle size close to 2 nm. But, the total loading of metal was very limited and did not exceed 10 wt%, which is a too low value for PEMFC application. Higher metal loadings could be obtained by slightly modifying the synthesis protocol [Pieck et al., 1996; Roman-Martinez et al. 2000, Vigier et al., 2004]. The anionic exchange method consists in the modification of the carbon surface with aqua regia as reactant to form surface hydroxyl groups. After acidification of the OH groups into OH_2^+ groups, an exchange reaction is performed with H_2PtCl_6 acid complex. The catalyst precursor is then reduced at 300°C under pure hydrogen flow to form metallic particles. By this way, Pt/C catalysts with loadings up to 30 wt%, which is suitable for PEMFC application, could be obtained, but the size distribution became multimodal [Coutanceau et al., 2008].

In the present book chapter, we will focus on the synthesis of platinum supported nanocatalysts by colloidal routes. Colloidal methods have been extensively developed in the literature for the synthesis of platinum nanoparticles. Most of them are based on the use of surfactants for stabilizing the colloidal suspension and controlling the particle size. Amongst these methods, nano-encapsulation methods [Hwang et al., 2007] are based on the use of reducing agent which will act as surfactant after metal salt reduction reaction. The "Bönnemann method" [Bönnemann & Brijoux, 1995] of metal colloid synthesis belongs to this family. The methods called "water in oil" microemulsion are based on the formation of water nanodroplets as microreactors, dispersed in a continuous oil phase, protected by a surfactant - cetyltrimethyl ammonium bromide (CTAB), sodium bis(2-ethylhexyl)sulphosuccinate (AOT), poly(ethylene glycol) dodecyl ethers such as Brij®30, etc. [Boutonnet et al., 1982; Ingelsten et al., 2001; Solla-Gullon et al., 2000], in which metal salts are dissolved. After addition of a reducing agent ($NaBH_4$ [Brimaud et al., 2007], N_2H_2 [Solla-Gullon et al., 2000], H_2 [Erikson et al., 2004]), platinum nanoparticles are formed and the surfactant is fixed on the metal surface, forming the colloidal solution. The first method involves a thermal treatment at 300°C under air to recover the Pt/C catalyst, whereas the second is carried out at room temperature. Other synthesis methods, such as polyol method [Liu et al., 2005; Oh et al. 2007 ; Lebègue et al., 2011] and instant method [Reetz & Lopez, 2004; Reetz et al., 2004; Devadas et al., 2011], activated by thermal treatment or microwave irradiation, do not necessitate the presence of an organic surfactant. In the case of polyol methods, the reducing and acid-base properties of the solvent allow obtaining a colloidal solution of metallic nanoparticles protected by glycolate species [Fievet et al., 1989; Oh et al., 2007; Grolleau et al;, 2010]. In the case of the instant method, metal oxide nanoparticles are formed and reduced using H_2 or $NaBH_4$ as reducing agents.

The metallic nanoparticle synthesis starts with the nucleation step [Lamer and Dinegar, 1950]. The nucleation step occurs out of thermodynamic equilibrium and involves the co-existence of the formed species with the reactant ones. The formation of the stable phase occurs via the local creation of solid seeds. Under these conditions, the seed surface area is very high in comparison with their volume, and surface tension is also very high. Therefore, to obtain stable platinum seeds, the energetic saving has to be higher than the energy involved for the solid-liquid interface formation; in other words, the seeds have to reach a critical size where the ratio volume/surface is high enough to stabilize them. Now, under thermodynamic equilibrium conditions, the equilibrium shape of the particle is given by the Wulf theorem [Henry, 2005]. In the case of a fcc structure material, such as platinum, the equilibrium shape is a truncated octahedron. Numerical simulations have shown that the transition between a non-crystal icosahedron structure and a fcc structure occurred for about 200 platinum atoms [Mottet et al., 2004]. Once platinum seeds are formed, the growth step can occur. Growth step can involve the collision of seeds due to Brownian agitation, and their fusion (decreasing the surface tension by increasing the size [Wang et al., 2000]); then little seeds can fusion with bigger particles participating to the so-called Otswald ripening [Wilson et al., 2006]. It is also possible that platinum ions are reduced on the surface of existing platinum particle leading to the particle growth trough a homoepitaxial mechanism [Park et al., 2007]. The growth step is then stopped due to ionic metal species depletion in the reaction mixture or to the adsorption of surfactant on the crystal surface. The growth rate, the growth mechanism, and the growth stop have obviously an influence on the particle morphology while the growth step leads to define the size, size distribution, the shape, the shape distribution and internal constraints of the particles [Shevchenko et al., 2003]. Reactions occurring in PEMFCs are known to be structure sensitive, therefore the preparation method of Pt/C catalysts, i.e. the history of the Pt/C catalysts, is expected to have a great importance on the catalyst activity, selectivity and stability towards oxygen reduction reaction and on the tolerance toward poison species contained in reformate gas, such as carbon monoxide.

2. Catalysts preparation

Different colloidal methods have been proposed in the present chapter for the synthesis of Pt/C catalysts. Some of them ("Bônnemann method" and "water in oil" microemulsion) were performed in the presence of organic surfactant, the first one involving a thermal treatment at 300°C, whereas the second one is carried out at room temperature. Two other synthesis methods presented here were carried out without addition of an organic surfactant, the polyol method involving an activation step by either temperature (ca. 200°C) or microwave irradiation, and the instant method involving low temperature activation (from 40 to 80°C).

2.1 Synthesis of Pt/C by the Bönnemann method

This method consists in the chemical reduction of metal ions using a molecule which after the reduction step will act as surfactant to protect metal particles [Bönnemann et al., 1991; Bönnemann et al., 1994]. In the present case, the tetraalkylammonium triethylborohydride reducing agent $(N(alk)_4)^+(Bet_3H)^-$ was first prepared by mixing tetraalkylbromide $(N(alk)_4)^+Br^-$ and triethylborohydrure $(Bet_3H)^-K^+$ in anhydrous tetrahydrofuran, according to the following equation:

$$[N(alk)_4)^+]Br^- + [Bet_3H^-]K^+ \rightarrow [N(alk)_4^+](Bet_3H)^- + KBr \qquad (1)$$

then, after KBr crystallization, filtration and removal, the reducing agent is conserved in THF at 0°C before use for the reduction of anhydrous $PtCl_2$ platinum salt. A given volume of $PtCl_2$ in anhydrous THF (1 g_{Pt} L^{-1}) [Grolleau et al., 2008; Sellin et al., 2009], are placed in a balloon and the reducing agent is dropwisely added to the mixture; the Pt^{2+} reduction reaction occurs according to the following equation:

$$PtCl_2 + 2 [N(C_nH_{2n+1})_4^+][B(C_2H_5)_3H^-] \rightarrow Pt([N(C_nH_{2n+1})_4^+]Cl^-)_2 + 2B(C_2H_5)_3 + H_2 \qquad (2)$$

A colloidal solution of platinum nanoparticles protected by a tetraalkyl chloride surfactant and dispersed in anhydrous THF is then obtained. Under these conditions, the particle growth is controlled by the presence of the surfactant which adsorbs on the platinum surface and prevents any growth of their size, as it is shown in Fig. 1a. Because the reducing agent acts as surfactant after the reduction process, the particle size appears very homogeneous and the size distribution is relatively narrow as it can be seen in the TEM image of a Pt[N(octyl)$_4$Cl]$_2$ colloid precursor in Fig. 1b. Once the reduction step is finished, a very stable colloid solution is obtained and can be kept for several months without any decantation of the platinum particles. This colloidal solution will be used for the impregnation of a carbon powder in order to obtain carbon supported platinum nanoparticles which will be further used as PEMFC catalysts. A given amount of carbon powder (Vulcan XC72 from Cabot inc.) is added to the platinum colloid solution (in our case in order to obtain a platinum loading of 40 wt%) and THF is completely removed by evaporation under ultrasonication. Then, the surfactant is removed by calcination of the remaining powder at 300°C under air atmosphere for 90 minutes [Grolleau et al., 2008; Sellin et al., 2009]. By this way, anchored platinum particles on the carbon support are obtained. A last step of catalytic powder washing with ultrapure water is carried out in order to remove chlorides and bromides coming from the synthesis procedure.

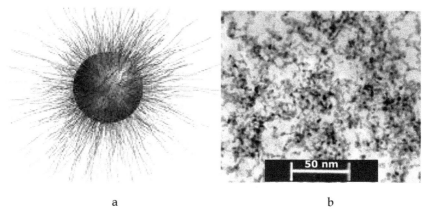

a   b

Fig. 1. **Synthesis by the Bönnemann method.** (a) scheme of a platinum particle protected by tetraalkyl chloride surfactant chains; (b) transmission microscopy image of a Pt[N(octyl)$_4$Cl]$_2$ colloid precursor.

The platinum loading is determined using thermogravimetric analysis under air atmosphere, and was evaluated to be ca. 37.5 wt%; such value is conform to the expected one (40 wt%) [Sellin et al., 2010]. The structural and morphological parameters of the Pt/C catalyst were determined by transmission electron microscopy (TEM), high resolution transmission electron microscopy (HRTEM), X-ray diffraction (XRD) and X-ray photoelectron spectroscopy (XPS). Both the mean particle size as determined by TEM (Fig. 2a) and the mean crystallite size as determined by XRD (Fig. 2c) were estimated to be ca. 2.5 nm, with a standard dispersion of 0.9 nm in the case of the crystallite size. The TEM size d_{TEM} was determined using the following equation,

$$d_{TEM} = \frac{\sum_{i=1}^{n} n_i d_i}{n} \tag{3}$$

where n is the total number of measured particles, n_i is the number of particles with a size d_i, whereas the crystallite size L_v was determined using the simple Sherrer equation:

$$L_v = \Phi \frac{\lambda}{FMWH \cos \theta} \tag{4}$$

Where L_v is the Sherrer length, φ is the shape factor (0.89 for spherical crystallite), λ the radiation wavelength (1.5406 Å), FWHM the full width at half-maximum, and θ the angle at the maximum intensity. Apparently L_v is extracted from the diffraction peak located close to $2\theta = 40°$, which corresponds to the (111) crystallographic plane of platinum. The diffraction pattern was analyzed by the method of Levenberg-Marquardt, using a Pearson VII fit by means of a computer refinement program (Fityk free software).

A calculation of the exhibited surface area has been performed and led to a value of 110 m^2 g^{-1}, assuming that Pt/C nanoparticles have a spherical shape (HRTEM image in Fig. 2b) and that the whole Pt surface is available. However, agglomerated platinum particles are visible on the TEM image in Fig. 2a. After counting on several tenths of images of different regions of the TEM grid, it has been estimated that 15% of the platinum entities were particle agglomerates with a size higher than 14 nm. This will certainly decrease the available platinum surface for electrochemical reaction. XRD characterization in Fig. 2c allowed estimating the value of the cell parameter (ca. 0.392 nm). Lattice defects like stacking faults and microstrains are common for metal nanoparticles, and could add substantially to the width of diffraction peaks [Vogel et al., 1998; Habrioux et al., 2009]. But in the present case, certainly due to the synthesis method of Pt/C catalysts, which involves a thermal treatment at 300°C for 90 min, no stacking fault was detected from HRTEM observations as shown in Fig. 2b as an example. Moreover, the lattice parameter evaluated from XRD measurements is ca. 0.392 mm, i.e., close to the value for platinum bulk. This seems to indicate that the contribution of microstrains is low. Therefore, by correlating the TEM, HRTEM and XRD data, it can be proposed that isolated Pt nanoparticles prepared via the Bönnemann method are à priori round shaped single crystals without or with few defects. XPS (X-ray photoelectron microscopy) measurements carried out on a Pt/C powder (Fig. 2d) indicated that platinum was only present under 0 and +II oxidation state. The peaks corresponding to Pt 4f7/2 are located at binding energies of 71.65 eV and 72.65 eV, corresponding to metallic Pt^0 and to Pt(OH)$_2$, whereas the Pt 4f7/2 peak related to PtO species should be located at binding energy higher than 73.5 eV.

2.2 Synthesis of Pt/C by the "water in oil" microemulsion method

The "water in oil" microemulsion is a ternary system currently composed of an aqueous phase dispersed in a continuous n-heptane phase. In order to obtain a translucent mixture thermodynamically stable of both these phases (aqueous and organic), the presence of a surfactant (Brij®30) is necessary. The surfactant will act as a separator between both phases leading to the formation of nano-droplets of the minority phase in the majority one. In the

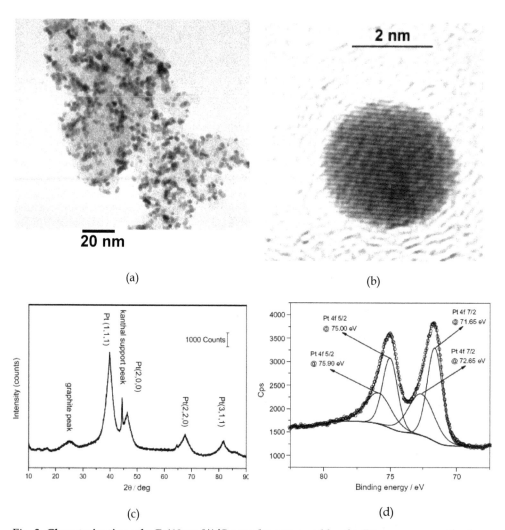

(a)   (b)

(c)   (d)

Fig. 2. **Charaterization of a Pt(40 wt.%)/C powder prepared by the Bönnemann method.** (a) TEM image, (b) HRTEM micrograph, (c) XRD pattern and (d) Pt 4f XPS core level spectrum recorded by XPS.

present case, the formation of water nano-droplets (where H_2PtCl_6 is dissolved) in the oil phase is the considered system, which will act as microreactor for the metal salt reduction to platinum nanoparticles. The reduction of metallic salt present in the microreactor is carried out by addition of a reducing agent ($NaBH_4$ for instance). Different chemical reactions were suggested in order to illustrate the reduction mechanism of H_2PtCl_6 by sodium borohydride; however, Ingelsten and coworker [Ingelsten et al., 2001] proposed the following one:

$$PtCl_6^{2-}(aq.) + BH_4^-(aq.) + 3H_2O \rightarrow Pt^0 + H_2BO_3^-(aq.) + 4H^+(aq.) + 6Cl^-(aq.) + 2H_2(g) \qquad (5)$$

The synthesis protocol is represented on the scheme in Fig. 3. The desired amount of H_2PtCl_6, $6H_2O$ (from Alfa Aesar) is dissolved in the appropriate water volume in order to obtain a concentration of 0.2 mol L^{-1} in Pt^{4+}. The ternary system is realized by mixing the aqueous solution in presence of Brij®30 (water/Brij®30 molecular ratio = 4) with n-heptane; the reduction reaction is carried out by adding a large excess of solid $NaBH_4$ (n_{NaBH4}/n_{Pt} =15).

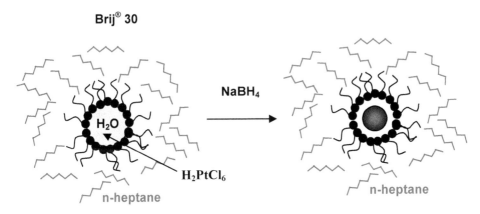

Fig. 3. **Synthesis by the "water in oil" microemulsion.** Scheme of the platinum nanoparticle synthesis via the "water in oil" microemulsion route.

The size of the microreactors was determined using a Laser DL135 granulometer from Cordouan Technologies; the hydrodynamic diameter was found to be ca. 16 nm; considering a size of 2 nm for a surfactant molecule, a microreactor with an internal diameter of 12 nm containing an aqueous solution of 0.2 mol L^{-1} H_2PtCl_6 should lead to platinum nanoparticles of ca. 1.5 nm.

After addition of the carbon powder (in a suitable amount to obtain a platinum loading of 40wt% assuming a total reduction of the platinum salt) and an abundant cleaning step with acetone, ethanol and water, the catalytic powder was characterized. TGA measurement indicated a platinum loading of ca. 37 wt%. TEM and XRD characterizations as shown in Fig. 4a and 4c, respectively, indicated that the mean platinum particle size was in the range of 3.0 nm, whereas the mean crystallite size was ca. 2.9 nm. It can be noted that TEM conical dark field clearly shows that the platinum particles obtained from "water in oil" microemulsion were composed by spherical crystallites (Fig. 4a). This observation is also confirmed by HRTEM of unsupported nanoparticles prepared by "water in oil" microemulsion on Fig. 4b.

In such synthesis route, the end of the growth process could be due to the depletion of metallic ions in microreactors. However, the mean particle size and crystallite size obtained from TEM and XRD measurements are significantly higher than that calculated from hydrodynamic diameter as determined by laser granulometry. Then, It seems that a second grain growth step is involved, which can be explained in terms of Brownian interparticles collision [Bradley, 1994] or Ostwald ripening [Peng et al., 1998].

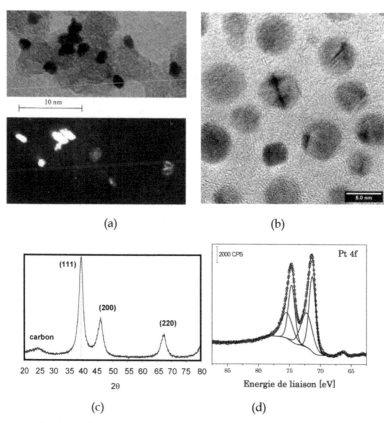

(a)   (b)

(c)   (d)

Fig. 4. **Characterization of a Pt(40 wt.%)/C powder prepared by the microemulsion route.**
(a) TEM (top) and TEM conical dark field (down) images, (b) HRTEM micrograph, (c) XRD
pattern and (d) Pt 4f XPS core level spectrum recorded.

Therefore, by correlating the TEM, HRTEM and XRD data, it can be proposed that isolated
Pt nanoparticles prepared via the microemulsion method are *a priori* polycrystalline. XPS
measurements carried out on a Pt/C powder (Fig. 4d) confirmed that platinum was also in
that case present under 0 and +II oxidation states. From the particle size determined by TEM
measurements, a real platinum surface area of ca. 55 m^2 g^{-1} could be calculated. It has also
been estimated that 23% of the platinum entities were particle agglomerates with a size
higher than 14 nm.

2.3 Synthesis of Pt/C by the polyol method

First described by Fievet et al. [Fievet et al., 1989], this synthesis method consists in the
reduction of a metallic salt by using the reductive properties of polyol which also acts as
solvent. Several polyol compounds were studied, particularly ethylene glycol, diethylene
glycol and glycerol [Larcher et al., 2000]. In the present book chapter, we will focus on
polyol synthesis methods using ethylene glycol as solvent and reducing agent.

Once the metallic salt is completely dissolved in an alkaline solution of ethylene glycol, intermediate phase of metal oxides or hydroxides are created. Then, the dehydration reaction of glycerol into acetaldehyde occurs, which allows the reduction reaction of metal oxides or hydroxides. The reaction mechanism is summarized in equations 6 to 8 [Larcher et al., 2000].

$$CH_2OH\text{-}CH_2OH \rightarrow CH_3\text{-}CHO + H_2 \qquad \text{dehydration reaction} \qquad (6)$$

$$CH_2OH\text{-}CH_2OH + 2OH^- \rightarrow CH_2O^-\text{-}CH_2O^- + H_2O \quad \text{acid-base reaction (pKa = 6.5)} \quad (7)$$

$$MO_2 + 4\,CH_3\text{-}CHO \rightarrow M + 2\,CH_3\text{-}CO\text{-}CO\text{-}CH_3 + 2\,H_2O \quad \text{reduction reaction} \qquad (8)$$

This reduction reaction under relatively soft conditions favours low grain growth kinetics. The grain growth process is stopped due to the combination of two phenomena: adsorption of reaction by-products on the metal surface (particularly glycolates formed in alkaline medium according to equation 7) and depletion of metal salt to be reduced. Moreover, glycolate ligands, which also act as surfactant, do interact weakly and adsorbs preferentially on certain crystalline surface domains of platinum particles, so that nanoparticles with surface orientations (facetted) can be created.

Because ethylene glycol (EG) is a soft reducing agent, the reduction reaction of $H_2PtCl_6,6H_2O$ (1 g_{Pt} L^{-1}) in alkaline EG solution (pH = 11), in presence of Vulcan XC72 carbon powder (in appropriate amount for a nominal Pt loading of 40 wt%) needs to be activated either by temperature (at reflux, ca. 200°C for 2 hours) or by microwave irradiation [Lebegue et al., 2010].

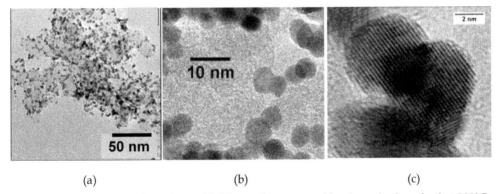

| (a) | (b) | (c) |

Fig. 5. Characterization of a Pt(40 wt.%)/C powder prepared by the polyol method at 200°C for 2 hours. (a) and (b) TEM images at different magnifications, and (c) HRTEM micrograph.

The TEM characterizations presented in Fig. 5 correspond to a Pt/C catalyst prepared by thermal treatment at reflux for 2 hours, leading to a platinum loading of ca. 36 wt%, as determined by TGA. The dispersion of platinum particle on the carbon support is relatively homogeneous, and the mean particle size was found to be ca. 4.0 nm. A majority of particles observed with higher magnification in Fig. 5b are clearly not round shaped, and facetted crystallites are formed. Concerning the crystalline structure of the Pt/C catalyst synthesized by the polyol method, XRD patterns have evidenced the classical fcc structure of platinum, from which the determined cell parameter was found to be close to that of bulk platinum (i.e. 0.392 nm) for a mean crystallite size of ca. 4 nm. The isolated nanoparticles would then correspond to single nanocrystals dispersed on the carbon substrate. This assumption is confirmed by the HRTEM image presented in Fig. 5c, where isolated particles appear really as single crystals.

2.4 Synthesis of Pt/C by the "instant method"

In order to obtain a high dispersion and prevent any undesirable nanoparticle agglomeration, a surfactant is commonly used in the chemical synthesis methods of platinum catalysts. This molecule controls the seed formation as well as its growth through the reduction of the metal salt. But its stabilizing effect depends on the pH and the temperature of the solution. As can be seen in the methods cited above, various molecules are proposed as surfactants for catalyst preparation. Reetz et al. [Reetz and Koch, 1998; Reetz and Koch, 1999] have shown that the use of a surfactant permits to produce metal oxides as intermediate species before their reduction to catalysts. Indeed, alkaline solutions such as NaOH, Na_2CO_3, or Li_2CO_3 are added to the platinum salts ($PtCl_4$ or H_2PtCl_6) in presence of water-soluble betaine stabilizer (3-(N,N-dimethyldodecylammonio)propane sulfonate [$R(CH_3)_2N^+(CH_3)_3SO_3^-$]). After 5 to 6 hours at 60°C, stable PtO_2 particles were obtained. The pH of the solution is one of the key parameter. Lithium carbonate permits to keep the pH close to 9, value at which chlorine are replaced by the hydroxide leading to PtO_2. The role of the surfactant in colloidal synthesis modes of Pt/C catalysts for fuel cells application is well demonstrated in literature. However, the synthesis methods of Pt/C catalysts described above require a heat treatment for removing this organic molecule. In order to avoid any use of surfactant and/or heat treatment, Reetz et al. [Reetz and Lopez, 2002, Reetz et al. 2004] developed a simple way called "instant method" for preparing PtO_x colloids free of surfactant. A catalysts support is used in place of the stabilizer. Therefore, the PtO_x colloids are immobilized on the support according to:

$$H_2PtCl_6 + C \xrightarrow{Li_2CO_3/H_2O} PtO_x/C$$

$$(9)$$

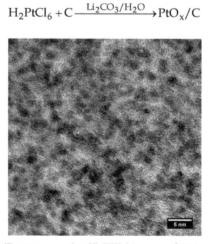

Fig. 6. PtO_x particles from "Instant method". TEM image of PtO_x colloids particles on carbon Vulcan XC72 (metal loading is 40%).

The pH of the solution is kept at 9 by the presence of lithium carbonate. The PtO_x colloid particles obtained in our group (Fig. 6) are small (1-2 nm). The reduction of the PtO_x colloids to Pt/C can be made by various reducing agents: hydrogen, as suggested by Reetz, sodium citrate or sodium borohydrate in a modified method. Fig. 7 shows high dispersion of Pt particles on carbon obtained by reducing PtOx colloids with an ice-cooled solution of sodium borohydrate.

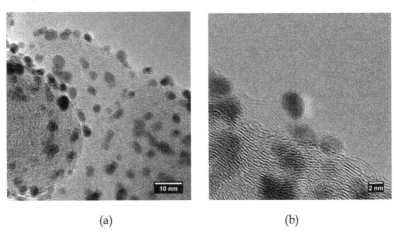

(a)   (b)

Fig. 7. Characterization of a Pt(40wt%)/C catalysts obtained by "instant method" at 40°C. (a) TEM image and (b) HRTEM image.

With the experimental conditions fixed above, the TEM characterizations show a mean Pt particle size about 3nm. The Thermogravimetry analysis confirmed the nominal Pt loading (38%). It is also observed that the particles size depends on the temperature of the colloidal solution and the metal loading. The higher dispersion and smaller crystallites were obtained with 40 wt% of Pt loading at 60°C.

3. Electrochemical characterization of the Pt/C catalysts

The working electrode was prepared by deposition of a catalytic ink on a glassy carbon disc (0.126 cm² geometric surface area) according to a method proposed by Gloaguen et al. [Gloaguen et al., 1994]. Catalytic powder was added to a mixture of 25 wt% (based on the powder content) Nafion® solution (5 wt.% from Aldrich) and ultra-pure water (MilliQ, Millipore, 18.2 MΩ cm). After ultrasonic homogenization of the Pt/C-Nafion® ink, a given volume was deposited from a syringe onto the fresh polished glassy carbon substrate yielding in the case of a Pt(40 wt.%)/C to a 60 µg cm⁻² catalytic powder loading (i.e. 24 µg$_{Pt}$ cm⁻²). The solvent was then evaporated in a stream of ultra-pure nitrogen at room temperature. By this way, a catalytic layer was obtained with a thickness lower than 1µm.

The electrochemical active surface area (EASA) of Pt/C catalysts is determined from cyclic voltammograms (CV) by integrating the charge in the hydrogen desorption region corrected from the double layer capacity contribution [Coutanceau et al., 2000; Grolleau et al., 2008]. For each catalyst, the average active surface area is determined by recording the CV on four different ink depositions. Measurements are carried out in a N_2-saturated electrolyte at a scan rate of 50 mV s⁻¹ after 10 voltammetric cycles performed between 0.05 V and 1.25 V vs. RHE in order to clean the platinum surface from remaining organic molecules coming from the synthesis and to obtain quasi-constant voltammograms [Maillard et al., 2007]. Fig. 8 displays a typical CV recorded on a commercially available e-Tek Pt(40wt%)/C catalyst in a 0.5 M H₂SO₄ electrolyte, as an example.

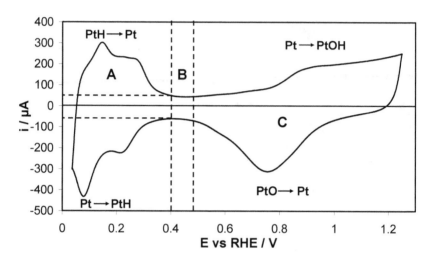

Fig. 8. Electrchemical characterization of a Pt(40wt%)/XC72 e-tek® catalyst.
Voltammogram recorded in N₂-saturated 0.5 M H₂SO₄ electrolyte (v=20 mV.s⁻¹; T = 25 °C,).

The potential region (A) corresponds to the hydrogen adsorption (i < 0) and desorption (i > 0), the potential region (B) corresponds to the double layer capacity and the potential region (C) corresponds to the platinum surface oxide formation (i > 0) and reduction (i < 0). In the hydrogen region (A), by analogy with studies performed on low index single crystals [Solla-Gullon et al., 2006; Solla-Gullon et al., 2008; Attard et al., 2001], the oxidation peaks located at ca. 0.12 V and ca. 0.26 V can be assigned to sites of (110) and (100) symmetries, respectively. Changes in the height and the width of these peaks are representative of different surface states of platinum nanoparticles. The electrochemical active surface area is usually estimated from the determination of the charge associated to hydrogen desorption in the potential range from 0.05 V to 0.4 V vs RHE [Biegler et al., 1971], after correction of the double layer capacitive contribution. It is then assumed that a hydrogen monolayer is formed at the platinum surface and that the adsorption stoechiometry is one hydrogen atom per platinum surface atom. The charge involved for the hydrogen desorption reaction is calculated using the following equation:

$$Q_{H_{des}} = \int i(t)dt = \frac{1}{v}\int i(E)dE \tag{10}$$

where $Q_{H_{des}}$ is the charge corresponding to the integrated current i(t) for a duration t, which corresponds to the integrated current in the potential range considered, i(E), divided by the potential scan rate, v.

From measurement on low index platinum single crystals, the charge involved during the hydrogen desorption could be determined [Markovic et al., 1997]:

- a Pt(111) surface exhibits $1.5 \; 10^{15}$ atom cm^{-2}, leading to an associated theoretical charge of 240 μC cm^{-2},
- a Pt(100) surface exhibits $1.3 \; 10^{15}$ atom cm^{-2}, leading to an associated theoretical charge of 225 μC cm^{-2},
- a Pt(110) surface exhibits $4.6 \; 10^{14}$ atom cm^{-2}, leading to an associated theoretical charge of 147 μC cm^{-2},

By weighting the different surface domains, a global charge of 210 μC cm^{-2} was found for the adsorption of a hydrogen monolayer on a polycrystalline platinum surface ($Q_{monolayer}$). This value is now generally accepted. From this value, the accessible platinum surface per mass unit, named electrochemically active surface area (EASA), can be calculated by using equation (11):

$$\text{EASA} = \frac{Q_{H_{des}}}{m_{Pt} \times Q_{monolayer}} = \frac{\frac{1}{v}\int i(E)dE}{m_{Pt} \times Q_{monolayer}} \tag{11}$$

where Q_{Hdes} is the charge determined in the hydrogen desorption region of the voltammogram recorded on the Pt/C catalyst in supporting electrolyte, $Q_{monolayer}$ is the charge related to the adsorption or desorption of a hydrogen monolayer on a polycrystalline Pt surface ($Q_{monolayer}$ = 210 μC cm^{-2}), i(E) is the current (in μA) recorded in the hydrogen desorption region, E is the electrode potential (in V vs RHE), v is the linear potential variation (in V s^{-1}) and m_{Pt} is the platinum loading on the electrode (in g). In the case of the catalyst used to record the voltammogram in Fig. 8, the measurement performed on five different electrodes prepared successively led to an EASA of 42 m^2 g_{Pt}^{-1}, with a standard deviation of 3.7 m^2 g_{Pt}^{-1}.

As it was shown in the previous part of this chapter, each synthesis method led to platinum particles having different structures and catalytic powders with different morphologies. Those changes will have an important effect on the structure of exhibited platinum surface and on the electrochemically active surface area, as it is shown in Fig. 9 where different Pt/C catalysts are compared with a commercial one (e-Tek catalyst).

The shape of the oxidation current peak in the hydrogen desorption region (i > 0) between 0.05 and 0.4 V vs RHE gives information on the catalyst surface. Considering for example the voltammogram of a Pt/C catalyst prepared by the polyol method, two well defined oxidation peaks are clearly visible, the first one centered at ca. 0.12 V and the second one at ca. 0.26 V were assigned to surface defects and to short range order (100) surface domains, respectively [Markovic et al., 1995; Attard et al., 2001]. Both those peaks are lower in intensity on the voltammogram recorded with a Pt/C catalyst prepared by the Bönnemann synthesis method. On the other hand, a third peak at ca. 0.2 V arises in this latter case, which was assigned to defective (110) planes [Gomez and Clavilier, 1993]. Commercial and "water in oil" microemulsion Pt/C catalysts exhibit the same general shape of the hydrogen region in the voltammograms, with a well-marked oxidation peak at ca. 0.12 V and a less-marked one at ca. 0.26 V. In the case of a Pt/C catalyst prepared via the "instant method" the three peaks are clearly visible as it can be seen in Fig. 10. This comparison indicates that the synthesis method of platinum nanoparticles, through the related Pt salt reduction kinetics and mode of growth stopping, has a great influence on the surface structure and the morphology of the catalyst, and can induce more or less surface defects, surface domains

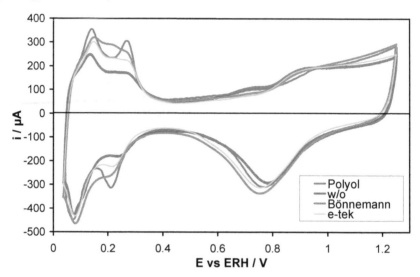

Fig. 9. **Electrochemical characterization of different Pt(40wt%)/XC72 catalysts.**
Voltammograms recorded in N_2-saturated 0.5 M H_2SO_4 electrolyte (v=20 mV.s^{-1}; T = 25°C,).

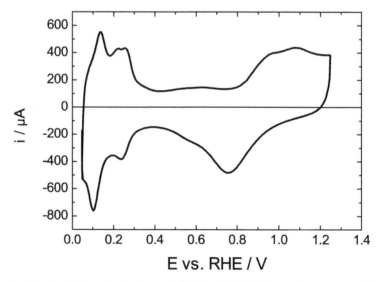

Fig. 10. Electrochemical characterization of a Pt(40wt%)/XC72 catalyst prepared via the "instant method". Voltammogram recorded in N_2-saturated 0.5 M H_2SO_4 electrolyte (v=20 mV.s^{-1}; T = 25°C,).

and surface heterogeneities. It is also likely that such changes in morphology and structure could lead to changes in the catalytic activity of the different Pt/C catalysts.

The electrochemical active surface area (EASA) of the different catalyst was also determined using equation (11) and was compared to the active surface area as determined from data obtained by TEM and XRD measurements (Table 1 summarizes the results). Assuming spherical particles of similar radius, an "electrochemical" mean particle size can be calculated from the following equations:

-   the geometric surface area of a spherical particle is:

$$S_{geom} = 4\pi r^2 \tag{12}$$

-   the volume of a spherical particle is:

$$V_p = \frac{4}{3}\pi r^3 \tag{13}$$

where r is the particle radius.

-   the electrochemical active surface area of the particle is:

$$EASA = \frac{S_{geom}}{V_p \times \rho} = \frac{3}{\rho \times r} = \frac{6}{\rho \times d} \tag{14}$$

where ρ is the platinum density (21 450 kg m^{-3}).

-   the electrochemical particle size d can be expressed as:

$$d = \frac{6}{\rho \times EASA} \tag{15}$$

	Bönnemann	Microemulsion w/o	Polyol	Instant Method	Commercial e-Tek
Pt loading (%)	37.5	37	36	38	40
Particle shape	Spherical	Spherical	spherical + facetted	Spherical	Spherical
d_{XRD} / nm	2.5	2.9	4.0	-	3.2
d_{TEM} / nm	2.5	3.0	4.0	2.8	3.5
$S_{real,TEM}$ / m^2 g^{-1}	110	91.7	68.8	98.3	78.6
EASA / m^2 g^{-1}	46	34	42	61	42
d / nm	6.0	7.6	6.5	4.5	6.5

Table 1. Structural and electrochemical characterization of different Pt(40 wt%)/C catalysts; EASA and d were determined from the voltammogram presented in Fig. 8, 9 and 10.

The values obtained from equation 15 are higher than that obtained by TEM or XRD, which seems to indicate that the whole surface of platinum particles is not used for electrochemical reactions. Considering the mean particle sizes determined by TEM for calculating the corresponding real surface areas ($S_{real,TEM}$), it appears that the electrochemical active surface area represents only between 37% (for the "water in oil" microemulsion" Pt/C catalyst) and 62% (for the Instant Method Pt/C catalyst) of the real surface areas as determined by TEM. This can be due to the non-utilization of the contact surface between platinum and carbon in the electrochemical reaction, to the aggregation of crystallites and to the presence of Nafion in the catalytic layer which can block some platinum adsorption sites. This discrepancy can also be due to the adsorption of species coming from the synthesis procedure, such as chloride (from used platinum salts), carbonate (for the instant method), traces of remaining surfactant, adsorbed ethylene glycol, etc. Considering the "water in oil" microemulsion catalyst, thermogravimetric analyses under air flow have shown that about 2 wt% Brij$_\circledR$30 remained on the catalyst surface, which could be desorbed from ca. 150°C (Fig. 11). The presence of traces of surfactant on the catalytic surface before calcination step, even with so low amount, can explain the very high difference between the electrochemical active surface area (34 m^2 g^{-1}) and that calculated from d_{TEM} (ca. 92 m^2 g^{-1}). The proportion of agglomerates with sizes higher than 14 nm (23 %) did not likely explain alone this very high difference. So, it can be proposed that a thermal treatment at ca. 200°C could clean the platinum particle surface from adsorbed surfactant. However, such treatment is expected to induce some changes in the platinum particle structure, and further on their electrocatalytic activity; nethertheless, it has been shown that the thermal treatment under air at 250°C of "water in oil" microemulsion Pt/C catalyst has no significant effect on the electrochemical active surface area (35 m^2 g^{-1}).

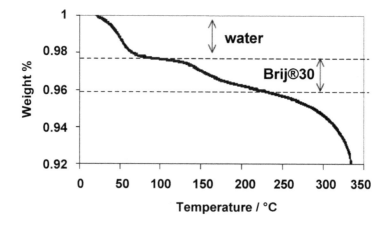

Fig. 11. TGA measurement of a Pt(40 wt%)/C catalyst prepared by the "water in oil" microemulsion method, recorded under air flow at 10°C min^{-1}.

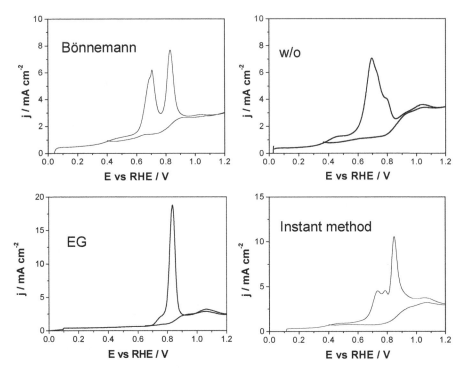

Fig. 12. **CO stripping measurements.** Voltammograms recorded on different Pt(40 wt %)/C in a deaerated H_2SO_4 0.5 M electrolyte at 20 mV s^{-1} (T = 293 K).

Interactions between platinum surfaces and carbon monoxide (CO) have been extensively studied in the few past decades [Iwasita, 2003]. CO is considered as a poisoning species for platinum-based anode catalysts. The strong affinity of CO for platinum and its sensitivity to surface structure originate its use in surface science as probe molecule for studying surface structure [Villegas and Weaver, 1994; Yoshimi et al., 1996] and eventually characterizing the electrocatalytic activity of platinum nanoparticles. On platinum nanoparticles, the most recognized CO electrooxidation mechanism is described by a Langmuir-Hinshelwood reaction mechanism, involving a water dissociative adsorption step:

$$H_2O \leftrightarrows OH_{ads} + H^+ + e^- \tag{16}$$

$$CO_{ads} + OH_{ads} \rightarrow CO_2 + H^+ + e^- \tag{17}$$

One of the main methods used for studying CO electrooxidation on platinum surface is the so-called CO-stripping voltammetry. Then, in order to get information on structure and morphology of nanoparticles, the complete surface coverage of platinum catalysts with CO is performed at 0.1 V vs. RHE for 5 min, then CO is removed from the bulk electrolyte by N_2 bubbling for 15 min maintaining the electrode under potential control at 0.1 V, at last CO stripping measurements are recorded by linearly varying the potential at 20 mV s^{-1} from 0 to 1.25 V vs RHE. Fig. 12 gives the recorded CO stripping voltammograms on the different synthesized catalysts. Clearly, the shape of the CO oxidation current feature is completely different according to the catalyst. The Bönnemann, w/o microemulsion and instant method catalysts, which are round shaped catalyst, lead to several oxidation peaks or shoulders, with a more or less important pre-peak from 0.4 to 0.6 V vs. RHE, whereas the polyol catalyst leads to a well defined single oxidation peak (and a small shoulder at lower potentials) and no pre-peak. The electrooxidation of carbon monoxide is often characterized by the existence of multiple oxidation peaks, which was explained either by a platinum particle size effect [Friedrich et al., 2000; Maillard et al., 2004], or by the presence of agglomerates [Maillard et al., 2005; Lopez-Cudero et al., 2010], or by the presence of grain boundaries [Plyasova et al., 2006] or again by surface crystallographic domain structures [Solla-Gullon et al., 2006; Solla-Gullon et al., 2008; Brimaud et al., 2008]. However, in the case of the catalysts presented here, the absence of an oxidation peak at ca. 0.7 V in the CO stripping voltammogram of the polyol catalyst could be interpreted by the absence of particle agglomeration according to other authors [Maillard et al., 2005; Lopez-Cudero et al., 2010], although TEM image in Fig. 5a clearly shows the presence of aggregates. Size effect could also hardly be proposed for the existence of the oxidation peak at ca. 0.85 V which was attributed by certain authors to CO oxidation on smaller particles [Friedrich et al., 2000; Maillard et al., 2004], as means particles size are in the same range for Bönnemann, w/o microemulsion and instant method catalysts. These experiments evidenced that the history of the Pt/C catalyst has a great importance on the electrochemical behaviour towards pre-adsorbed CO layer oxidation reaction, which is a well know surface structure sensitive reaction. However, based on current data and knowledge concerning this reaction, it is yet very difficult and too early to rely the electrochemical behavior to a given structure or morphology of a Pt/C catalyst.

4. Electrocatalytic activity towards the oxygen reduction reaction in acid medium

The oxygen reduction reaction is the limiting process in proton exchange fuel cells [Ralph and Hogarth, 2002]; this very irreversible reaction involves indeed the exchange of several electrons, and as a consequence, occurs with a high overvoltage. For this reason it has been extensively studied for the last few decades.

The reduction of dioxygen dissolved at saturation in 0.5 M H_2SO_4 supporting electrolyte was investigated at the Pt(40 wt%)/C deposited on a glassy carbon rotating disc electrode for several rotation rates Ω (from 0 to 3600 round per minute). The current vs potential curves were recorded during a slow voltammetric sweep (sweep rate of 2 mV s^{-1}) between 1.1 and 0.2 V vs RHE, so that the recorded curves are quasi-stationary. As a typical example, the current density (referred to the geometric surface area A_g) vs potential curves are shown in Fig. 13, for a platinum catalyst prepared by the polyol method.

A more general analysis of data was carried out by separating the contribution of the diffusion of molecular dioxygen from that of the surface processes involved in the oxygen reduction reaction using the $1/j$ vs $1/\Omega^{-1/2}$ Koutecky–Levich plots. For that purpose, a detailed reaction mechanism, involving the formation of H_2O_2, is written as follows [Tarasevich et al. 1983]:

$$O_{2,sol} \rightarrow O_{2,surf} \quad \text{diffusion in the bulk electrolyte (diffusion coefficient D)} \tag{18}$$

$$O_{2,surf} \rightarrow O_{2,cata} \quad \text{diffusion in the catalytic film (limiting current density } j_l^{film}) \tag{19}$$

$$O_{2,cata} \rightarrow O_{2,ads} \quad \text{adsorption process (limiting current density } j_l^{ads}) \tag{20}$$

$$O_{2,ads} + 2e^- \rightarrow [O_{2,ads}]^{2-} \text{ electron transfer rds (exchange current density } j_0, \text{ Tafel slope b)} \tag{21}$$

$$[O_{2,ads}]^{2-} + H^+ \rightarrow [HO_{2,ads}]^- \tag{22}$$

$$[HO_{2,ads}]^- + H^+ \rightarrow [H_2O_2]_{ads} \rightarrow H_2O_2 \tag{23a}$$

$$\text{or } [HO_{2,ads}]^- + H^+ + e^- \rightarrow H_2O_2 \tag{23b}$$

followed by $H_2O_2 + 2 H^+ + 2 e^- \rightarrow 2 H_2O$

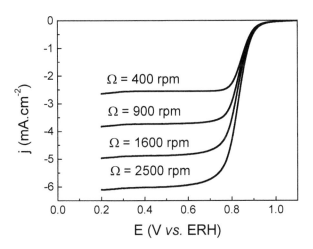

Fig. 13. **ORR study at a Pt/C catalyst prepared by a polyol method.** j(E) polarization curves at different electrode rotation rates Ω in revolution per minute(rpm) recorded in a O_2-saturated 0.5 M H_2SO_4 electrolyte. (T=20°C, scan rate = 2mVs^{-1}).

Assuming that the reaction is first order with respect to oxygen and that the first electron transfer is the rate determining step, Koutecky–Levich plots can be drawn from these polarization curves, as it is shown in Fig. 14, using Equations (24) and (25) [Schmidt et al., 1999; Coutanceau et al., 2000]:

$$\frac{1}{j} = \frac{1}{j_0(\theta/\theta_e)e^{\eta/b}} + \frac{1}{j_l^{ads}} + \frac{1}{j_l^{film}} + \frac{1}{j_l^{diff}} \tag{24}$$

where j_l^{film} and j_l^{ads} correspond to the diffusion limiting current density in the catalytic film and to the adsorption limiting current density of dioxygen, respectively; $\eta = E-E_{eq}$ is the overpotential; b is the Tafel slope; j_0 is the exchange current density; θ and θ_e are the coverage of platinum surface by species coming from oxygen adsorption at potential E and at the equilibrium potential E_{eq} (1.185V vs RHE), respectively, and j_l^{diff} is the diffusion limiting current density which can be calculated from the Levich law:

$$j_l^{diff} = 0.2nF\left(D_{O_2}\right)^{2/3} v^{-1/6} C_{O_2} \Omega^{-1/2} \tag{25}$$

with n the number of electron exchanged per oxygen molecule, F the Faraday constant (96,500 C mol^{-1}), D_{O2} the coefficient diffusion of oxygen molecular in 0.5 M H$_2$SO$_4$ (2.1×10^{-5} cm^2 s^{-1}), v the kinematic viscosity of the electrolyte (1.07×10^{-2} cm^2 s^{-1}), C$_{O2}$ the bulk concentration of oxygen in a saturated electrolytic solution (1.03×10^{-3} mol L^{-1}) [Jakobs et al., 1985]. The coefficient 0.2 is used when the working electrode rotation rate Ω is expressed in revolutions per minute (rpm) [Bard et al., 2001; Zagal et al., 1980].

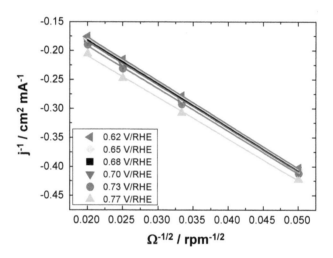

Fig. 14. **ORR study at a Pt/C catalyst prepared by a polyol method.** Koutecky–Levich plots at different potentials determined from j(E) curves presented in Fig. 12.

The film diffusion limiting current density and the adsorption limiting current density are both independent on disk electrode rotation rates and applied potential (E), thus it is impossible to dissociate them. Then, the current density j can be expressed as follows:

$$\frac{1}{j} = \frac{1}{j_0(\theta/\theta_e)e^{\eta/b}} + \frac{1}{j_l} + \frac{1}{j_l^{diff}} \quad \text{with} \quad \frac{1}{j_l} = \frac{1}{j_l^{ads}} + \frac{1}{j_l^{film}} \tag{26}$$

Assuming that in the considered potential range the adsorption process of oxygen is more rapid than the charge transfer step, i. e. $\theta \approx \theta_e$ for the whole electrode potential range [Coutanceau et al., 2001]. The kinetic current density j$_k$ can be expressed as:

$$\frac{1}{j_k} = \frac{1}{j_0 e^{\eta/b}} + \frac{1}{j_l} \tag{27}$$

Therefore, the slopes of the Koutecky-Levich straight lines will lead to the determination of the number of exchanged moles of electron per mole of oxygen, whereas their intercept with the Y-axis will lead to the determination of the kinetic current values as a function of potential. Equation 25 indicates that when the overpotential η tends to ∞ then $1/j_k$ tends to $1/j_l$, so that one can obtain the limiting current density j_l by extrapolating equation 25 at high η (Fig. 15a). This allowed to transform equation 25 as follows and to access the Tafel slope b (Fig. 15b) as well as the exchange current density j_0:

$$\eta = E - E_{eq} = -b\left(\ln\frac{j_l}{j_0} + \ln\left|\frac{j_k}{j_l - j_k}\right| \right) \qquad (28)$$

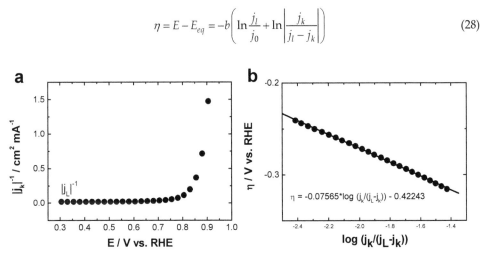

Fig. 15. **ORR study at a Pt/C catalyst prepared by a polyol method.** Plot of $1/j_k$ as a function of the electrode potential and (d) related Tafel plot determined from Koutecky-Levich curves presented in Fig. 13.

	Bönnemann	Microemulsion w/o	Microemulsion w/o heat treated @ 250°C	Polyol	Commercial e-tek
Pt loading (%)	37.5	37	-	36	40
Particle shape	Spherical	Spherical	Spherical	Spherical +facetted	Spherical
EASA / m² g⁻¹	46	34	35	42	42
j_k @ 0.95 V / mA cm⁻² j_{kw} @ 0.95 V / µA cm$_{Pt}^{-2}$	0.82 22	0.17 6	0.81 22	0.74 22	0.014 0.4

Table 2. **ORR kinetics data.** Values determined from RDE measurements recorded on different Pt(40wt%)/C catalysts in H_2SO_4 electrolyte according to the mathematical treatment of the Koutecky-Levich equation presented above.

Equation 26 not only allows determining the exchange current density j_0 but also the kinetic current densities j_k at a given potential for the different catalysts. An electrode potential of 0.95V vs. RHE for the determination of the diffusion corrected kinetic current density j_k is close to that of a cathode working in a PEMFC. So, the kinetic current at a potential convenient with that of a working electrode and the limiting current in the catalytic film can be determined; both parameters allow for well characterizing the catalytic film and selecting the best catalyst toward oxygen reduction reaction. However, in order to compare the catalytic activity, the kinetic current density at 0.95 V vs RHE may also be assessed by the active surface area expressed in μA cm_{Pt}^{-2}. Table 3 gives the kinetics parameters obtained for the different platinum catalysts prepared via the different methods presented previously.

Although platinum materials synthesized by the Bönnemann method, the "water in oil" microemulsion way and the polyol route, display a fcc structure with a cell parameter very close to that of bulk platinum (0.392 nm), these results indicate an important dependence of the catalytic activity on the catalyst synthesis method. However, due to the invariance of the cell parameter for all prepared catalyst, it is unlikely that the change in catalytic activity of the platinum material is related to geometric effect (change in Pt-Pt interatomic distance).

The catalyst prepared by the "water in oil" microemulsion leads to the lower activity. TEM images (Fig. 4) clearly show that platinum species are polycrystalline in the case of the w/o microemulsion Pt/C catalyst. A thermal treatment under air at 250°C of a w/o microemulsion Pt(40wt%)/C material allowed obtaining a catalyst leading to a kinetic current density of 0.81 mA cm^{-2} (corresponding to a j_{kw} of 22 μA cm^{-2}), whereas no significant change was observed by TEM and XRD measurements. Such improvement could be due to the removal of remaining surfactant as it was proposed in Fig. 11. However, no significant increase of the active surface area was recorded (35 m^2 g^{-1} after the thermal treatment against 34 m^2 g^{-1} for the fresh catalyst). Then, the low activity of the w/o microemulsion Pt/C catalyst can be attributed to a higher surface defect concentrations, i. e. higher density of low coordinated platinum atoms, which is not in favour of the activation of the oxygen reduction reaction [Kinoshita, 1990]. Indeed, conversely to the catalysts prepared by the Polyol route and the Bönnemann route, the growth stopping in the microemulsion synthesis is mainly due to the platinum cation depletion rather than to blocking of the platinum surface by surfactant adsorption. Moreover, no crystalline reconstruction occurs due to the absence of thermal treatment. Indeed it is known that a thermal treatment could improve the catalytic performance of PEM fuel cell catalysts for oxygen reduction reaction [Bezerra et al., 2007]. The residual constrains decrease with the thermal treatment, which leads to non negligible changes in the electronic effect.

The catalyst prepared by the polyol method display a high activity (j_k = 0.74 mA cm^{-2} and j_{kw} = 22 μA cm$_{Pt}^{-2}$), almost reaching that of the catalyst prepared by the Bönemann method (j_k = 0.82 mA cm^{-2} and j_{kw} = 22 μA cm$_{Pt}^{-2}$), although a higher mean particle size (and mean crystallite size) than that determined for the Bönnemann Pt/C, with ca. 4.0 nm against ca. 2.5 nm. This high activity of the Polyol Pt/C catalyst may be related to a size effect, a mean particle size of ca. 4.0 nm could be more favourable for the oxygen reduction reaction

[Kinoshita, 1990]. It could also be related to the presence of facets on the platinum surface, which could also favour the ORR. Indeed, facets display less low-coordinated surface atoms than edges or corners, which could be blocked by strongly adsorbed oxygen containing species difficult to reduce.

5. Conclusion

This book chapter intended to point out the importance of the Pt/C history on the electrocatalytic activity. For this purpose, different colloidal methods have been developed for the synthesis of platinum-based catalysts for Proton Exchange Membrane Fuel Cell (PEMFC) applications. Such methods are expected to produce metallic nano-sized particles with a narrow size distribution. However, the germination process, the grain growth and the metallic particles stabilization steps leading to Pt-surfactant stabilized colloids were different according to the synthesis route. For example, synthesis performed in organic or aqueous media could sometimes require thermal treatment (Bönnemann method) or not (Instant method). The end of the grain growth mechanism could occur either from depletion of metallic salts in surfactant stabilized nanoreactors ("water in oil" microemulsion method) or from the blocking of the platinum particle surface by adsorption of a surfactant (Bönnemann and polyol methods). All these mechanisms led to reach different catalyst morphologies and structures. The different methods of metal nanocatalyst synthesis via colloidal routes were first described, particularly Bönnemann method, "water in oil" microemulsion, polyol methods with thermal or microwave activation and instant method, in which different mechanisms are involved for the metal nanoparticle formation. Because the reactions occurring in Fuel Cells are known to be sensitive to the catalyst structure, the present book chapter has then focused on the relationship between synthesis methods, catalyst structure and morphology, crystallite microstructure, electrochemical properties and electrocatalytic behaviour of the materials.

The influence on the composite material structure and morphology has been discussed on the basis of microscopy observations (TEM, HRTEM) and X ray measurements (XRD, XPS). The characterization of the catalyst structure (mean sizes and size distribution of isolated particles), of the composite material morphology (mean size and size distribution of aggregates), and the platinum microstructure (crystallinity, crystallite sizes), etc. has also been performed and discussed.

Then, the electrochemical active surface areas, the behavior toward pre-adsorbed CO saturating layer oxidation and the activity toward the oxygen reduction reaction (ORR) have been compared using electrochemical methods (Cyclic voltammetry, rotating disc and ring disc electrodes). In the case of ORR studies, the number of exchanged electrons and the kinetic current densities at 0.95 V vs RHE, as determined from mathematical treatment of the Koutecky-Levich equation, have been used to compare the activity and selectivity of the different Pt/C catalysts. An important dependence of the catalytic activity on the catalyst synthesis method has been evidenced. However, due to the invariance of the cell parameter for all prepared catalyst, it is unlikely that the change in catalytic activity of the platinum material is related to geometric effect (change in Pt-Pt interatomic distance). But, the surface structure, presence of surface domains (facets) or of surface defects, seems also to have an important effect on the catalytic activity. It also seems that a thermal treatment, leading to crystallite reconstruction and decrease of residual strains, could lead to an enhancement of the catalytic activity toward orr.

Although the final objective, which was to correlate the activity of Pt/C catalysts to their morphology and structure, is not fully achieved, this book chapter shows unambiguously that the microstructure, macrostructure and morphology of a Pt/C catalyst has a great effect on its electrochemical behaviour toward molecules having energetic interest in fuel cell technology.

6. References

Attard, G. A. Gillies, J. E. Harris, C. A. Jenkins, D. J. Johnston, P. Price, M. A. Watson, D. J. Wells, P. B. (2001). Electrochemical evaluation of the morphology and enantioselectivity of Pt/graphite. *Appl. Catal. A : Gen.*, Vol. 222, N° 1-2, (December 2001) pp. 393-405.

Ayyadurai, S. M. Choi, Y-S. Ganesan, P. Kumaraguru, S. P. Popov. B. N. (2007). Novel PEMFC Cathodes Prepared by Pulse Deposition. *J. Electrochem. Soc.*, Vol. 154, N° 10 (August 2007) pp. B1063-B1073.

Bard, A. J. Faulkner, L. R.. in: Electrochemical Methods: Fundamentals and Applications, 2nd ed., John Wiley & Sons Inc., New York, 2001,.

Bezerra, C. W. B. Zhang, L. Liu, H. Lee, K. Marques, A. L. B. Marques, E. P. Wang, H. Zhang, J. (2007). A review of heat-treatment effects on activity and stability of PEM fuel cell catalysts for oxygen reduction reaction. *J. Power Sources*, Vol. 173, N° 2 (November 2007) pp. 891-908.

Biegler, T. Rand, D.A.J. Woods, R. (1971). Limiting Oxygen Coverage on Platinized Platinum; Relevance to Determination of Real Platinum Area by Hydrogen Adsorption. J. Electroanal. Chem., Vol. 29, N° 2, (February 1971) pp. 269-277.

Billy, E. Maillard, F. Morin A. Guetaz, L. Emieux, F. Thurier, C. Doppelt, P. Donet, S. Mailley, S. (2010). Impact of ultra-low Pt loadings on the performance of anode/cathode in a proton-exchange membrane fuel cell. *J. Power Sources*, Vol. 195, N° 9, (May 2010), pp. 2737-2746.

Bönnemann, H. Brijoux, W. (1995). The preparation, characterization and application of organosols of early transition metals. NanoStructured Materials, Vol. 5, N° 2, (February 1995) pp. 135-140.

Bönnemann, H. Brijoux, W. Brinkmann, R. Dinjus, E. Joussen, T. Korall, B. (1991). Formation of Colloidal Transition Metals in Organic Phases and Their Application in Catalysis. *Angew. Chem. Int. Engl.*, Vol. 30, N° 10, (October 1991) 1312-1314.

Bönnemann, H. Brijoux, W. Brinkmann, R. Fretzen, R. Joussen, T. Köppler, R. Korall, B. Neiteler, P. Richter, J. (1994). Preparation, characterization, and application of fine metal particles and metal colloids using hydrotriorganoborates. *J. Mol. Catal.*, Vol. 86, N° 1-3, (January 1994) 129-177.

Boulmer Leborgne, C. Benzerga, R. Scuderi, D. Perriere, J. Albert, O. Etchepare, J. Millon, E. (2006). Femtosecond laser beam in interaction with materials for thin film deposition. *SPIE Proceeding on High Power Laser Ablation VI*, Vol. 6261, Taos-New Mexico (USA), May 2006.

Boutonnet, M. Kizling, J. Stenius, P. Maire, G. (1982). The preparation of monodisperse colloidal metal particles from microemulsions. *Colloid and Surfaces*, Vol. 5, N° 3, (November 1982) pp. 209–225.

Bradley, J. S. (1994). The Chemistry of Transition Metal Colloids, In: *Clusters and Colloids, From Theory to Applications*, G. Schmid, pp. 459–544, Springer, ISBN 9783527290437, Weinheim, Germany.

Brault, P. Caillard, A. Thomam, A. L. Mathias, J. Charles, C. Boswell, R. W. Escribano, S. Durand, J. Sauvage, T. (2004). Plasma sputtering deposition of platinum into porous fuel cell electrode. *J. Phys. D : Appl. Phys.*, Vol. 37, N° 24, (December 2004), pp. 3419-3423.

Brimaud, S. Coutanceau, C. Garnier, E. Léger, J.-M. Gérard, F. Pronier, S. Leoni, M. (2007). Influence of surfactant removal by chemical or thermal methods on structure and electroactivity of Pt/C catalysts prepared by water-in-oil microemulsion. *J. Electroanal. Chem.*, Vol. 602, N° 2, (April 2007) pp. 226-236.

Brimaud, S. Pronier, S. Coutanceau, C. Léger, J.-M. (2008). New findings on CO electrooxidation at platinum nanoparticle surfaces. *Electrochem. Comm.*, Vol. 10, N° 11, (November 2008), pp. 1703-1707.

Caillard, A. Charles, C. Boswell, R. Brault, P. Coutanceau, C. (2007). Plasma based platinum nanoaggregates deposited on carbon nanofibers improve fuel cell efficiency. *Appl. Phys. Lett.*, Vol. 90, N° 22, (May 2007), pp. 223119-1-223119-3.

Chen, S. Kucernak. A. (2003). Electrodeposition of Platinum on Nanometer-Sized Carbon Electrodes. *J. Phys. Chem. B*, Vol. 107, N° 33, (August 2003) pp. 8392-8402.

Cho, Y-H. Yoo, S. J. Cho, Y-H. Park, H-S. Park, I-S. Lee, J.K. Sung, Y-E. Enhanced performance and improved interfacial properties of polymer electrolyte membrane fuel cells fabricated using sputter-deposited Pt thin layers. *Electrochim. Acta*, Vol. 53, N° 21,(September 2008) pp. 6111-6116.

Coutanceau, C. Croissant, M. J. Napporn, T. Lamy, C. (200). Electrocatalytic reduction of dioxygen at platinum particles dispersed in a polyaniline film. *Electrochim. Acta*, Vol. 46, N° 4, (December 2000) pp. 579-588.

Coutanceau, C. Rakotondrainibe, A. Lima, A. Garnier, E. Pronier, S. Léger, J. M. Lamy, C. (2004). Preparation of Pt-Ru bimetallic anodes by galvanostatic pulse electrodeposition : characterization and application to the direct methanol fuel cell. *J. Appl. Electrochem.*, Vol. 34, N° 1, (January 2004), pp. 61-66.

Coutanceau, C. Brimaud, S. Dubau, L. Lamy, C. Léger, J.-M. Rousseau, S. Vigier, F. (2008). Review of different methods for developing nanoelelectrocatalysts for the oxidation of organic compounds. *Electrochim. Acta*, Vol. 53, N° 23, (October 2008) pp. 6865-6880.

Devadas, A. Baranton, S. Napporn, T. W. Coutanceau, C. (2011). Tailoring of RuO_2 nanoparticles by microwave assisted "Instant method" for energy storage applications. *J. Power Sources*, Vol. 196, N° 8, (April 2011) pp. 4044-4053.

Eriksson, S. Nylén, U. Rojas, S. Boutonnet, M. (2004). Preparation of catalysts from microemulsions and their applications in heterogeneous catalysis. *Appl. Catal. A: Gen.*, Vol. 265. N° 2, (July 2004) pp. 207-219.

Fievet, F. Lagier, J. P. Blin, B. Beaudoin, B. Figlarz, M. (1989). Homogeneous and heterogeneous nucleations in the polyol process for the preparation of micron and submicron size metal particles. *Solid State Ionics*, Vol. 32-33, N° 1, (February-March 1989) pp. 198-205.

Friedrich, K. A. Henglein, F. Stimming, U. Unkauf, W. (2000). Size dependence of the CO monolayer oxidation on nanosized Pt particles supported on gold. *Electrochim. Acta*, Vol. 45, N° 20, (Juin 2000) pp. 3283-3293.

Gloagen, F. Andolfatto, N. Durand, R. Ozil, P. (1994). Kinetic study of electrochemical reactions at catalyst-recast ionomer interfaces from thin active layer modeling. *J. Appl. Electrochem.*, Vol. 24, N° 9, (September 1994) pp. 863-869. ISSN 0021-891X.

Gómez, R. Clavilier, J. (1993). Electrochemical behaviour of platinum surfaces containing (110) sites and the problem of the third oxidation peak . *J. Electroanal. Chem.*, Vol. 354, N° 1-2, (August 1993) pp. 189-208.

Grolleau, C. Coutanceau, C. Pierre, F. Leger, J. M. (2010). Optimization of a surfactant free polyol method for the synthesis of Platinum-Cobalt electrocatalysts using Taguchi design of experiments. *J. Power Sources*, Vol. 195, N° 6, (March 2010) pp. 1569-1576.

Grolleau, C. Coutanceau, C. Pierre, F. Léger. J. M. (2008). Effect of potential cycling on structure and activity of Pt nanoparticles dispersed on different carbon supports. *Electrochim. Acta*, Vol 53, N° 24, (October 2008) pp. 7157-7165.

Habrioux, A. Vogel, W. Guinel, M. Guetaz, L. Servat, K. Kokoh, B. Alonso-Vante, N. (2009). Structural and electrochemical studies of Au–Pt nanoalloys. *Phys. Chem. Chem. Phys.*, Vol. 11, N° 18, (May 2009), pp. 3573-3579. ISSN 1463-9076.

Henry, C. R. (2005). Morphology of supported nanoparticles. Progress in Surf. Sci. Vol. 80, N° 3-4, pp. 92–116.

Hwang B. J. Kumar, S. M. S. Chen, C-H. Monalisa. Cheng, M-Y. Liu, D-G. Lee J-F. (2007). An Investigation of Structure-Catalytic Activity Relationship for Pt-Co/C Bimetallic Nanoparticles toward the Oxygen Reduction Reaction. *J. Phys. Chem. C* , Vol. 111, N° 42,(October 2007) pp. 15267-15276.

Ingelsten, H. H. Ggwe, R. Palmqvist, A. Skoglundh, M. Svanberg, C. Hlmberg, K. Shah, D. O. J. (2001). Kinetics of the Formation of Nano-Sized Platinum Particles in Water-in-Oil Microemulsions. *J. Colloid Interface Sci.*, Vol. 241, N° 1, (Septembre 2001) pp. 104–111.

Iwasita, T. (2003). Methanol and CO électrooxydation. in *Handbook of Fuel Cells – Fundamentals, Technology and Applications*, Vielstich, W., Gasteiger, H. A., Lamm, A. John Wiley & Sons, Ltd.; New York, 2003, Vol. 2: Electrocatalysis, pp. 603-624.

Jakobs, R. C. M. Janssen, L. J. J. Barendrecht, E. (1985). Oxygen reduction at polypyrrole electrodes – I. Theory and evaluation of the rrde experiments. *Electrochim. Acta*, Vol. 30, N° 8, (August 1985) pp. 1085-1091.

Kinoshita, K. (1990). Particle size effects for oxygen reduction on highly dispersed platinum in acid electrolytes. J. Electrochem Soc., Vol. 137, N° 3, (March 1990) pp. 845-848.

Larcher, D. Patrice. R. (2000). Preparation of Metallic Powders and Alloys in Polyol Media: A Thermodynamic Approach. *J. Solid State Chem.*, Vol. 154, N° 2, (Novembre 2000) pp. 405-411.

Lebègue, E. Baranton, S. Coutanceau, C. (2011). Polyol synthesis of nanosized Pt/C electrocatalysts assisted by pulse microwaves activation. *J. Power Sources*, Vol. 196, N° 3, (February 2011) pp. 920-927.

Liu, Z. Gan, L. M. Hong, L. Chen, W. Lee, J. Y. (2005). Carbon supported Pt nanoparticles for PEMFC. *J. Power Sources*, Vol. 139, N° 1-2, (January 2005) 73-78 : ""

Lopez-Cudero, A. Solla-Gullon, J. Herrero, E. Aldaz, A. Feliu, J. M. (2010). CO electrooxidation on carbon supported platinum nanoparticles: Effect of aggregation. *J. Electroanal. Chem.*, Vol. 644, N° 2 (June 2010) pp. 117-126.

Maillard, F. Savinova, E. R. Simonov, P. A. Zaikovskii, V. I.. Stimming, U. (2004). Infrared Spectroscopic Study of CO Adsorption and Electro-oxidation on Carbon-Supported Pt Nanoparticles: Interparticle versus Intraparticle Heterogeneity. *J. Phys. Chem. B*, Vol. 108, N° 46, (November 2004) pp. 17893-17904.

Maillard, F. Schreier, S. Hanzlik, M. Savinova, E. R. Weinkauf, S. Stimming, U. (2005). Influence of particle agglomeration on the catalytic activity of carbon-supported Pt nanoparticles in CO monolayer oxidation. *Phys. Chem. Chem. Phys.*, Vol. 7, N° 2, (January 2005) pp. 385-393.

Maillard, F. Savinova, E. R. Stimming, U. (2007). CO monolayer oxidation on Pt nanoparticles: Further insights into the particle size effects. *J. Electroanal. Chem.*, Vol. 599, N° 2, (January 2007) pp. 221-232.

Markovic, N. M. Gasteiger, H. A. Ross, P. N. Oxygen Reduction on Platinum Low-Index Single- Crystal Surfaces in Sulfuric Acid Solution: Rotating Ring - Pt(hkl) Disk Studies. J. Phys. Chem., Vol. 99, N° 11, (March 1995) pp. 3411-3415.

Markovic, N. M. Grgur, B. N. Ross, P. N. (1997). Temperature-Dependent Hydrogen Electrochemistry on Platinum Low-Index Single-Crystal Surfaces in Acid Solutions. *J. Phys. Chem. B*, Vol. 101, N° 27, (July 1997) pp. 5405-5413.

Mottet, C. Gonialowski, J. Baletto, F. Ferrando, R. Tréglia, G. Modeling free and supported metallic nanoclusters: structure and dynamics. *Phase Transitions*, Vol. 77, N° 1-2, (January-February 2004) pp. 101–113. ISSN 0141-1594.

Oh, H-SOh, . J-G. Hong, Y-G. Kim, H. (2007). Investigation of carbon-supported Pt nanocatalyst preparation by the polyol process for fuel cell applications. *Electrochim. Acta*, Vol. 52, N° 25, (September 2007) pp. 7278–7285.

Park, J. Joo, J. Kwon, S. G. Jang, Y. Hyeon, T. (2007). Synthesis of monodisperse spherical nanoparticules. *Angewandte Chem. – Int. Ed.*, Vol. 46, N°25, (June 2007) pp. 4630-4660. ISSN 1521 3773.

Peng, X. Wickham, J. Alivisatos, A. P. (1998). Kinetics of II-VI and III-V Colloidal Semiconductor Nanocrystal Growth: "Focusing" of Size Distributions. *J. Am. Chem. Soc.*, Vol. 120, N° 21, (June 1998) pp. 5343-5344.

Perrière, J. Millon , E. Chamarro , M. Morcrette, M. Andreazza, C. (2001). Formation of GaAs nanocrystals by laser ablation. *Appl. Phys. Lett.*, Vol. 78, N° 19, (May 2001) 2949-1-2949-3.

Pieck, C. L. Marecot, P. Barbier, J. (1996). Effect of Pt-Re interaction on sulfur adsorption and coke deposition. *Appl. Catal. A: Gen.*, Vol. 145. N° 1-2, (October 1996) pp. 323-334.

Plyasova, L. M. Molina, I. Y. Gavrilov, A. N. Cherepanova, S. V. Cherstiouk, O. V. Rudina, N. A. Savinova, E. R. Tsirlina, G. A. (2006). Electrodeposited platinum revisited: Tuning nanostructure via the deposition potential. *Electrochim. Acta*, Vol. 51, N° 21 (June 2006) pp. 4477-4488.

Ralph, T. R. Hogarth, M. P. (2002). Catalysts for low temperature fuel cell. Part I: the cathode challenges. *Platinum Metal Rev.*, Vol. 46, N° 1 (January 2002) pp. 3-14

Richard, D. Gallezot, P. (1987). Preparation of highly dispersed carbon supported platinum catalysts. In: *Preparation of Catalysts IV*. B. Delmon, P. Grange, P.A. Jacobs, G. Poncelet, pp. , Elsevier Science Publishers B.V., Amsterdam.

Roman-Martinez, C. Cazorla-Amoros, D. Yamashita, H. de Miguel, S. Scelza, O. A. (2000). XAFS Study of Dried and Reduced PtSn/C Catalysts: Nature and Structure of the Catalytically Active Phase. *Langmuir*, Vol. 16, N° 3, (February 2000) pp. 1123-1131.

Reetz, M. T. Koch, M. G. Patent application DE-A 19852547.8, 13.11.1998.

Reetz, M. T. Koch. M. G. (1999). Water-Soluble Colloidal Adams Catalyst: Preparation and Use in Catalysis. *J. Am. Chem. Soc.*, Vol. 121, N° 34 (September 1999) pp. 7933–7934.

Reetz, M. T. Lopez, M. Patent Application DE-A 102 11701.2, 16.03.2002.

Reetz, M. T. Schulenburg, H. Lopez, M. Splienthoff, B. Tesche, B. (2004). Platinum-Nanoparticles on Different Types of Carbon Supports: Correlation of Electrocatalytic Activity with Carrier Morphology. *Chimia*, 58, N°12, (December 2004), pp. 896-899.

Sellin, R. Grolleau, C. Coutanceau, C. Léger, J.-M. Arrii-Clacens, S. Pronier, S. Clacens, J.-M. (2009). Effects of Temperature and Atmosphere on Carbon-Supported Platinum Fuel Cell Catalysts. *J. Phys. Chem. C,* Vol. 113, N° 52, (December 2009) pp. 21735-21744.

Sellin, R. Clacens, J-M. Coutanceau, C. (2010). A thermogravimetric analysis/mass spectroscopy study of the thermal and chemical stability of carbon in the Pt/C catalyst system. *Carbon,* Vol. 48, N° 8, (July 2010) pp. 2244-2254.

Shevchenko, E. V. Talapin, D. V. Schnablegger, H. Kornowski, A. Festin, O. Svedlindh, P. Haase, M. Weller H. (2003). Study of nucleation and growth in the organometallic synthesis of magnetic alloy nanocrystals : the role of nucleation rate in size control of $CoPt_3$ nanocrystals. *J. Am. Chem. Soc.,* Vol. 125, N° 30, (July 2003) pp. 9090-9101.

Schmidt, T. J. Gasteiger, H. A. Behm, R. J. (1999). Rotating Disk Electrode Measurements on the CO Tolerance of a High-Surface Area Pt/Vulcan Carbon Fuel Cell Catalyst. *J. Electrochem. Soc.,* Vol. 146, N° 4, (April 1999) pp. 1296-1304.

Solla-Gullon, J. Montiel, V. Aldaz, A. Clavilier. J. (2000). Electrochemical characterisation of platinum nanoparticles prepared by microemulsion: how to clean them without loss of crystalline surface structure. *J. Electroanal. Chem.,* Vol. 491, N° 1-2,(September 2000) pp. 69-77.

Solla-Gullón, J. Vidal-Iglesias, F. J. Herrero, E. Feliu, J. M. Aldaz, A. (2006). CO monolayer oxidation on semi-spherical and preferentially oriented (1 0 0) and (1 1 1) platinum nanoparticles. *Electrochem. Comm.,* Vol. 8, N° 1, (January 2006) pp. 189–194.

Solla-Gullón, J. Rodríguez, P. Herrero, E. Aldaz, A. Feliu, J. M. (2008). Surface characterization of platinum electrodes. *Phys. Chem. Chem. Phys.,* Vol. 10, N° 10, (March 2008) pp. 1359-1373. ISSN 1463-9076.

Tarasevich, M. R. Sadkowski, A. Yeager, E. (1983). Oxygen Electrochemistry. In *Kinetics and Mechanisms of Electrode Processes*; Conway, in: B.E. Conway, J.O'M. Bockris, E. Yeager, S.U.M. Khan (Eds.), Comprehensive Treatise of Electrochemistry, vol. 7, Plenum Press, New York, 1983, p. 301-398.

Vigier, F. Coutanceau, C. Perrard, A. Belgsir, E. M. Lamy, C. (2008). Development of anode catalyst for Direct Ethanol Fuel Cell. *J. Appl. Electrochem.,* Vol. 34, N° 4, (April 2004) pp. 439-446.

Villegas, I.. Weaver, M. J. Carbon monoxide adlayer structures on platinum (111) electrodes: A synergy between in situ scanning tunneling microscopy and infrared spectroscopy (1994). *J. Chem. Phys.,* Vol. 101, N° 2, (July 1994) pp. 1648-1660.

Vogel, W. Bradley, J. Vollmer, O. Abraham, I. (1998). Transition from Five-Fold Symmetric to Twinned FCC Gold Particles by Thermally Induced Growth. *J. Phys. Chem. B,* Vol. 102, N° 52, (December 1998) pp. 10853-10859.

Wang, J. Xue, J. M. Wan, D. M. Gan, B. K. (2000). Mechanically activating nucleation and growth of complex perovskites. *J. Solid State Chem.,* Vol. 154, N° 2, (November 2000) pp. 321-328.

Wilson, G. J. Matijasevitch, A. S. Mitchell, D. R. G. Schulz, J. C. Will, G. D. (2006). Modification of $TiO2$ for enhanced surface properties : finite Ostwald ripening by a microwave hydrothermal process. *Langmuir,* Vol. 22, N° 5, (February 2006) pp. 2016-2027.

Yoshimi, K. Song M. -B. Ito, M. (1996). Carbon monoxide oxidation on a Pt(111) electrode studied by in-situ IRAS and STM: coadsorption of CO with water on Pt(111). *Surf. Sci.,* Vol. 368, N° 1-3 (December 1996) pp. 389-395.

Zagal, J. Bindra, P. Yeager, E. (1980). A Mechanistic Study of O_2 Reduction on Water Soluble Phthalocyanines Adsorbed on Graphite Electrodes . *J. Electrochem. Soc.,* VoL 127, N° 7, (July 1980) pp. 1506-1517.

9

Bulk Nanocrystalline Thermoelectrics Based on Bi-Sb-Te Solid Solution

L.P. Bulat[1], D.A. Pshenai-Severin[2], V.V. Karatayev[3],
V.B. Osvenskii[3], Yu.N. Parkhomenko[3,5], M. Lavrentev[3],
A. Sorokin[3], V.D. Blank[4], G.I. Pivovarov[4],
V.T. Bublik[5] and N.Yu. Tabachkova[5]

1. Introduction

Thermoelectric energy conversion represents one of ways of direct conversion of the thermal energy to the electric energy. The thermoelectric converters – a thermoelectric power generator or a thermoelectric cooler are solid-state devices, therefore it possesses following advantages: environmentally cleanliness, simplicity of management and convenience of designing, high reliability and possibility of longtime operation without service, absence of moving parts, absence of noise, vibration and electromagnetic noise, compactness of modules, independence of space orientation, ability to carry out of considerable mechanical overloads. Despite obvious advantages of thermoelectric conversion it has the important lack – rather small value of the efficiency η; in the best cases η = (5 – 8) %. Therefore thermoelectric generators are used today, as a rule, only in «small power», where it is impossible or is economically inexpedient to bring usual electric mains: for power supply of space missions, at gas and oil pipelines, for power supply of sea navigating systems, etc. Nevertheless, the thermoelectric method of utilization of waste heat from units of cars and vessels is unique technically possible. It appears the thermoelectric generators can save up to 7 % of automobile fuel. The thermoelectric coolers become economically justified at enough small cooling power Qc as a rule no more than 10 – 100 W. However the thermoelectric coolers are widely applied in the most different areas: domestic refrigerators, water-chillers, picnic-boxes; coolers for medicine and biology, for scientific and laboratory equipment, refrigeration systems for transport facilities. Very important area connects with strong up-to-date requirements for the thermal management of micro- and optoelectronics elements, including microprocessors and integrated circuit; the requirements have essentially increased owing to the increase in their speed and miniaturization. And the desired value of local heat removal from concrete spots of chips can be realized only by means of the thermoelectric cooling.

[1]National Research University ITMO, St. Petersburg, Russia
[2]Ioffe Physical Technical Institute, St Petersburg, Russia
[3]GIREDMET Ltd., Moscow, Russia
[4]Technological Institute of Superhard and New Carbon Materials, Troitsk, Russia
[5]National University of Science and Technology "MISIS", Moscow, Russia

It is known, that the efficiency of thermoelectric generators and the coefficient of performance (COP) of thermoelectric coolers are defined by the dimensionless parameter of a thermoelectric material $ZT = \dfrac{\sigma \alpha^2}{\kappa} T$, where T – is the absolute temperature, Z – thermoelectric figure of merit, σ, κ, α – accordingly electric conductivity, thermal conductivity and Seebeck coefficient (thermoelectric power) of a used material.

Today the best commercial thermoelectric materials (thermoelectrics) has the efficiency ZT =1.0. Let us underline that ZT increased from 0.75 only to 1.0 during lust five decades. Obviously, competitiveness of thermoelectric generators and coolers will rise if it will be possible to increase the figure of merit. Thus the thermoelectric generation and cooling can provide a weighty contribution to a decision of the problem of utilization of renewed energy sources, a recycling of a low potential heat, and maintenance of storage of a foodstuff. Recently the increasing attention is involved to the thermoelectric refrigeration and the electric power generation as to environmentally clean methods. It is caused by several reasons. The main of them is caused by new scientific results on improve of the thermoelectric figure of merit. Important results in the development of highly effective nanostructured thermoelectric materials have been published last decade; see for example the reviews (Dresselhaus et al., 2007; Minnich et al., 2009; Dmitriev & Zvyagin, 2010; Lan et al., 2010). The thermoelectric efficiency ZT = 2.4 has been reached at T = 300K in the p-type semiconductor Bi_2Te_3/Sb_2Te_3 in superlattices with quantum wells (Venkatasubramanian et al, 2001); the estimation indicates that the value ZT ~ 3.5 has been received at T = 575 K in nanostructured n-type PbSeTe/PbTe with quantum dots (Harman et al., 2000, 2005). It is possible also to include thermotunnel elements (thermal diodes) to nanostructured thermoelectrics in which there exists the electron tunneling through a narrow vacuum or air gap (Tavkhelidze et al., 2002). The efficiency $ZT=1.7$ at the room temperature was received experimentally in thermoelements with cold junctions, consisting from the semiconductor branches of p-type $Bi_{0.5}Sb_{1.5}Te_3$ and n-type $Bi_2Te_{2.9}Se_{0.1}$ (Ghoshal et al., 2002a, 2002b). Let us note also the important results received in a set of papers, for example (Shakouri & Bowers, 1997), which specify in perspectives of the use of the emission nanostructures for creation of effective thermoelectric energy converters and coolers. The nanostructuring gives a new way of improvement of the thermoelectric efficiency because the governance of the sizes of nanostructured elements is a new important parameter for influence on the thermoelectric properties of a material.

Unfortunately the best values of ZT that were specified in nanostructures based on superlattices with quantum wells and quantum dots have not been reproduced in one laboratory of the world. On the other hand fabrication of such superlattices uses very expensive technologies; therefore industrial manufacture of such nanostructures is very problematic from the economical point of view. Good values of the thermoelectric figure of merit in thermotunnel devices and in thermoelements with point contacts also have not been reproduced. Therefore the special interest represents a creation of thermoelectric nanostructures by means of an adaptable to streamlined production and a cheap technique. An example of such technology is fabrication of bulk nanostructured thermoelectric samples by ball milling of initial materials with subsequent hot pressing (Poudel et al., 2008; Bublik et al., 2009; Bulat et al., 2008a, 2008b, 2009, 2011b; Minnich et al., 2009; Lan et al., 2010), spark plasma sintering (Bublik et al., 2010a, 2010b) or extrusion (Vasilevskiy et al., 2010). In Ref.

(Poudel et al., 2008) the value ZT = 1.4 have been received at T=100⁰C and ZT = 1.2 at the room temperature in such bulk nanostructured thermoelectrics fabricated from the solid solutions based on p-type Bi-Sb-Te.

Thus from the set forth above it can be concluded that the following reliable preconditions of the obtaining of the high thermoelectric figure of merit in the nanostructured thermoelectrics take place: (a) the experimental results specify a possibility for the achievement of a high thermoelectric figure of merit in nanostructured thermoelectrics of various types; (b) in particular some experimental results confirm the possibility of the obtaining of high figure of merit in bulk nanostructured semiconductors. Experimental and theoretical results that obtained by the authors during lust few years on investigation of bulk nanocrystalline thermoelectrics based on Bi-Sb-Te solid solution including nanocomposites are summarized, systematized and analyzed in the present chapter.

2. Experiment

2.1 Fabrication of bulk nanocrystalline thermoelectrics

Two stages should be executed for preparation of bulk nanocrystalline materials. At first a powder from nanoparticles should be fabricated, and then it should be consolidated into a bulk sample. A crystalline thermoelectric material with high thermoelectric efficiency should be chosen as an initial material for the nanopowder preparation. In our case the solid solution based on p-type $Bi_xSb_{2-x}Te_3$ was selected as the initial material (Bublik et al., 2009, 2010a; Bulat et al., 2008a, 2008b, 2009a, 2009b, 2011b). It was grown up by zone melting method; and the dimensionless figure of merit ZT = 1.0 was measured along the C axis at the room temperature in primary samples. The initial crystalline material was grinded and purifying. The mechanoactivation process (the ball milling) is the most convenient and cheap way for fabrication of a nanopowder. We used the high-speed planetary mill AGO-2U to achieve the further superthin crushing and to prepare the nanopowder. Other types of mills: the Activator 2S, Retsch PM 400 also were applied at different stages of the nanopowder preparation. The processing of the powders fabrication in the mill is made by steel spheres which were collided with acceleration up to 90g. Tightly closed containers of the mill rotate in flowing water that protects a material from a warming up. It is necessary to provide absolute absence of the oxidation of the nanopowder. Therefore all operations were spent in the boxing filled with argon.

The duration of the mechanoactivation processing was varied from 30 min till 2 hours. The diffraction analysis has shown that the main sizes of nanoparticles of the powder are 8-10 nm. The following methods of pressing for fabrication of compact samples from highly active ultradisperse powder have been used (Bublik et al., 2009, 2010a; Bulat et al., 2009a): cold pressing of powders with the subsequent sintering in inert gas; sintering in graphite compression moulds; sintering in steel compression moulds (at more high pressure in comparison with the previous variant). Hot pressing of the nanopowder was made under the pressure in the range from 35 MPa to 3.3 GPa in the range of temperatures from 250 to 490 °C. To prevent the oxidation of nanoparticles all basic operations are made in the atmosphere of argon. As a result, series of compact p-type $Bi_xSb_{2-x}Te_3$ samples were produced. The method of spark plasma sintering (SPS) with the equipment SPS–511S for preparation of bulk nanostructure was also used (Bublik et al., 2010b, 2010c).

2.2 Methods of experimental investigation

For investigation of thermoelectric properties it is necessary to know values of four material parameters: Seebeck coefficient (thermoelectric power), electric conductivity, heat conductivity and thermoelectric figure of merit. The heat conductivity measurement of small samples is the most difficult because all traditional techniques of direct measurement are based on passing of a calibrated thermal flow through a sample; but a thermal flow measurement with sufficient accuracy and consideration of all losses on small samples is very complicated. Therefore we had been used the Harman method (Bublik et al., 2009, 2010a) which allows to fulfill the measurement of thermoelectric figure of merit Z directly by measurement of only electric parameters, not mentioning about thermal flows. Besides, the technique allows receiving in the same cycle of the measurement the values of Seebeck coefficient and the electric conductivity also. Then the value of heat conductivity can be calculated using the known value of Z. A mathematical model for calculations of thermoelectric parameters on the Harman method measurement has been developed, and processing of results of the measurement was carried out under specially developed soft.

For determining of the speed of longitudinal acoustic waves and for subsequent calculation of modules of elasticity and the modulus of dilatation the modified echo-pulse method with application of focusing system of an acoustic microscope has been used. The mode of ultrashort probing impulses has been utilized; it has given the opportunity to register separate signals caused by the reflexion of the impulse from walls of a sample. The microstructure of samples was investigated at metallographic sections made on grinding-and-polishing machine "Struers". Microhardness was measured on microhardness gauge PMT-3M by the method of cave-in of diamond tips. The microscope Olympus BX51 was used for the metallographic analysis.

The working capacity and reliability of thermoelectric devises are substantially caused by their strength characteristics. The strength at the extension occupies a special place among them. However for investigated materials the method of direct test for the extension is the extremely inexact for some reasons. Therefore the method of diametrical compression of disk or cylindrical samples was used; the advantage of the method consists that the extension pressure destruction begins inside a sample instead of its surface. Determination of the density of samples was made by the method of hydrostatic weighing. Laboratory analytical electronic scales "KERN", model 770-60 were used. The option "Sartorius" was applied for determination of the weight of a solid in a liquid.

The X-ray diffractometer methods were used for investigation of structure of nanopowder and bulk samples. The phase analysis was carried out by the method of X-rays diffraction with the diffractometer Bruker D8, equipped by the scintillation detector Bruker. The lattice constant of a solid solution of a thermoelectric material was determined by shooting of a diffractograms in the standard symmetric scheme of reflexion. A composition of the solid solution on the basis of measurements of the lattice constant was estimated. Values of nanograins were estimated by sizes of coherent dispersion areas (CDA) determined by the method of X-ray diffractometry on broadening of diffraction maxima. Calculation of CDA and estimation of microdeformation were spent by means of Outset program. The received values of CDA size were compared with the data received by a method of high resolution transmission electron microscopy (HRTEM). The structure of a sample was analyzed by means of creation of return polar figures. They were carried out by shooting of diffractograms in the standard symmetric scheme of reflexion. The following microscope equipment was used: the scanning electron microscope JSM-6480LV with the option for the energy-dispersive analysis INCA DRY; transmission electron microscope JEM 2100 with ultrahigh resolution and X-ray photoelectron spectroscope.

2.3 Structure and mechanical properties of nanopowder

Different types of nanopowder from $Bi_xSb_{2-x}Te_3$ solid solutions with different value of x were prepared with the following duration of the mechanoactivation process (the ball milling): 15 min, 30 min, 60 min and 120 min. For each type of powder the X-ray diffractograms, a distribution of CDA size and HRTEM images were received. A typical diffractogram of $Bi_{0.5}Sb_{1.5}Te_3$ powder is presented in (Bulat et al., 2009b). Examples of distribution of CDA size of the nanopowder prepared from $Bi_{0.5}Sb_{1.5}Te_3$ during 60 min ball milling and the correspondent HRTEM image are shown in Fig.1.

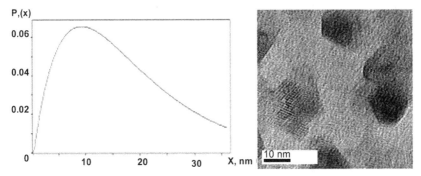

Fig. 1. Distribution of CDA size and TEM image of nanoparticles for $Bi_{0.5}Sb_{1.5}Te_3$ (60 min ball milling)

The nanopowder is the single-phase solid solution of $Bi_{0.5}Sb_{1.5}Te_3$ at each duration of the mechanoactivation process. Microdeformations of nanoparticles did not reveal. The electron microscopic data on the average size of nanoparticles confirm the calculation of CDA size determined by broadening of the diffraction maxima. In particular for 2 hours of ball milling the average value of nanoparticles was 8.5 nm and the greatest size was 35 nm. The distribution of sizes is homogeneous enough. An insignificant increase of the average size of CDA in comparison with the powder passed processing during 60 min is observed. A monotonous reduction of the average size of CDA was observed according to increase of the duration of the ball milling processing from 15 min to 60 min, and the size distribution of nanoparticles became more homogeneous. However the further increase of duration of milling leads any more to a reduction but to an increase of the average CDA size. Microstrains have been found out in the powder after 120 min ball milling. The mean-square microstrains are equal to 0.144%. So small particles cannot contain a dislocation therefore the presence of microstrains in the powder can be connected with heterogeneity of a structure of the solid solution, arising at long processing.

The lattice constants of the solid solution for all duration of processing are: a=0.4284 nm and c=3.0440 nm.

2.4 Structure and mechanical properties of bulk nanocrystalline samples

Taking into account set forth above the 60 min duration of the milling have been chosen for prepare of bulk samples. Therefore the average size of nanoparticles in the starting powder from $Bi_{0,5}Sb_{1,5}Te_3$ was equal to 8 – 10 nm. The cold and hot pressing and SPS method were applied for preparation of the bulk nanostructures.

The cold pressure under 1.5 GPa was carried out within 60 min (without sintering). The correspondent diffractogram shows that the sample is single-phase one, and it does not contain an exudation of moisture of another phase. The diffraction peaks belong to threefold solid solution of $Bi_{0,5}Sb_{1,5}Te_3$ with the lattice constants: a=0.4284 nm and c=3.0440 nm. The diffraction streaks remained strongly blurring as well as in a powder after the milling. The distribution of CDA size after the cold pressure illustrates Fig.2. The average size of CDA after the cold pressure is equal to 12 nm. As follows from Fig.2 at the cold pressing CDA size does not increase practically, and also uniformity of the nanoparticles size distribution increases. Microdeformations did not reveal.

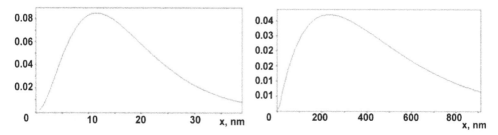

Fig. 2. Distribution of CDA size for bulk $Bi_{0,5}Sb_{1,5}Te_3$ after cold pressure; (a) – without sintering, (b) – after sintering

The investigation has shown that the sintering after cold pressing at temperatures ~ 300⁰C and above leads to two effects: (a) to occurrence of the second phase (tellurium) and (b) to origination of microdeformations. For example after the cold pressing a sample was sintered at the temperature 350⁰C within 25 min in argon atmosphere. It contains two phases $Bi_{0,5}Sb_{1,5}Te_3$ and tellurium. The lattice constants of the solid solution a=0.4296 nm and c=3.0447 nm are increased in comparison with the lattice constants in the initial sample before sintering. Root-mean-square microdeformation was equal to 0,086 %. Increase of the lattice constants of the solid solution as well as occurrence of microdeformations are apparently results of the excretion of tellurium from the solid solution. Thus as it was marked above microdeformations can be caused by heterogeneity of the solid solution arising at the raised temperature.

Fig.2 shows also the distribution of CDA size for the cold pressed bulk sample of $Bi_{0,5}Sb_{1,5}Te_3$ with the subsequent sintering. The average size of CDA is equal to 230 nm and the maximum size – 900 nm. Character of the size distribution in comparison with the cold pressing without the sintering has not changed, but the curve was displaced towards the big size. The increase of the CDA size testifies that there have passed processes of recrystallization during the sintering at 350 ^{0}C.

The SEM images of the sample received by cold pressing are characterized by high porosity (Fig.3). Pores are dark formations as they do not reflect electrons. The size of pores reaches 5 μm. It is visible at a big resolution that pores are not spherical; and coagulation of small pores takes place, therefore a facet is formed. The pores are faceted as a result of the diffusion processes during the sintering.

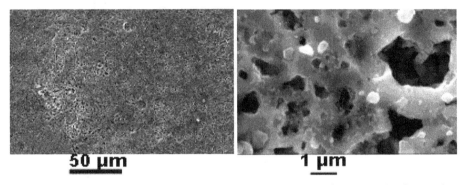

Fig. 3. The SEM images of the bulk $Bi_{0,5}Sb_{1,5}Te_3$ (cold pressing with sintering). The resolution is 500 and 10000 accordantly

The CDA size and structure of samples fabricated by the hot pressing are defined by three factors: a temperature of pressing, duration of stand-up under the loading and a value of pressure. But in any case the hot pressing as like as the cold pressing with the subsequent sintering leads to an occurrence of the second phase and to microdeformations. The increase of the CDA size due to the processes of recrystallization also takes part at the hot pressing. Fig.4 shows a typical distribution of CDA size for the hot pressed bulk sample; it was pressed during 20 min at 0.2 GPa and 289 ^{0}C. The sample contains two phases: $Bi_{0,5}Sb_{1,5}Te_3$ and tellurium. The lattice constants are increased in comparison with the lattice constants in the initial nanopowder. Root-mean-square microdeformation was equal to 0,055 %, the average CDA size ~ 85 nm, the biggest CDA has the size ~ 300 nm.

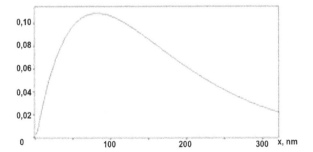

Fig. 4. Distribution of CDA sizes of the bulk $Bi_{0,5}Sb_{1,5}Te_3$ (hot pressing)

The SEM images of the surfaces of this sample are shown at Fig.5. A relief specified to different speed of dissolution arises after chemical polishing. The observable elements of structure: consertal formation, micropores, cracks, allocation of a second phase, microdeformations of CDA, connected with a method of hot pressing, are the factors reducing thermoelectric properties of the sample, first of all, the electric conductivity.

Fig. 5. The SEM images of the sample of bulk $Bi_{0,5}Sb_{1,5}Te_3$ (hot pressing). The resolution is 500 and 10000 accordantly

The samples for SPS were prepared by the cold pressure from the nanopowder at the room temperature. Then the SPS processing in a graphite press mould was made by passing of a pulse electric current under the pressure 50 MPa and the temperature 250-400 ^0C up to achievement of the hundred-percent density (from theoretical value for the given material). The correspondent SEM immerges can be seen in Fig.6.

At the temperatures of sintering 350 ^0C and 400 ^0C the grains grow and facet; that testifies about the active process of recrystallization. For samples that were sintered at a lower temperature the finely divided structure is typical, fragments of the fracture surface are not faceted, i.e. the grains have not recrystallized yet.

The average density of the samples fabricated from $Bi_{0,5}Sb_{1,5}Te$ solid solution by cold and hot pressing (plus 4 mass % of Te) are presented in table 1; the accuracy is ±0.02 g/cm^3. We see that increase of the temperature and increase of the pressure lead to gain of the density almost to the density of initial samples. However according to the ultrasonic microscopy microdefects in the form of separate cracks are found out even in fabricated at the high temperature and pressure samples. Such defects can lead to the decrease of the strength of nanostructured samples and can lead to reduce of the density. The values of elastic modules are presented in table 2.

Initial samples	Cold pressing at 1.5 GPa without sintering	Cold pressing at 1.5 GPa with subsequent sintering at 350°C	Hot pressing at 35 MPa and 470°C	Hot pressing at 250 MPa and 490°C
6.71	5.02	5.60	6.62	6.69
6.45	5.00	5.67	6.41	6.64
6.69	5.12	5.86	6.48	6.70

Table 1. Density of samples fabricated under different modes, g/cm^3

The change of the elasticity module E (Young's module) can be caused by changes of the concentration and sizes of defects in the type of micro- and sub-microcracks, formed at consolidation of nanostructured materials. It proves to be true according to the direct supervision of the samples structure (Fig.6).

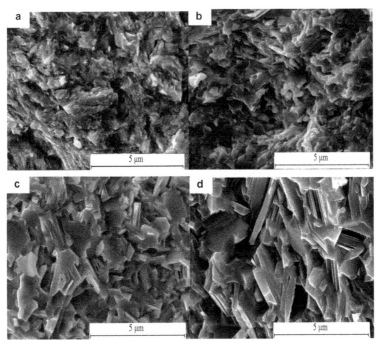

Fig. 6. SEM fractographs of sintered $Bi_{0,4}Sb_{1,6}Te_3$ samples at the pressure 50 MPa. SPS temperatures: a) 240 °C, b) 300 °C, c) 350°C, d) 400 °C.

Mode of sample fabrication	Elastic modules						
	V_L	V_T	ρV_L^2	ρV_T^2	B	E	Σ
Initial sample	3,26	1,77	71,84	21,1	43,7	54,5	0,292
Cold pressing at 1.5 GPa with subsequent sintering at 350°C	2,94	1,75	58,17	20,61	30,7	50,5	0,226
Hot pressing at 350 MPa and 470°C	3,46	2,16	80,09	31,21	38,48	73,65 73,71	0,181
Hot pressing at 250 MPa and 490°C	2,64	1,55	41,96	14,46	22,68	66,19	0,237
Hot pressing at 250 MPa and 490°C	2,72	1,59	44,54	15,22	24,25	69,36	0,240

Table 2. Elastic modules of samples fabricated under different modes, GPa

The received by SPS method samples were strong mechanically at all temperatures of sintering. Pores were absent. Results of the samples testing on diametrical compression are presented in Table 3.

Samples cut out from an ingot	Samples after cold pressing and sintering	Samples after hot pressing at 350 MPa and 470°C
6.8	20.3	27.3
11.4	15.0	31.0
4.5	22.6	29.5
	19.7	

Table 3. Compressive at diametrical strength (MPa)

2.5 Structure of bulk nanocomposites

Thermoelectric properties of materials with nanocrystalline structure should depend essentially on the size of nanograins (CDA) in a bulk sample. In turn the size of grain is defined by a number of factors: a temperature of the sintering or the hot pressing, duration of the hot pressing, value of the pressure, composition of materials, including presence of nano-adding of a second phase in a composite material. It is possible to ascertain that fabrication of initial nanopowder is less complex technological problem than maintain of the bulk nanostructure during the hot pressing that is caused by growth of initial nanoparticles due to recrystallization.

We see from Sec.2.4 that that the samples sintered after the cold as like as the hot pressure leads to increase of CDA size (or nanograins' size). In general nanostructured material is nonequilibrium by its nature, therefore thermal influences (at a manufacturing or an operation) are usually accompanied with the recrystallization of a compact material and a degradation of its properties. A possible way to reduce the average size of nanograins can be an inclusion of nanoparticles from another chemical composition, it means fabrication of nanocomposites. To investigate the relative change of nanograin size we added another nanoparticle-phase to the same solid solution matrix. They were added before the mechanical activation process. Three types of the extra nanoparticles were used for fabrication of nanocomposites: (a) MoS_2 with a laminated structure; (b) fullerene C_{60}, and (b) thermally expanded graphite (TEG). Values of nanograins was estimated by sizes of coherent dispersion areas (CDA) determined by the method of X-ray diffractometry on broadening of diffraction maxima. The received values of CDA sizes were compared to the data, obtained by the method of high resolution transmission electron microscopy (HRTEM). Both methods have shown a good consent of results at least at the size of grains up to several tens in nanometer. Larger grains also will consist from CDA with various crystallographic orientations which still influence on physical properties.

The content of MoS_2 was varied from 0.1 to 0.4 mass %. Only the peaks belonging to the triple solid solution $Bi_{0,4}Sb_{1,6}Te_3$ can be seen in the X-ray diffractogram of such a nanocomposite. The lattice constant of the nanocomposite does not change. Such a situation is repeated regardless of the pressing temperature. TEM study shows that MoS_2 nanoparticles are situated at the grain boundary, and do not dissolve in the matrix (Fig.7).

The MoS_2 particles have sizes ~ 20nm, and they have a crystalline structure. The introduction of MoS_2 greatly reduces the average size of nanograins and makes their size distribution more uniform. The maximum size of the nanograins decreases from 180 nm (in the solid solution without additives) to 80 nm (at a content of 0.1 mass % of MoS_2). The increase of the contents of MoS_2 up to 0.4 mass % leads to a future reduction of the average size of the nanograins and also leads to a more uniform size distribution. The toughness of the sample of the same composition was not less than 150 MPa. The addition of 0.1 mass % of MoS_2 brought about an increase of the toughness of 20-30 %.

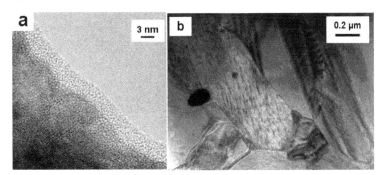

Fig. 7. (a) - HRTEM image of $Bi_{0.4}Sb_{1.6}Te_3$ nanoparticle covered by levels of MoS_2; (b) - image of the $Bi_{0.4}Sb_{1.6}Te_3$ sample fabricated at 350 MPa and 350⁰C with 0.1 mass % of MoS_2

The fullerene C_{60} (1,5 mass %) or the thermally expanded graphite (TEG) (0.1 mass %) were added to the micropowder from initial crystalline material $Bi_{0.5}Sb_{1.5}Te_3$ of p-type. Then the mechanoactivation processing was made at different temperatures under pressure 350 MPa during 20 min in the argon atmosphere. The nanopowder received without carbon additives represented 100÷300 nm units consisting in turn from nanoparticles. The average CDA size was 8÷10 nm.

The mechanoactivation of $Bi_{0.5}Sb_{1.5}Te_3$ samples in the presence of TEG was accompanied by a stratification of the graphite and a formation of flakes with the size of few nanometers; layers from the graphite flakes cover the semiconductor nanoparticles. Fig.8 shows the size and a the configuration of carbon layers. The received layered covers on the semiconductor nanoparticles had as the ordered (similarly as layer of graphite on a surface) and the disorder structure. Let us notice that formation in the same process of the ordered and the disordered carbon covers is undesirable as they have different type of conductivity. This factor can cause a bad reproducibility of the properties of thermoelectric nanocomposites. Unlike TEG the fullerenes possess strongly pronounced electrophilic properties; therefore it would be interesting to track a combination of this form of carbon with the semiconductors' nanoparticles. The state of the interface «semiconductor – C_{60} – semiconductor» can make an essential impact on the transport properties at the expense of change of an electronic condition in thin layers of nanoparticles without chemical doping (Bulat et al., (2006). Nanoparticles from the semiconductor covered by layers with disorder structure from C_{60} molecules have been received by the mechanoactivated processing of $Bi_{0.5}Sb_{1.5}Te_3$ together with the fullerenes. The typical structure of such particle is shown at Fig. 9.

It has been determined, that at mechanoactivation processing of $Bi_{0.5}Sb_{1.5}Te_3$ solid solution the additive of nanocarbon do not influence to the average size of CDA; in all cases it was 8÷10 nm. However the application of nanocarbon has allowed to reduce essentially disorder of CDA size, and to reduce in 1.5÷2 times a share of concerning large (more then

30 nm) particles. Apparently the received carbon covers effectively break the recrystallization of nanoparticles: the CDA size have decreased in 1.7÷1.9 times at the temperature 400÷450°C.

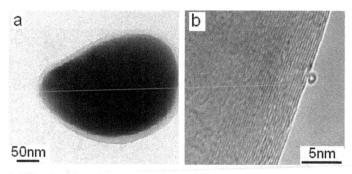

Fig. 8. Carbon covers from mechanoactivated TEG on the surface of $Bi_{0.5}Sb_{1.5}Te_3$ nanoparticles: (a) – ordered and (b) - disordered structure

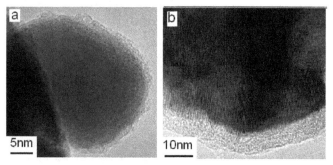

Fig. 9. HRTEM images of a semiconductor particle in a cover from molecules C_{60}: (a) - $Bi_{0.5}Sb_{1.5}Te_3$; (b) - $Bi_{0.4}Sb_{1.6}Te_3$

The samples cut out from an initial ingot have shown a considerable disorder of the strength σ_p=0.5÷2.5 MPa. The nanostructures samples from $Bi_{0.5}Sb_{1.5}Te_3$ had the strength σ_p=18.5÷20 MPa, and for $Bi_{0.5}Sb_{1.5}Te_3$ samples with TEG and with C_{60} the value of strength 26.3 and 31.0 MPa accordingly have been received.

Fig. 10 shows the consolidated data of the influence of different factors on the temperature dependencies of the average nanograins size for p-type nanocrystalline samples that were fabricated under different pressing mode (Bublik et al., 2009; Bulat et al., 2010b). We see that the main factors allowing slow growth of nanograins as a result of recrystallization are the reduction of the temperature and of the duration of the process, the increase of pressure, as well as the addition of small amount of additives (like MoS_2, TEG or fullerenes). In the case of additives the accidental particles in a nanocomposite settle down on borders of particles of the basic solid solution creating the structure like "core – cover". Let us underline that the CDA size coincides with the size of grains revealed on the SEM image of the break of surface in a compact sample at the sizes of grains to several tens nm. In the larger grains ~ 1–2 μm CDA are a part of the internal structure of a grain.

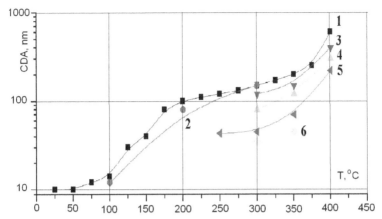

Fig. 10. Temperature dependence of the average size of nanograins of samples fabricated under different modes of pressing. 1 – in situ heating of $Bi_{0,4}Sb_{1,6}Te_3$ samples pressed at 25⁰C and 1.5GPa in the thermocamera of the diffractometer; 2 – vacuum annealing at different temperatures of $Bi_{0,4}Sb_{1,6}Te_3$ samples pressed at 25⁰C and 1.5GPa; 3 – $Bi_{0,4}Sb_{1,6}Te_3$ samples hot pressed at 35MPa; 4 – $Bi_{0,4}Sb_{1,6}Te_3$ samples hot pressed at 350MPa; 5 – $Bi_{0,4}Sb_{1,6}Te_3$ samples plus 0.1 mass% MoS_2 pressed at 350MPa; 6 – $Bi_{0,4}Sb_{1,6}Te_3$ samples plus 0.4 mass % MoS_2 pressed at 350MPa

As properties of a material to a great extent depend on its structure in micro- and nanoscale, the comparative analysis of the structure of the bulk samples received by SPS methods and traditional hot pressing has been carried out; these results are presented at Fig. 11, 12. The analysis of the received results shows that unlike the method of hot pressing the SPS method allows to receive at rather low pressure 50 MPa mechanically strong well sintered nanostructured materials. It does not contain pores even at temperatures more low then 300°C. The explanation of this result is that in SPS process the high density of allocated energy in contact zones between the powder particles causes the very strong local warming up (up to fusion of a grain blanket) whereas the basic volume of a the material remain at lower temperature. The CDA size for both methods of consolidating up to temperatures

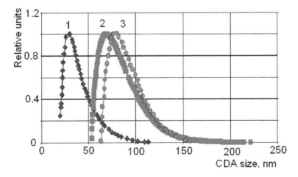

Fig. 11. Relative quantitative portion of different CDA size in nanostructured sintered $Bi_{0,4}Sb_{1,6}Te_3$. SPS temperatures: 1 - 240 °C; 2 - 300 °C; 3 - 350 °C; pressure 50 MPa

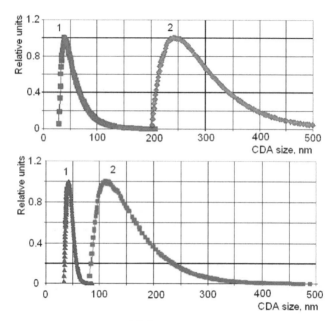

Fig. 12. Relative quantitative portion of different CDA size in hot pressed nanostructured $Bi_{0.4}Sb_{1.6}Te_3$ (350 MPa; 1-300°C &, 2- 400°C). a) $Bi_{0.4}Sb_{1.6}Te_3$, b) $Bi_{0.4}Sb_{1.6}Te_3$+0,1 mass % MoS_2

350°C are comparable, whereas at sintering temperatures 400°C and above (that in practice corresponds to temperatures of hot pressing) the CDA size in SPS method increases much less.

2.6 Thermoelectric properties of bulk nanostructures and nanocomposites

The main transport properties of nanocrystalline materials fabricated under different conditions were investigated. Dependences of the transport properties on average nanograins size were also analyzed.

The temperature dependences of the thermoelectric parameters of typical hot pressed nanostructured p-$Bi_{0.3}Sb_{1.7}Te_3$ sample was published and discussed in Ref. (Bulat et al., 2010b). The correspondent maximum value ZT=1.12 takes place at the temperatures ~ 350÷375 K. The same maximum efficiency ZT=1.1 in the same nanostructured material at the same temperature was measured in Ref. (Vasilevskiy et al., 2010) but the extrusion instead of hot pressing for consolidation of samples was used here.

Let as consider more in detail our investigation of thermoelectric properties of samples fabricated by the SPS method. Thermoelectric properties were studied depending on sintering temperature on samples of p-type $Bi_{0.5}Sb_{1.5}Te_3$ and $Bi_{0.4}Sb_{1.6}Te_3$. All samples have been received by SPS method at pressure 50 MPa, temperature from 250 to 500 °C, the duration of sintering was 5 min (for $Bi_{0.4}Sb_{1.6}Te_3$) and 20 min (for $Bi_{0.5}Sb_{1.5}Te_3$). The samples sintered at 300 °C have the maximum value of thermoelectric power. A small distinction exists between values of thermoelectric power for samples of various compositions. The electric conductivity increases with the raise of sintering temperature. The similar dependence is observed also for the heat conductivity; increase of the heat conductivity with rise of the temperature sintering is caused by increase in the electronic heat conductivity which is proportional to the electric conductivity.

It follows from experimental results that the samples received by SPS method at sintering temperature 450 ^{0}C have the greatest value of the efficiency ZT. The samples fabricated from the 0.5 μm powder have lower efficiency ZT in comparison with the nanostructured material. It was established that the pressure 50 MPa is the optimum one for obtain the high thermoelectric efficiency. Samples of $Bi_{0.4}Sb_{1.6}Te_3$ composition obtained at the sintering temperatures ~ 350 ^{0}C have higher ZT than $Bi_{0.5}Sb_{1.5}Te_3$ samples. The temperature dependence of thermoelectric parameters in nanostructured samples $Bi_{0.4}Sb_{1.6}Te_3$ received at sintering temperature 400 ^{0}C and pressure 50 MPa is presented at Fig. 13. Peak efficiency is reached at 90 ^{0}C and makes ZT=1.22.

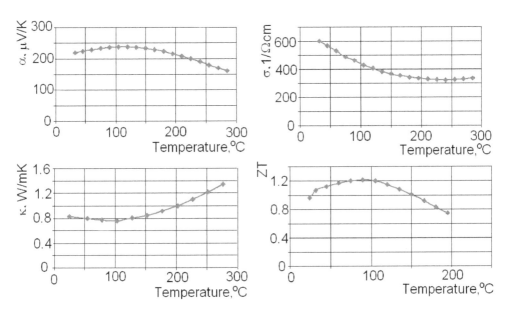

Fig. 13. Thermoelectric properties of sintered bulk nanostructured materials $Bi_{0.4}Sb_{1.6}Te_3$ as a function of measurement temperature: a) electrical conductivity, b) Seebeck coefficient, c) thermal conductivity, d) figure of merit, ZT.

Some dependences of measured thermoelectric coefficients in bulk nanostructured materials on the grain size for solid solution $Bi_xSb_{2-x}Te_3$ (Bulat et al., 2010c, 2011a) will be presented in Sec. 3.2 and 3.3.

3. Theory

Three mechanisms that can improve the thermoelectric efficiency are studied theoretically and compared with the experiment in Sec.3.

3.1 Electron tunneling

One of the possible mechanisms of electric transport in nanostructured materials is the tunneling of charge carriers through the intergrain barriers. This effect is similar to the thermionic or field emission thought the vacuum gap. The studies of the thermionic emission applied to the field of energy conversion began in the 1960th (Anselm, 1951). Though the efficiency of thermionic generators can reach 20% their working temperatures are about 1000K because of the large values of work function in metals and semiconductors.

To use thermionic devices at lower temperatures one should decrease the work function, e.g. by applying high electric field (Fleming & Henderson, 1940; Murphy & Good, 1956), by using special cathode coatings that can decrease the work function down to 0.8eV (Sommer, 1980) or by utilizing the tunneling effect through the thin vacuum gap (Hishinuma et al., 2001; Tavkhelidze et al., 2002). As was shown by Mahan (Mahan, 1994) to use thermionic devices for refrigeration at room temperature one needs to decrease the work function down to the values of 0.3-0.4eV that are not available at the present time. But in the case of nanoscale tunneling junction the tunneling probability increases and the noticeable cooling power can be reached even at the work functions of about 0.8eV (Hishinuma et al., 2001). One of the possible cooling applications of such device that consisted of metallic tip over the semiconducting plate was described in (Ghoshal, 2002b). Alternatively Schottky barriers or semiconductor heterostructures can be used instead of vacuum barriers (Mahan & Woods, 1998; Mahan et al., 1998). In such structures the barrier energy height can be as low as 0.1eV but the phonon thermal conductivity of semiconducting barrier will increase the total thermal conductivity and produce negative influence on the figure of merit.

In this section the bulk nanostructured material that consists of grains separated by tunneling junctions is considered. The influence of the charge carrier tunneling on the thermoelectric figure of merit of such material is theoretically investigated. The shape of nanoparticles is modelled by two truncated cones with the same base (Fig. 14) that allows one to perform calculations in an analytical way (Bulat & Pshenai-Severin, 2010a). The calculations of the heat flow inside the nanoparticles take into account the difference between the electron and phonon temperatures in the limiting case of vacuum gap when phonons cannot tunnel through the barriers.

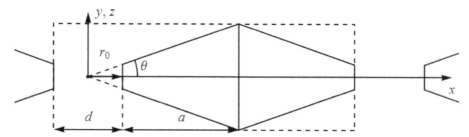

Fig. 14. The cross section of nanoparticle modeled by two truncated cones with the same base. $2a$ is the size of nanoparticle, 2θ is the cone aperture angle, the radius r_0 determines the size of truncated part, d is the tunneling junction width. Dashed rectangle represents the single period of the whole structure.

In order to estimate the thermoelectric figure of merit of such material the transport coefficients were calculated. The use of the linear approximation for tunneling coefficients in bulk nanostructured materials seems to be quite reasonable. Indeed if the typical size of samples are several millimeters and the grain size is about 10-40 nm then the voltage drop on single junction is 10^5 times smaller than on the whole sample. So even at several hundreds of volts bias on the sample the voltage drop on the single junction will be $\Delta V \sim 10^{-3} V$. Similarly the temperature difference on single junction is about 10^{-3}K at the total temperature difference of 100K. So at the room temperature if the barrier energy height $\varepsilon_b \sim 0.1$eV than $|k_0 \Delta T|$ and $|q_0 \Delta V|$ are much less than both the thermal energy $k_0 T$ and barrier height ε_b. In this case one can use linear transport coefficient theory for the tunneling junction. The total current density through the tunneling junction is determined by the difference of emission currents from two electrodes (Burstein, 1969)

$$j_x = \sum_{k, v_x > 0} 2(q_0 v_x) D(\varepsilon_x) \left[f_0\left(\frac{\varepsilon - \mu}{k_0 T} \right) - f_0\left(\frac{\varepsilon - (\mu + q_0 \Delta V)}{k_0 (T + \Delta T)} \right) \right]. \tag{1}$$

In this expression x-axis is directed at the right angle to the junction cross section (Fig. 14), ε is the total energy of electron with the wave vector k, μ is the chemical potential of the left electrode, f_0 is the Fermi-Dirac distribution function, v_x and ε_x are velocity and kinetic energy of electron corresponding to its motion along x direction and $D(\varepsilon_x)$ is the tunneling probability. In order to obtain linear transport coefficients the expression for the total current density (1) was linearized with respect to small voltage ΔV and temperature ΔT differences. Finally the barrier electric conductivity can be obtained as $\sigma_b = -j_x / \Delta V$ (Bulat & Pshenai-Severin, 2010a)

$$\sigma_b = \frac{q_0^2 m k_0 T}{2 \pi^2 \hbar^3} \int_0^\infty D(\varepsilon_x^*) f_0(\varepsilon_x^* - \mu^*) d\varepsilon_x^*, \tag{2}$$

where energy ε_x^* and chemical potential μ^* with asterisks are measured in $k_0 T$ units and the effective mass of electron m is assumed to be the same inside nanoparticle and barrier. The Seebeck coefficient can be obtained from the zero current condition $\alpha_b = -(\Delta V / \Delta T)_{j_x = 0}$ and it was expressed as $\alpha_b = \beta_b / \sigma_b$, where (Bulat & Pshenai-Severin, 2010a)

$$\beta_b = \frac{k_0}{q_0} \frac{q_0^2 m k_0 T}{2 \pi^2 \hbar^3} \int_0^\infty D(\varepsilon_x^*) \left[u f_0(u) + \ln\left(1 + e^{-u}\right) \right] d\varepsilon_x^*, \tag{3}$$

and the following notation was introduced $u = \varepsilon_x^* - \mu^*$.

The expression for electronic heat flow through the junction can be obtained from (1) after replacing $q_0 v_x$ with $(\varepsilon - \mu) v_x$. The value of barrier thermal conductivity measured at zero current κ_b can be expressed through the thermal conductivity at zero voltage drop $\kappa_{b, \Delta V = 0}$ as (Bartkowiak & Mahan, 1999)

$$\kappa_b = \kappa_{b,\Delta V=0} - \alpha_b{}^2 \sigma_b T \ , \tag{4}$$

where (Bulat & Pshenai-Severin, 2010a)

$$\kappa_{b,\Delta V=0} = \frac{m k_0{}^3 T^2}{2\pi^2 \hbar^3} \int_0^\infty D(\varepsilon_x^*) \left[u^2 f_0(u) + 2u \ln\left(1 + e^{-u}\right) - 2 Li_2\left(-e^{-u}\right) \right] d\varepsilon_x^* , \tag{5}$$

and the dilogarithm function is denoted as $Li_2(x)$. It is worth to note that the barrier electrical and thermal conductivities are determined with respect to voltage and temperature difference instead of their gradients as in the bulk case. Hence for the case of comparison with the bulk values it is more convenient to use the values of $\sigma_b d$ and $\kappa_b d$.

In the present calculations the intergrain barrier shape was assumed to be rectangular. In linear operating region the change of the tunneling barrier shape under applied field can be neglected. So the well-known expression for tunneling probability of rectangular barrier was used

$$D(\varepsilon_x) = \left(1 + \frac{\left(k_x{}^2 + k_b{}^2\right)^2}{4 k_x{}^2 k_b{}^2} \sinh^2\left(k_b d\right) \right)^{-1} , \tag{6}$$

where $k_x = \sqrt{2m\varepsilon_x}/\hbar$ and $k_b = \sqrt{2m(\varepsilon_b - \varepsilon_x)}/\hbar$. Note that if $\varepsilon_x > \varepsilon_b$ the wave vector became pure imaginary $k_b = i|k_b|$ and hyperbolic sine should be changed to $sin(|k_b|d)/i$. Very often instead of exact expression for tunneling probability the WKB approximation is used $D_{WKB}(\varepsilon_x) = exp(-2k_b d)$. In WKB approximation tunneling probability for $\varepsilon_x > \varepsilon_b$ is equal to unity. In the following the values of tunneling transport coefficients calculated using these two approximations will be compared.

Fig. 15-17 show the dependencies of barrier electrical conductivity, Seebeck coefficient and Lorenz number on the size of tunneling junction for different barrier heights 0.4 and 0.8eV. In these estimations the effective mass was equal to $m = 0.7 m_0$ that corresponds to the typical hole effective mass in Bi_2Te_3 (Goltsman et al., 1972). Doping impurity concentration was equal to 10^{19}cm^{-3} for chemical potential close to the band edge. The curves plotted using exact expression for tunneling probability (6) and obtained in WKB approximation illustrate noticeable difference of two approaches. For metallic electrodes electron energies close to the Fermi level are important and WKB approximation can be used for small tunneling probabilities $D(\varepsilon_x) < 1/e$ as was stated in (Stratton, 1962). For semiconducting electrodes when the charge carrier energies are close to the band edge the preexponential factor can also be important because it approaches zero for small carrier energies. When the tunneling junction width becomes larger these difference decreases because the contribution of small energy carriers is less important.

As can be seen from Fig. 15 at the junction thicknesses smaller than 2nm the Seebeck coefficient can reach the values of about 300-350μV/K and slowly varies with the barrier thickness. The electrical conductivity is small (Fig. 16) and decreases exponentially with the barrier thickness. At larger junction thicknesses $d>2nm$ the thermionic emission becomes more important than tunneling. Charge carriers with small energies are filtered out of the current that leads to the sharp increase in Seebeck coefficient and Lorenz factor (Fig. 15, 17). The dimensionless figure of merit for single junction is rather high $ZT \approx 3-4$.

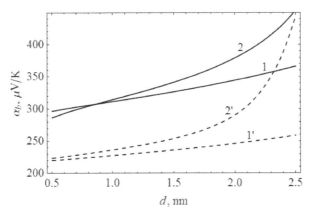

Fig. 15. The dependence of the barrier Seebeck coefficient on the tunneling junction size at room temperature for $\varepsilon_b = 0.8eV$ (1, 1′) and 0.4eV (2, 2′) calculated using exact expression for tunneling probability (1, 2) and WKB approximation (1′, 2′).

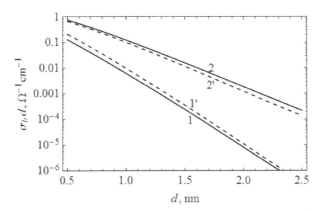

Fig. 16. The dependence of the barrier electrical conductivity on the tunneling junction size (see Fig. 15 for notation).

In order to calculate effective transport coefficients in the whole structure charge and heat flow inside nanoparticles should be taken into account. As the approximation of zero phonon thermal conductivity of the barrier is considered the heat flow through the junction is only due to charge carriers. Hence the equations for the heat flow should take into account the differences in electron T_e and phonon T_p temperatures (Ghoshal, 2002b;

Bartkowiak & Mahan, 1999). The electron-phonon scattering inside nanoparticle leads to the equilibrating of their temperatures on the length scale l_c which is called cooling length. The general solution for conical geometry was obtained in (Ghoshal, 2002b) where the limiting case $l_c \ll a$ was analyzed. In the considered materials based on bismuth antimony telluride solid solutions the values of cooling length are 66nm for Bi_2Te_2 and 156nm for Sb_2Te_3 (da Silva & Kariany, 2004). So for nanoparticle size of 10-20nm the limit of $l_c \gg a$ can be considered. For this case the heat transfer equations were solved for each of two truncated cones representing the nanoparticle in (Bulat & Pshenai-Severin, 2010a). As a result the equations for total resistance and thermal conductance of nanoparticle were obtained (Bulat & Pshenai-Severin, 2010a)

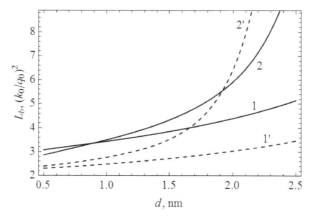

Fig. 17. The dependence of the barrier Lorenz number on the tunneling junction size (see Fig. 2 for notation).

$$R_n = \left(\sigma \gamma \Omega r_0^2 \right)^{-1}, \quad K_n = \kappa_e \xi \gamma \Omega r_0^2, \tag{7}$$

where $\gamma = (r_0 + a) / r_0 L_n$, $L_n = 2a$ is the total length of nanoparticle in the x-axis direction, r_0 determines the size of truncated part (Fig. 14), $\Omega = 4\pi \sin^2 \theta / 2$ is the cone solid angle and

$$\xi = 1 + \frac{\kappa_p}{\kappa_e + \kappa_p} \frac{(a / l_c)^2}{3 \gamma L_n}. \tag{8}$$

From these equations it can be seen that in the limit of the small nanoparticle size compared to the cooling length the electrical resistance does not change due to the difference in T_e and T_p. The correction to the thermal conductance due to this effect is only second order of magnitude with respect to small parameter a / l_c and K_n is determined mainly by electronic contribution.

It is interesting to note that the transition to the layered geometry can be obtained if $r_0 \to \infty$ and $\theta \to 0$ in such a way that the area Ωr_0^2 is constant. In this limit $\gamma L_n = 1$ and from (7) it is easy to get corresponding equation for the layered system (Anatychuk & Bulat, 2001).

Though in real nanostructured material the size on nanoparticles and their positions are randomly distributed here for estimations of effective transport coefficient the material is modeled as an ordered set of primary cells outlined by dashed lines on the Fig. 14. In this case the total current flow is directed along x-axis and effective transport coefficients can be calculated based on equations for layered medium (see, e.g., Snarsky et al., 1997). The effective transport coefficients for the present case were calculated in (Bulat & Pshenai-Severin, 2010a). The thermal conductivity can be obtained as a series connection of barrier and nanoparticle thermal conductivities

$$\kappa_{eff} = \frac{\kappa_b \, \kappa_e \, \xi\gamma}{\kappa_b + \kappa_e \, \xi\gamma} \gamma_t \, , \tag{9}$$

where geometric factor $\gamma_t = r_0^2 \, (d + 2a) / (r_0 + a)^2$ was introduced. The effective Seebeck coefficient can be obtain as a sum of Seebeck coefficients of barrier and nanoparticle taking into account corresponding temperature differences on each part

$$\alpha_{eff} = \frac{\alpha_n \, \kappa_b + \alpha_b \, \kappa_e \, \xi\gamma}{\kappa_b + \kappa_e \, \xi\gamma} \, . \tag{10}$$

In calculations of electrical conductivity the average sample temperature T_{av} is assumed to be constant. But due to Peltier effect the temperatures of neighboring contacts are different. So in the equation for effective electrical conductivity in addition to common expression for series resistance the factor due to Peltier effect induced thermopower should be taken into account

$$\sigma_{eff} = \frac{\sigma_b \, \sigma\gamma\gamma_t}{\sigma_b + \sigma\gamma} \left(1 + \frac{(\alpha_b - \alpha_n)^2}{\kappa_b + \kappa_e \, \xi\gamma} \frac{\sigma_b \, \sigma\gamma T_{av}}{\sigma_b + \sigma\gamma} \right)^{-1} , \tag{11}$$

where T_{av} is the average temperature of sample.

The effective figure of merit of bulk nanostructured material can be calculated using equation (9)-(11) as $Z_{eff} = \alpha_{eff}^2 \sigma_{eff} / \kappa_{eff}$. For estimations the typical room temperature parameter for Bi_2Te_3 from (Goltsman et al., 1972) were used: α_n=200 μV/K, σ=830 $\Omega^{-1}cm^{-1}$ and κ_p=1 W/m K. Though Bi_2Te_3 is anisotropic material nanocrystals inside the sample are randomly oriented. So for the estimations the values of thermal and electrical conductivities were average over all directions.

On Fig. 18 the dependencies of effective electrical conductivities on the tunneling junction thickness are plotted. It is interesting to note that effective transport coefficients are independent of cone aperture angle θ because only cross-section areas depend on it and these dependences are canceled out. For larger r_0 the electrical conductivity increases due to the increase of the smaller cross-section of the cone. As was noted above in the limit of large r_0 the transport coefficients approach the values for layered geometry (compare i, i' and i'' with L_i on Fig. 18 for i=1, 2). For considered parameter range the electrical conductivity of the tunneling junction is much less than the usual values in semiconductors. Hence the effective electrical conductivity is determined mainly by barrier part but it is related to the total period of the structure. For example for layered geometry $\sigma_{eff} \approx (\sigma_b d)(L_n + d) / d > \sigma_b d$ (compare L_i with B_i on the Fig. 18 for i=1, 2). For the case of conical geometry the factor γ_t should be taken into account that diminishes σ_{eff} for small r_0.

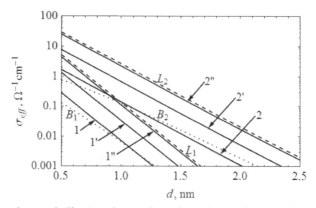

Fig. 18. The dependence of effective electrical conductivity on the tunneling junction thickness for barrier height $\varepsilon_b = 0.8$ (1, 1', 1'') and 0.4eV (2, 2', 2'') for cone-shaped nanoparticles. The same dependences for single barrier and layer structure are plotted as B_1, B_2 and L_1, L_2 correspondingly. The ratio r_0/a=0.3 (1,2), 1 (1', 2') and 10 (1'',2''); $2a$=20nm.

On Fig. 19 the dependence of effective thermoelectric figure of merit on the tunneling junction thickness is plotted. The estimations showed that in the absence of the phonon thermal conductivity in the barrier for all considered ranges of tunneling junction parameters (see Fig.16, 17) the barrier thermal conductivity is much smaller than the thermal conductivity of nanoparticle. Hence relatively high values of the effective Seebeck coefficient are determined mainly by large α_b (Fig. 15) and the ratio $\sigma_{eff} / \kappa_{eff}$ in Z_{eff} is determined by the effective Lorenz number that has usual values for small d and begins to increase with the increase of d (see Fig. 17). So in the present case the large values of $Z_{eff} T \approx 2.5-4$ are determined by large barrier Seebeck coefficient and the decrease of $Z_{eff} T$ for larger d is due to the increase of barrier Lorenz number. Simple estimations of the effect of phonon thermal conductivity of the barrier performed in one-temperature approximation showed that to increase the thermoelectric figure of merit compared to initial semiconducting material the phonon barrier thermal conductivity $\kappa_{b,ph}$ should be about 4 time smaller than the electronic contribution.

To conclude this section it can be said that the thermoelectric figure of merit of the structures with tunneling junctions can be quite large $Z_{eff} T \approx 2.5-4$ if the barrier phonon thermal conductivity is negligible. These large values are determined by the large values of the barrier Seebeck coefficient and greatly reduced in the presence of $\kappa_{b,ph}$. In addition irregularities in the tunneling junction width or the size of nanoparticles can also lead to the decrease of the figure of merit in real structure. The comparison with the experimental data from Sec.2.6 and Ref. (Poudel et al., 2008; Bulat et al., 2010b; Vasilevskiy et al., 2010) showed that the increase of the figure of merit in these materials is hardly connected with the tunneling effect because of the large difference between measured electrical conductivity (of the order of $1000\ \Omega^{-1}\,cm^{-1}$) and estimated values that are much less than $100\ \Omega^{-1}\,cm^{-1}$.

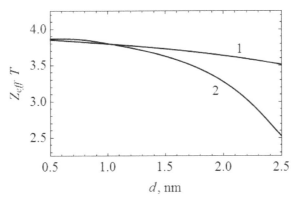

Fig. 19. The dependence of effective thermoelectric figure of merit on the tunneling junction thickness for barrier height $\varepsilon_b = 0.8$ (1) and 0.4eV (2).

3.2 Boundary scattering

In this section the influence of boundary scattering on the thermal conductivity of bulk nanostructured materials obtained by ball-milling with subsequent hot pressing is considered following (Bulat et al., 2010c). These materials are polycrystalline with small grain sizes in the range from 10 nm to several hundreds on nanometers depending on the temperature of hot pressing. The most common way to estimate the influence of grain boundary scattering on the thermal and electrical conductivities is to include additional scattering mechanism with the mean free path equal to the grain size L_n. The theory of this effect applied to thermal conductivity in different polycrystalline solid solutions was described in (Goldsmid et al., 1995) but due to relatively large grain size considered there it was predicted that the effect of boundary scattering on thermal conductivity in bismuth telluride alloys is negligible.

Usually the grain boundary effect is considered to be important for thermal conductivity only at low temperatures when the probability of phonon-phonon scattering decreases. But it is related mainly to pure single crystals (Goldsmid et al., 1995). In solid solutions at high temperatures the contribution of short wavelength phonons to the thermal conductivity is reduced due to the point defect. So in solid solutions the contribution of long wavelength phonons to thermal conductivity is relatively more important than in pure crystals. This contribution can be effectively reduced by introducing boundary scattering.

The estimations performed here are based on Debye model for acoustic phonons with linear spectrum up to Debye frequency ω_D. The following scattering mechanisms are taken into account. In pure single crystals the most important scattering mechanism at room temperature is phonon-phonon umklapp scattering with the relaxation time $\tau_U = A_U / \omega^2$. The thermal conductivity in this case can be written as (Goldsmid et al., 1995)

$$\kappa_0 = \frac{1}{3} c_V \, v_D \, \overline{l_0} \, , \tag{12}$$

where $c_V = 3 k_0 N_V$ is the heat capacity associated with acoustic modes of the crystal containing N_V primary cells, v_D is the mode averaged Debye speed of sound and $\overline{l_0}$ is the mean free path associated with the umklapp scattering. Knowing the value of thermal conductivity in single crystal κ_0 the mean free path $\overline{l_0}$ and the constant A_U can be obtained.

When the second component is added forming solid solution the thermal conductivity κ_s becomes less than κ_0 due to the point defect scattering $\tau_p = A_p / \omega^4$. The constant A_p can be deduced from experimental value of κ_s assuming that A_U is the same as in initial single crystal. Finally the boundary scattering is described by frequency independent relaxation time $\tau_b = L_n / v_D$. In the simplified treatment (Goldsmid et al., 1995) it was proposed to divide the total range of phonon frequencies into three parts. For each part of the spectrum only the most important relaxation time is considered: τ_b, τ_U and τ_p for lower, medium and high frequency parts correspondingly. Then the simple equation for phonon thermal conductivity in polycrystalline material was obtained (Goldsmid et al., 1995)

$$\kappa_{ph} = \kappa_s - \frac{2}{3} \kappa_0 \sqrt{\frac{\overline{l_0}}{3 L_n}} . \tag{13}$$

In order to compare the values of κ_{ph} with experiment for nanostructured material based on p-$Bi_xSb_{1-x}Te_3$ (Bulat et al., 2010c) the hole contribution should be subtracted from experimental values of thermal conductivity. So the proper estimations of electrical conductivity and hole thermal conductivity are necessary. The electrical conductivity in initial solid solution is anisotropic but after ball-milling and hot pressing the samples became isotropic on average. To take the anisotropy into account it was assumed that it is connected mainly with the anisotropy of effective masses and the relaxation time is a scalar. Then using the effective medium theory for average electrical conductivity the effective mass of conductivity in polycrystalline material can be expressed as (Bulat et al., 2010c)

$$m_c = \frac{4 m_{c11}}{1 + \sqrt{1 + 8 m_{c11} / m_{c33}}} , \tag{14}$$

where m_{cii} are effective conductivity masses along main crystalline directions ($i = 1, 2, 3$).

The boundary scattering of holes was taken into account using relaxation time in the form $\tau_{b,h} = L_n / v$. The relaxation time energy dependence for acoustic scattering is $\tau_a \sim \varepsilon^{-1/2}$. It is the same as that for point defect scattering or alloy scattering in solid solution. It appears that this energy dependence is the same also for boundary scattering of holes. So the change of mobility in nanostructured material can be describe as (Bulat et al., 2010c)

$$u = \frac{L_n / l_s}{1 + L_n / l_s} u_s , \tag{15}$$

where l_s and u_s are the mean free path and mobility in initial solid solution. The Lorenz number and the Seebeck coefficient in this case are the same as for acoustical scattering due to the same energy dependencies of the relaxation times.

The experimental values of electrical conductivity together with estimations based on equation (15) are shown on Fig. 20. In the initial solid solution $Bi_{0.4}Sb_{1.6}Te_3$ the values of electrical conductivity in the cleavage plane and the Seebeck coefficient were equal to $1000\ \Omega^{-1}cm^{-1}$ and $195\ \mu V/K$ correspondingly. In $Bi_{0.3}Sb_{1.7}Te_3$ these values were equal to $1387\ \Omega^{-1}cm^{-1}$ and $187\ \mu V/K$. The experimental values of mobility in $Bi_{0.3}Sb_{1.7}Te_3$ were 15% higher than in $Bi_{0.4}Sb_{1.6}Te_3$. The values of effective masses were taken from two different sources (Luk'yanova et al., 2010) and (Stordeur et al., 1988). The effective masses of the density of state per one ellipsoid m_{d1} and of conductivity m_c obtained using (14) were equal to $0.069m_0$ and $0.054m_0$ (Luk'yanova et al., 2010) and $0.305m_0$ and $0.186m_0$ (Stordeur et al., 1988). Due to the wide spread of the effective mass values the estimations of the mean free path in the initial solid solution $Bi_{0.4}Sb_{1.6}Te_3$ were quite different $l_a=23$ nm and 4nm correspondingly. This is reflected on the Fig. 20 where the effect of boundary scattering is more prominent for the estimations with larger l_a (compare curves 1 and 1').

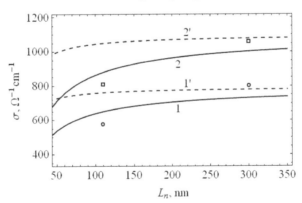

Fig. 20. The dependence of electrical conductivity of bulk nanostructured materials on the grain size for $Bi_{0.4}Sb_{1.6}Te_3$ (circles– experimental data; 1, 1' – estimations) and for $Bi_{0.3}Sb_{1.7}Te_3$ (squares – experimental data; 2, 2' – estimations). Estimations use effective mass values from (Luk'yanova et al., 2010) - 1, 2 and from (Stordeur et al., 1988) – 1', 2'.

For estimations of the influence of boundary scattering on the phonon thermal conductivity the following material parameters were used. The lattice thermal conductivity in Sb_2Te_3 in the cleavage plane at room temperature is equal to $\kappa_{0,11} = 1.9\ W/m\ K$ (Goltsman et al., 1972) and the anisotropy of the thermal conductivity is equal to 2.38 (Madelung et al., 1998). Averaging similar to (14) gives the thermal conductivity $\kappa_0 = 1.47\ W/m\ K$. In $Bi_{0.4}Sb_{1.6}Te_3$ solid solution the thermal conductivity in the cleavage plane is 1.2 W/m K (Goltsman et al., 1972) that after averaging using anisotropy value of 2.22 (Madelung et al., 1998) gives

$\kappa_s = 0.94$ W/m K. The Debye temperature in Sb_2Te_3 is about 160K, Debye velocity was estimated as $v_D = 3.6 \cdot 10^5$ cm/s and the heat capacity at room temperature is close to usual value 24.9 J/mol K (Goltsman et al., 1972). This data allowed estimating the average mean free path in the pure crystal as $\overline{l_0} = 4.7$ nm.

The comparison of the estimated thermal conductivity of nanostructured material with the experimental data is presented on Fig. 21. The electronic contribution to the thermal conductivity was subtracted using Lorenz factor calculated as described above. The results of estimations are quite well correlate with the experimental data. This allows one to conclude that the boundary scattering is important mechanism of reduction of phonon thermal conductivity in bulk nanostructured materials. The estimation of the decrease of the lattice thermal conductivity due to boundary scattering is shown on Fig. 22. It can be seen that the decrease can reach the values of 30-40% at the grain size of about 10-20 nm.

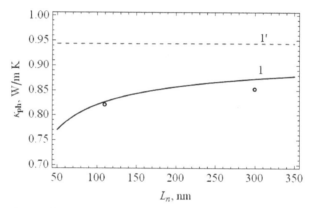

Fig. 21. The dependence of phonon thermal conductivity of bulk nanostructured materials on the grain size for $Bi_{0.4}Sb_{1.6}Te_3$ (circles – experimental data; 1 – estimations; 1' – the value in initial solid solution).

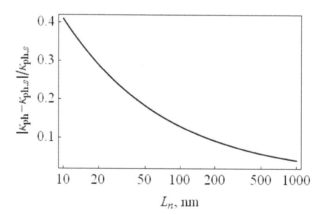

Fig. 22. The dependence of the relative decrease of lattice thermal conductivity on the grain size in nanostructured material based on $Bi_{0.4}Sb_{1.6}Te_3$ solid solution.

3.3 Energy filtering

In the previous section the influence of the boundary scattering on the electrical conductivity of nanostructured material was considered. The boundary scattering was described using constant mean free path equal to the size of grains L_n. As was noticed above the energy dependence of the relaxation time for boundary scattering in this approximation is the same as for acoustical scattering. So the Lorenz factor and the Seebeck coefficient should not differ from that in initial solid solution if the concentration remains the same. On the other hand the experimental data (Bulat et al., 2011a) showed the increase of the Seebeck coefficient in the samples with smaller grain size. To describe this effect the more detailed study of the scattering process was performed. The energy dependence of the probability of carrier scattering on the potential barrier at the grain boundary was taken into account. As the charge carriers with smaller energy scatter more intensively their contribution to the electrical and heat current decrease. This energy filtering can lead to the increase of the Seebeck coefficient if the energy relaxation length l_ε is much greater than the momentum mean free path l_p (Moizhes & Nemchinsky, 1998). At the temperatures much higher than Debye temperature the estimations for typical parameters of semiconductors (Moizhes & Nemchinsky, 1998) showed that as $l_p \sim 50\,\mathrm{nm}$ the energy relaxation length is about 500 nm that is much greater than the grain size considered in the present section.

There are several approaches that take into account the influence of the energy filtering on the transport coefficient. In (Ravich, 1995) the scattering on the single barrier was considered. In (Popescu et al., 2009) the exact expression for scattering probability was used but it was not taken into account that it should depend on the part of the kinetic energy corresponding to the motion normal to the boundary rather than the total energy. In (Mayadas & Shatzkes, 1970; Gridchin et al., 2005) the boundary scattering in polycrystalline thin films was considered but the relaxation time was anisotropic.

In the bulk nanostructured samples considered in this section the electrical conductivity appears to be isotropic due to random grain orientation and the following approach for calculation of relaxation time was used (Bulat et al., 2011a). In this approach the boundary scattering is modeled through the specular scattering on the randomly oriented planes representing grain boundaries and the inter plane distance is equal to the grain size L_n. The estimations of the mean free path in the previous section gave $l_p \sim 20\,\mathrm{nm}$. So the grain size is greater than the mean free path. In this case the multiple scattering can be taken into account through the summing up the probabilities of scattering rather than the matrix elements. In isotropic polycrystalline material with random grain orientation the summation of the probability of multiply scattering leads to the averaging over the boundary plane orientations. The total number of planes was estimated as $3L/L_n$, where L is the characteristic sample size. Due to the conservation lows only two final states in the individual scattering act are possible, namely forward scattering and reflection. In the relaxation time calculation only the second type gives contribution. For the probability of reflection the exact expression is used $W_r(k_n) = 1 - D(k_n)$ where tunneling probability $D(k_n)$ is defined by equation (6) and k_n is the wave vector normal to the grain boundary. As the number of incident electrons on the unit area of the boundary in one second is equal to the density of electron flow $j_i = \hbar k_n / mL$, the number of reflections in the unit time is equal to $j_i W_r(k_n)$. Finally the relaxation time can be calculated as

$$\tau_b^{-1} = \sum_n \frac{\hbar k_n}{mL} W_r(k_n) \frac{-\Delta k_n k}{k^2}, \qquad (16)$$

where the summation over n takes into account all possible boundary orientations. The summation can be replaced with the integration over polar and azimuthal angles θ and φ determining the direction of normal vector n. Then the expression for relaxation time due to boundary scattering can be obtained in the following form (Bulat et al., 2011a)

$$\tau_b^{-1} = \frac{6\hbar k}{mL_n} \int_0^1 W_r(k\chi)\chi^3 \, d\chi, \qquad (17)$$

where $\chi = \cos\theta$.

The experimental data and theoretical estimations for electrical conductivity and Seebeck coefficient in the bulk nanostructured materials based on Bi_2Te_3-Sb_2Te_3 solid solutions are presented on Fig. 23, 24. The scattering on the grain boundaries including energy filtering (17) and the scattering on acoustic phonons were taken into account. Because the exact account of anisotropy in Bi_2Te_3 based materials is complicated in equation (17) the density of state effective mass m_{d1} was used. The unknown parameters in calculations were the width d and the energy height ε_b of the intergrain barrier. The estimations showed that quite good agreement with the experimental data can be obtained at the reasonable values of these parameters equal to $d = 5\,nm$ and $\varepsilon_b = 1.5 k_0 T$. The other parameters were the same as for the estimations of boundary scattering in the constant mean free path approximation discussed in the previous section. In order to check the applicability of relaxation time

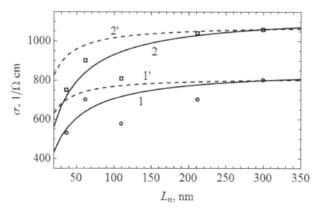

Fig. 23. The dependence of electrical conductivity of bulk nanostructured materials on the grain size for $Bi_{0.4}Sb_{1.6}Te_3$ (circles– experimental data; 1, 1' – estimations) and for $Bi_{0.3}Sb_{1.7}Te_3$ (squares – experimental data; 2, 2' – estimations). Estimations use effective mass values from (Luk'yanova et al., 2010) - 1, 2 and from (Stordeur et al., 1988) – 1', 2'.

approximation the estimation of typical values of the relaxation time were made. It is known that relaxation time approximation is applicable if $\tau > \hbar / k_0 T$ and if the temperature difference on the length of mean free path is small compared to average temperature. If the temperature difference on the sample with $L \sim 0.1$ cm is 100K, then the temperature difference of the length of the order of mean free path about 10 nm is 10^{-3}K. The estimations gave $\tau \sim 10^{-13}$ s and $\hbar / k_0 T = 2.5 \cdot 10^{-14}$ s at room temperature so the both criteria are well satisfied.

Finally the conclusion can be made that the energy filtering effect quite well describes the change of both electrical conductivity and the Seebeck coefficient in nanostructured materials. The estimations showed that in the bulk nanostructured materials based on $Bi_xSb_{1-x}Te_3$ the increase of the Seebeck coefficient due to this effect can reach 10-20% at the grain size of 20-30nm. If the lattice thermal conductivity decrease is the same as that for electrical conductivity this can give the 20-40% increase in the figure of merit.

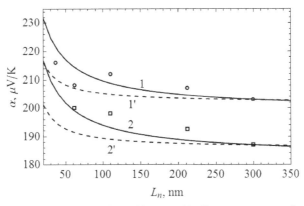

Fig. 24. The dependence of the Seebeck coefficient of bulk nanostructured materials on the grain size (see Fig. 23 for notation).

4. Conclusion

The nanopowder p-Bi-Sb-Te with particles ~ 10 nm were fabricated by the mechanical activation method (ball milling) using different technological modes. Cold and hot pressing at different conditions and also SPS process were used for consolidation of the powder into a bulk nanostructure and nanocomposites.

Nanoparticles keep composition of initial solid solution $Bi_xSb_{2-x}Te_3$. The change of the hot pressing temperature did not result in the change of phase composition and of the lattice parameter of samples. The main factors allowing slowing-down of the growth of nanograins as a result of recrystallization are the reduction of the temperature and of the duration of the pressing, the increase of the pressure, as well as addition of small value additives (like MoS_2, thermally expanded graphite or fullerenes). The SPS processing is also an effective way for reduction of the CDA (or nanograins) size, and as consequence for improvement of the figure of merit. The best value of the efficiency ZT=1.12 (at the temperature ~ 350÷375 K) was measured in the hot pressed bulk nanostructures $Bi_{0.3}Sb_{1.7}Te_3$, while it was reached ZT=1.22 (at 360 K) in the bulk nanostructure $Bi_{0.4}Sb_{1.6}Te_3$ fabricated by SPS method.

The theoretical dependence of the electric and heat conductivities and the thermoelectric power as the function of nanograins size L_n in $Bi_xSb_{2-x}Te_3$ bulk nanostructure are quite accurately correlates with the experimental data (see Sec. 3.2 & 3.3). It means that the phonons and holes scattering on the nanograin boundaries takes place. And the intensity of the scattering increases with reduction of L_n, that results in simultaneous decrease both phonons heat conductivity and electric conductivity. Our study shows that reduction of CDA size really can lead to improvement of the thermoelectric figure of merit.

Some theoretical results (Sec.3) on investigation of mechanisms of the thermoelectric efficiency improvement in bulk nanocrystalline semiconductors based on $Bi_xSb_{2-x}Te_3$ are summarized in Table 4.

Mechanism of improvement Z	Ways of realization	Probable value of increasing Z
Additional phonon scattering	$L_n < (10 - 20)$ nm	$(15 - 25)\%$
Tunneling of carriers	$L_n < (10 - 20)$ nm Vacuum gaps between nanograins $\sim 1 - 2$ nm	ZT - up to 3,0 - 3,5
Energy filtering of carriers	$L_n < (20 - 30)$ nm Decrease of electrical conductivity and lattice thermal conductivity compensate each other	$(20 - 40)\%$

Table 4. Comparison of mechanisms of the figure of merit improvement

The increase of the thermoelectric power by 10-20 % at L_n =20 - 30 nm can lead to significant (20 - 40 %) increase of the thermoelectric efficiency provided that the reduction of the electric conductivity and the lattice heat conductivity compensate each other. In the investigated samples the full indemnification does not occur, however the thermoelectric efficiency nevertheless managed to be increased up to the values ZT =1.1 - 1.2 (see Sec.2.6).

Table 4 shows that it is necessary to provide the small nanograin size (be more exact – CDA size) ~ 10 – 20 nm for realization of all three mechanisms of the figure of merit improvement. It is difficult to create such nanostructure technologically; the reason is the growth of the initial nanoparticles due to the recrystallization processes. However technological conditions have been determined (see Sec.2.5) for fabrication of the bulk nanostructures and nanocomposites based on $Bi_xSb_{2-x}Te_3$ solid solution from nanopowder by hot pressing and SPS methods which have given the reliable opportunity to obtain the CDA sizes L_n ~ 40 nm.

Fabrication of the vacuum gap ~ 1 – 2 nm between the nanograins for realization of the tunneling mechanism of the improvement of the figure of merit and the cutting off the phonons transport hardly will be possible by the technology of ball milling with the hot pressing or SPS process. Moreover, the electronic microscopy research has not found out any gaps between grains (or CDA) in studied nanostructures - no vacuum, no oxide (Sec.

2.4, 2.5). The accomplishment of all listed in Table 4 requirements to the structure of the nano-thermoelectrics based on $Bi_xSb_{2-x}Te_3$ solid solution should provide the increase of ZT up to 3,5 at the room temperatures. If the vacuum gaps ~ (1 - 2) nm between the grains can not be created technologically, but if the bulk nanostructure with the grain sizes ~ (10 - 20) nm can be realized, the increase of ZT up to 1.5 can be expected.

5. Acknowledgment

The work was supported by the Ministry of Education and Science of the Russian Federation, contract № 16.523.11.3002 and partially by the grant of the President of the Russian Federation № MK-7419.2010.2.

6. References

Anatychuk L.I. & Bulat L.P. (2001). *Semiconductors in Extreme Temperature Conditions*. Nauka, ISBN 5-02-024960-2, St. Petersburg, Russia (In Russian)

Anselm A.I. (1951). *Thermionic Vacuum Thermoelement*. Academy of Science of USSR, Leningrad, USSA (In Russian)

Bartkowiak M. & Mahan G.D. (1999). Boundary effects in thin-film thermoelectrics. *MRS Proceedings*, Vol.545, pp.265-272, ISSN 1946-4274

Bublik V.T., Bulat L.P., Karataev V.V., Maronchuk I.I., Osvenskii V.B., Pivovarov G. I., Pshenai-Severin D.A., Sagalova T.B. & Tabachkova N.Yu. (2009). Connection between Properties and Composition Parameters of Thermoelectric Material Based on Chalcogenides of Bismuth and Antimony. *Materials for electronic technics*, No.4, pp.61-64, ISSN: 1609-3597 (In Russian)

Bublik V.T., Bulat L.P., Karataev V.V., Maronchuk I.I., Osvenskii V.B., Pivovarov G. I., Pshenai-Severin D.A. & Tabachkova N.Yu. (2010a). Possibilities of nanostructured state maintenance at Thermoelectric Material Based on Bismuth and Antimony Chalcogenides. News of Universities, Physics, No.3-2, pp. 37-41, ISSN 0021-3411 (In Russian)

Bublik V.T., Dashevsky Z.M., Drabkin I.A., Karataev V.V., Kasian V.A., Osvenskii V.B., Pivovarov G. I., Pshenai-Severin D.A., Tabachkova N.Yu. & Bohmsteon N. (2010b). Transport Properties in 10-300 K Temperature Range for Nanostructured p-$Bi_{0,5}Sb_{1,5}Te_3$ Obtained by Spark Plasma Sintering. *Thermoelectrics and their Application*, pp.47-52, ISBN 978-5-86763-272-4, Ioffe PTI, St. Petersburg, Russia (In Russian)

Bublik V.T., Dashevsky Z.M., Drabkin I.A., Karataev V.V., Kasian V.A., Lavrentiev M.G., Osvenskii V.B., Pivovarov G. I., Sorokin A.I., Tabachkova N.Yu. & Bohmsteon N. (2010c). Thermoelectric Properties of Nanostructured p-$Bi_{0,5}Sb_{1,5}Te_3$ Obtained by SPS Method. *Thermoelectrics and their Application*, pp.53-57, ISBN 978-5-86763-272-4, Ioffe PTI, St. Petersburg, Russia (In Russian)

Bulat L.P., Pivovarov G.I., Snarskii A.A. (2006), Thermoelectrics based on fullerenes. In: *Thermoelectrics and their Application*, pp.36-40, ISBN 5-86763-185-0, Ioffe PTI, St. Petersburg, Russia (In Russian)

Bulat L.P., Osvensky V.B., Pivovarov G.I., Snarskii A.A., Tatyanin E.V. & Tay A.A.O. (2008a), On the effective kinetic coefficients of thermoelectric nanocomposites.

Proceedings of 6th European Conference on Thermoelectrics, pp. I2-1 – I2-6. Paris, France, July 2-4, 2008

Bulat L.P., Drabkin I.A., Pivovarov G.I., Osvensky V.B., (2008b). On Thermoelectric Properties of Materials with Nanocrystalline Structure. *Journal of Thermoelectricity,* No.4, pp. 27-31, ISSN 1607-8829

Bulat L.P., Bublik V.T., Drabkin I.A., Karatayev V.V., Osvensky V.B., Pivovarov G.I., Pshenai-Severin D.A., Tatyanin E.V. & Tabachkova N.Yu. (2009a), Bulk Nanostructured Thermoelectrics Based on Bismuth Telluride. *Journal of Thermoelectricity,* No.3, pp. 67-72, ISSN 1607-8829

Bulat L.P., Bublik V.T., Karatayev V.V., Osvensky V.B., Pivovarov G.I. (2009b). Methods of Investigation of Mechanical Properties and Structure of Nanomaterials for Thermoelectric Coolers. Bulletin of the International Academy of Refrigeration, No.3, pp.4-7, ISSN 1606-4313 (In Russian)

Bulat L.P. & Pshenai-Severin D. A. (2010a), Effect of Tunneling on the Thermoelectric Efficiency of Bulk Nanostructured Materials, *Physics of the Solid State,* Vol.52, No. 3, pp.485–492, ISSN 1063-7834

Bulat L. P., Bublik V.T., Drabkin I.A., Karataev V.V., Osvenskii V.B., Parkhomenko Yu.N., Pivovarov G. I., Pshenai-Severin D.A. & Tabachkova N.Yu. (2010b), Bulk Nanostructured Polycrystalline p-Bi-Sb-Te Thermoelectrics Obtained by Mechanical Activation Method with Hot Pressing, *Journal of Electronic Materials,* Vol.39, No.9, (September, 2010), pp.1650-1653, ISSN 0361-5235

Bulat L. P., Drabkin I.A., Karataev V.V., Osvenskii V.B. & Pshenai-Severin D.A. (2010c), Effect of Boundary Scattering on the Thermal Conductivity of a Nanostructured Semiconductor Material Based on the $Bi_xSb_{2-x}Te_3$ Solid Solution, *Physics of the Solid State,* Vol. 52, No. 9, pp. 1836–1841, ISSN 1063-7834

Bulat L. P., Drabkin I.A., Karataev V.V., Osvenskii V.B., Parkhomenko Yu. N., Pshenai-Severin D. A., Pivovarov G. I., & Tabachkova N. Yu. (2011a), Energy Filtration of Charge Carriers in a Nanostructured Material Based on Bismuth Telluride, *Physics of the Solid State,* Vol. 53, No. 1, pp. 29–34, ISSN 1063-7834

Bulat L.P., Pshenai-Severin D.A., Drabkin I.A., Karatayev V.V., Osvensky V.B., Parkhomenko Yu.N., Blank V.D., Pivovarov G.I., Bublik V.T. & Tabachkova N.Yu. (2011b), Mechanisms of Increasing of Thermoelectric Efficiency in Bulk Nanostructured Polycrystals. *Journal of Thermoelectricity,* No.1, pp. 14-19, ISSN 1607-8829

Burstein E. (ed.) (1969). *Tunneling Phenomena in Solids.* New York, Plenum Press, 579 p., ISBN: 0306303620

da Silva L.W. & Kariany M. (2004). Micro-thermoelectric cooler: interfacial effects on thermal and electrical transport. *International Journal of Heat and Mass Transfer,* Vol.47, No.10-11, pp.2417-2435, ISSN 0017-9310

Dmitriev A.V., Zvyagin I.P. (2010). Current trends in the physics of thermoelectric materials. *Uspekhi Fizicheskikh Nauk,* Vol.180, No.8, pp. 821- 838, ISSN 0042-1294

Drabble J.R. & Wolfe R. (1956). Anisotropic galvanomagnetic effects in semiconductors. *Proc. Phys. Soc. (London),* Vol.B69, No.2, pp.1101-1108, ISSN 0370-1328

Dresselhaus M.S., Chen G., Tang M.Y., Yang R., Lee H., Wang D., Ren Z., Fleurial J.-P. & Gogna P. (2007). New Directions for Low-Dimensional Thermoelectric Materials. *Adv. Mater.* Vol.19, No.8, (April, 2007), pp.1043–1053, ISSN 1521-4095

Fleming G.M. & Henderson J.E. (1940). The energy losses attending field current and thermionic emission of electrons from metals. *Phys. Rev.* Vol.58, No.10, pp.887-894, ISSN 1943-2879

Ghoshal, U., Ghoshal, S., McDowell, C., Shi, L., Cordes, S. & Farinelli, M. (2002a). Enhanced thermoelectric cooling at cold junction interfaces. *Appl. Phys. Letters,* Vol.80, No.16, pp. 3006-3008, ISSN 0003-6951

Ghoshal U. (2002b). Design and Characterization of Cold Point Thermoelectric Coolers. *Proceedings of XXI International Conf. on Thermoelectrics,* pp. 540-543, ISBN: 0-7803-7683-8, Long Beach, California, USA, August 26-29, 2002.

Goldsmid H.J., Lyon H.B. & Volckmann E.H. (1995). A simplified theory of phonon boundary scattering in solid solutions. *Proc. of the XIV Int. Conf. on Thermoelectrics,* pp.16-19, ISBN: 5-86763-081-1, St. Petersburg, Russia, June 27-30, 1995.

Goltsman B.M., Kudinov B.A. & Smirnov I.A. (1972). *Semiconductor Thermoelectric Materials Based on Bi_2Te_3.* Moskow, Nauka, 320 p., (In Russian)

Gridchin V.A., Lyubimskii V.M. & Moiseev A.G. (2005). Scattering of charge carriers at the boundaries of crystallites in films of polycrystalline silicon. *Semiconductors,* Vol.39, No.2, pp.192-197, ISSN 1063-7826

Harman, T. C., P. J. Taylor, Spears D. L. & Walsh M. P. (2000). Thermoelectric quantum-dot superlattices with high ZT. *Journal of Electronic Materials,* Vol.29, No.1, pp. L1-L2, ISSN 0361-5235

Harman, T. C., M. P. Walsh, Lafarge B.E. & Turner G.W. (2005). Nanostructured thermoelectric materials. *Journal of Electronic Materials,* Vol.34, No.5, pp. L19-L22, ISSN 0361-5235

Hishinuma Y., Geballe T.H., Moyzhes B.Y. & Kenny T.W. (2001). Refrigeration by combined tunneling and thermionic emission in vacuum: Use of nanometer scale design. *Appl. Phys. Lett.,* Vol.78, No.17, pp.2572-2574, ISSN 0003-6951

Lan Y., Minnich A.J., Chen G. & Ren Z. (2010). Enhancement of Thermoelectric Figure-of-Merit by a Bulk Nanostructuring Approach. *Advanced Functional Materials,* Vol.20, No.3, pp. 357-376, ISSN 1616-301X

Madelung O., Rössler U. & Schulz M. (ed.). (1998). $(Bi_{(1-x)}Sb_{(x)})_2Te_3$ physical properties, In: *Landolt-Börnstein - Group III Condensed Matter. Numerical Data and Functional Relationships in Science and Technology. Vol. 41C: Non-Tetrahedrally Bonded Elements and Binary Compounds I.* Berlin, Springer-Verlag, ISBN: 978-3-540-64583-2

Luk'yanova L.N., Kutasov V.A., Konstantinov P.P. & Popov V.V. (2010). Thermoelectric figure-of-merit in p-type bismuth- and antimony-chalcogenide-based solid solutions above room temperature. *Physics of the Solid State,* Vol.52, No.8, pp.1599-1605, ISSN 1063-7834

Mahan G.D. (1994). Thermionic refrigeration. *J. Appl. Phys.,* Vol.76, No.7, pp. 4362-4366, ISSN 0021-8979

Mahan G. D. & Woods L. M. (1998). Multilayer Thermionic Refrigeration. *Phys. Rev. Lett.* Vol.80, No.18, (4 May 1998) pp.4016-4019, ISSN 0031-9007

Mahan, G. D., Sofo, J. O. Bartkowiak, M. (1998). Multilayer thermionic refrigerator and generator. *J. Appl. Phys.* Vol.83, No.9, pp. 4683-4689, ISSN 0021-8979

Mayadas A. F. & Shatzkes M. (1970). Electrical-resistivity model for polycrystalline films: the case of arbitrary reflection at external surfaces. *Phys. Rev. B,* Vol.1, No.4, pp.1382-1389, ISSN 1098-0121

Minnich A.J., Dresselhaus M.S., Ren Z.F., & Chen G. (2009). Bulk nanostructured thermoelectric materials: current research and future prospects, *Energy and Environmental Science*, Vol. 2, No.5, pp. 466-479, *ISSN* 1754-5692

Moizhes B.Ya. & Nemchinsky V. (1998). Thermoelectric figure of merit of metal-semiconductor barrier structure based on energy relaxation length. *Appl. Phys. Lett.*, Vol.73, No.13, pp.1895-1897, ISSN 0003-6951

Murphy E.L. & Good R.H. (1956). Thermionic emission, field emission, and the transition region. *Phys. Rev.*, Vol.102, pp.1464-1473, ISSN 1943-2879

Popescu A., Woods L.M., Martin J. & Nolas G.S. (2009). Model of transport properties of thermoelectric nanocomposite materials. *Phys. Rev. B*, Vol.79, No.20, pp.205302-205308, ISSN 1098-0121

Poudel Bed, Hao Qing, Ma Yi, Lan Yucheng, Minnich Austin, Yu Bo, Yan Xiao, Wang Dezhi, Muto Andrew, Vashaee Daryoosh, Chen Xiaoyuan, Liu Junming, Dresselhaus Mildred S., Chen Gang, Ren Zhifeng (2008). High-Thermoelectric Performance of Nanostructured Bismuth Antimony Telluride Bulk Alloys. *Science*, Vol. 320, No. 5876, (2 May 2008), pp. 634-638, ISSN 0272-4634.

Ravich Yu.I. (1995). Selective scattering in thermoelectric materials. In: *CRC Handbook of Thermoelectrics. Ed. by Rowe D.M.* CRC Press, N.Y., pp.67-73, ISBN 0-8493-0146-7

Shakouri Ali & Bowers J. E. (1997). Heterostructure integrated thermionic coolers. *Appl. Phys. Lett.* Vol.71, No.9, pp.1234-1236, ISSN 1077-3118

Snarsky A. A., Palti A.M. & Ascheulov A.A. (1997). Anisotropic thermoelements. Review. *Fizika i Tekhnika Poluprovodnikov*, Vol.31, No.11, pp.1281-1298, (In Russian)

Sommer A. H. (1980). *Photoemissive Materials*. Krieger, New York, 270p., ISBN: 0898740096

Stordeur M., Stölzer M., Sobotta H. & Riede V. (1988). Investigation of the valence band structure of thermoelectric $(Bi_{1-x}Sb_x)_2Te_3$ single crystals. *phys. stat. sol. b*, Vol.150, No.1, pp.165–176, ISSN 0370-1972

Stratton R. (1962). Theory of field emission from semiconductors. *Phys. Ref.* Vol.125, No.1, pp.67-82, ISSN 1943-2879

Tavkhelidze, A., Skhiladze, G., Bibilashvili, A., Tsakadze, L., Jangadze, L., Taliashvili, Z., Cox, I. & Berishvili, Z. (2002) Electron Tunneling Through Large Area Vacuum Gap – Preliminary Results. *Proceedings of XXI International Conf. on Thermoelectrics*, pp. 435- 438, ISBN: 0-7803-7683-8, Long Beach, California, USA, August 26-29, 2002

Vasilevskiy, D., Dawood, M. S., Masse J.-P., Turenne, S. & Masut R. A. Generation of Nanosized Particles during Mechanical Alloying and Their Evolution through the Hot Extrusion Process in Bismuth-Telluride-Based Alloys. *Journal of Electronic Materials*, Vol.39, No.9, (September, 2010), pp. 1890-1896, ISSN 0361-5235

Venkatasubramanian R., Siivola E., Colpitts T. & O'Quinn B. (2001). Thin-film thermoelectric devices with high room-temperature figures of merit. *Nature*, Vol. 413 (11 October 2001), pp. 597-602, *ISSN* 0028-0836

10

Self-Assembling Siloxane Nanoparticles with Three Phases

Masatoshi Iji
NEC Corporation
Japan

1. Introduction

Inorganic nanometer-sized particles (nanoparticles) are attracting attention as reinforcing fillers for use in polymer-nanoparticle composites (nanocomposites) because they improve key characteristics of these composites at a relatively low content (Hussain et al., 2006, Jordan et al., 2005). For example, clay nanoparticles (Usuki et al, 2005), metal oxide nanoparticles such as silica (Rosso et al 2006) and titania nanoparticles (Zelikman et al., 2006), and carbon nanotubes (Moniruzzaman & Winey, 2006) increase the mechanical properties, especially the elasticity modulus, of many kinds of polymer composites. However, currently available nanoparticles insufficiently improve the tenacity (elongation at breaking point) of nanocomposites, which is necessary if nanocomposites are to be used in durable products such as electronic equipments and automobiles. This is mainly because these nanoparticles lack rubber-like elasticity although they have an affinity for a polymer matrix. Adding a typical elastomer such as rubber or plasticizer (e.g., a long chain alkyl ester) with a high affinity for a polymer matrix, however, reduces the breaking strength and elasticity modulus of the composite due to their lack of rigidity (Li & Turng, 2006, Shibata et al., 2006).

Nanoparticles with multiple-phases, a high-density phase (core) with rigidity and outside phases with rubber-like elasticity and affinity for matrix polymers, should improve the tenacity of nanocomposites without degrading their breaking strength. Metal oxide nanoparticles are typically formed by hydrolysis and condensation reactions of organic metal compounds, mainly silicon alkoxides (Ha & Cho, 2000, Chujo & Saegusa, 1992, Tamaki & Chujo,1999, Li et al., 2001, Kim et al., 2003) and perhydropolysilazane (Yamano, & Kozuka, 2009), in solvents or polymers, i.e., a sol-gel method. However, the formation of multiple-phased nanoparticles using these organic metal compounds is difficult due to their limited chemical structures. Moreover, the use of conventional surface treatment agents such as organic metal alkoxides to uniformly form multiple phases on nanoparticles while preventing their coagulation is practically difficult because such treatment is an extremely complex process.

Poly L-lactic acid (PLLA), a representative mass-produced biopolymer made of biomass (starch), is attractive for use in environmentally sensitive applications because its use prevents petroleum exhaust and reduces plastic waste due to its biodegradability after disposal. Although it has a relatively high breaking strength, its tenacity is extremely inadequate for it to be used in a variety of applications including durable products like those mentioned above because of its stiff structure, which is due to the hard crystalline region. Studies on PLLA nanocomposites have focused on the usual nanoparticles such as clay and calcium carbonate ones (Li & Turng, 2006, Petersson & Oksman, 2006, Jiang et al.,2007). However, to the best of the authors' knowledge, the use of multi-phased nanoparticles to increase the tenacity of PLLA has not been reported.

In this chapter, self-assembling siloxane nanoparticles with three phases that improve the tenacity of PLLA are reported. The particles consist of a high-density siloxane phase (plural cores), an elastomeric silicone phase, and a caprolactone oligomer phase. Self-assembly by aggregation and condensation of an organosiloxane with three units forms each phase. Testing showed that the use of these nanoparticles increases the tenacity (breaking strain) of PLLA while maintaining its relatively high breaking strength (Iji, 2011).

2. Self-assembling siloxane nano-particle with three phases

2.1 Preparation of siloxane nano-particles with three phases

Figure 1 illustrates the process used to prepare the organosiloxane with three units and the assumed self-assembly of the nanoparticles with three phases through aggregation and condensation of the organosiloxane, which consists of three units: isocyanatepropyltrimethoxysilane (IPTS), polymethylpropyloxysiloxane (PMPS), and a caprolactone oligomer (CLO).

IPTS was selected to form the high-density siloxane phase (plural cores) because it contains methoxy groups, which are highly polar and reactive, at a high molecular ratio, and thus preferentially aggregates and condensates, producing a rigidly cross-linked (high-density) siloxane network that forms more than one core in the particle. PMPS was selected to form the elastomeric silicone phase with an appropriate (nanometer) size around the cores because it contains propyloxy groups, which have moderate polarity and reactivity, at a low molecular ratio in an adequate-length siloxane chain (siloxane number: 8.0). Its use produces a loosely cross-linked (relatively low-density) siloxane network after the core formation. The CLO was selected to form the outside phase due to its low polarity compared with those of IPTS and PMPS and its high affinity for the PLLA matrix (it is highly soluble in melted PLLA).

The organosiloxane was synthesized by mixing PMPS with CLO at a molecular ratio of 1:1 and then mixing the resulting compound with IPTS at a molecular ratio of 1:1 (Figure 1 (A)). After the binding reactions had ceased, the unreacted IPTS, PMPS and CLO were removed using a column packed with polystyrene particles. The reactions were confirmed by hydrogen-nuclear magnetic resonance (H-NMR) analysis of the functional groups in the resulting compounds. Moreover, we performed gel permeation chromatography (GPC) analysis to determine the molecular weight, element analysis of carbon, hydrogen, and nitrogen, and Fourier transform infrared spectroscopy (FT-IR) to determine major bonds. These results showed that the IPTS, PMPS, and CLO were combined in almost equal molecular proportions as we intend. The CLO and IPTS were probably randomly located on the PMPS in the final compound but apart to some-extent because of the difference in their polarities. The reference organosiloxanes with two units (IPTS+PMPS or PMPS+CLO) were prepared and determined based on these methods.

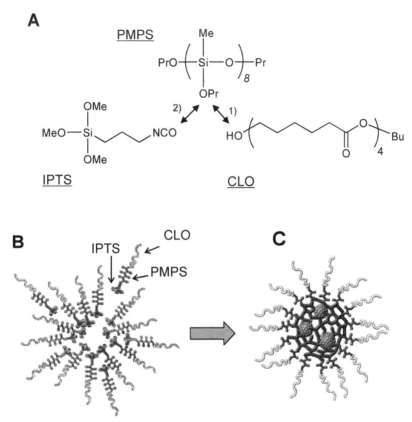

Fig. 1. Preparation of organosiloxane with three units (A) and self-assembly of three-phased nanoparticles through aggregation (B) and condensation (C) of organosiloxane.

We confirmed that the organosiloxane with three units dissolved in tetrahydrofuran (THF) aggregates and forms nanoparticles due to condensation by using water and ammonia as a base catalyst. To avoid confusing the formation of the high-density siloxane cores with the formation of the elastomeric silicone phase around them, we initiated two-step condensation of the organosiloxane by taking advantage of the higher reactivity of the methoxy groups in the IPTS unit than that of the propyloxy groups in the PMPS unit. The first step was core formation through hydrolysis and condensation of the methoxy groups in the IPTS unit at room temperature for 24 hours in THF. The second step was elastomeric silicone phase formation through hydrolysis and condensation of the propyloxy groups in the PMPS unit by heating at a high temperature (180°C) for 20 minutes after replacing the THF with dimethyl sulfoxide, which has a higher boiling point (189°C). These stepwise reactions were ascertained by H-NMR analysis for the methoxy, propyloxy and silanol. After the first step, the methoxy groups in the IPTS were almost completely hydrolyzed and condensed, forming siloxane, while the propyloxy groups in the PMPS unit did not hydrolyzed. After the second step, the propyloxy groups were almost completely hydrolyzed and condensed. These detail methods and results were shown in our paper (Iji, 2011).

2.2 Structures of siloxane nano-particles

As shown in Figure 2, the formed nanoparticles were observed with a scanning electron microscope (SEM). The size distribution, measured by light scattering analysis, indicated a relatively narrow size range, with an average diameter of 13 nm.

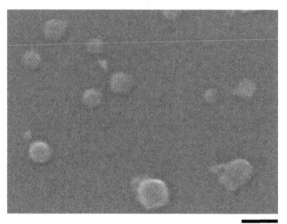

20nm

Fig. 2. Observation of three-phased nanoparticles using scanning electron microscope.

The inside structures of the nanoparticles were investigated by scanning electron microscopy and energy dispersive X-ray (SEM-EDX) analysis. Figure 3 shows representative results for the amounts (intensities) of silicon and carbon that were detected along cross-sections of the nanoparticles. The amount of silicon originating from the IPTS and PMPS units was remarkably higher at several points around the center. This indicates that the nanoparticle had plural high-density siloxane cores formed mainly from the IPTS unit and that, around the cores, there was a relatively low-density siloxane phase formed mainly from the PMPS unit. The amount of carbon, which originated from all the units, especially the CLO unit, did not significantly vary throughout the particle. This indicates that there was a relatively high concentration of carbon in the outer layer of the particle, meaning that the CLO unit formed the outside phase fairly well.

Figure 4 shows the thermo-gravity analysis of the nanoparticle formed by the organosiloxane with IPTS, PMPS, and CLO, the reference particle formed by the organosiloxane with PMPS and CLO (through the second step described above), and these organosiloxanes. The results indicated that the nanoparticle showed higher thermo-degradation resistance than the reference particle, which suggests the formation of the high density cross-linking phase to retard the thermo-degradation; we considered the phase to be the cores in the nanoparticle. Furthermore, these particles showed considerably higher thermo-degradation resistances than those of the organosiloxanes. These suggest that not only the high- density cores derived from IPTS but also the relatively low density cross-linking silicone phase derived from PMTS were formed.

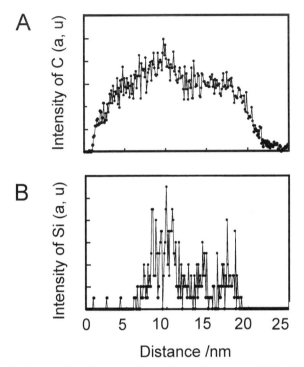

Fig. 3. Analysis of silicon and carbon in a three-phased nanoparticle by SEM-EDX:
intensities of silicon (A) and carbon (B) detected along cross-section of the nanoparticle.

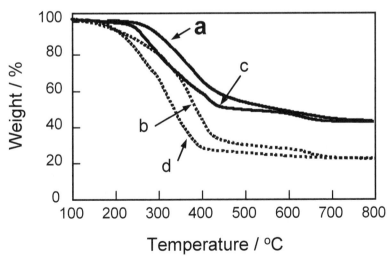

Fig. 4. Thermo-gravimetric analysis of nanoparticles and organosiloxanes (in nitrogen by
heating at 10 centigrade per minute)
(a) Three-phased nanoparticle, (b) Organosiloxane consisting of IPTS, PMPS and CLO to

form (a), (c) Two-phased particle consisting of elastomeric silicone and CLO phases, (d) Organosiloxane consisting of PMPS and CLO to form (c)

From these results, it seems reasonable that three-phased nanoparticles can be formed by self-assembly of the organosiloxane with three units, as shown in Figure 1 (B, C). The IPTS unit mainly performs the aggregation of the organosiloxane. The aggregated IPTS unit mainly forms the high-density cross-linked siloxane phase (plural cores) through preferential hydrolysis and condensation reactions of its methoxy groups. After the core formation, the PMPS unit mainly forms the middle phase, the relatively low-density cross-linked siloxane network (elastomeric silicone) around the cores through hydrolysis and condensation reactions of its propyloxy groups. The CLO unit mainly forms the outside phase of the nanoparticles after the organosiloxane has aggregated because of its position and relatively low polarity.

2.3 Characteristics of nanocomposites consisting of siloxane nano-particles and polylactic acid

Using these three-phased siloxane nanoparticles at 5wt%, we prepared a molded PLLA nanocomposite to measure its tenacity by flexural and tensile testing. After the first step (core formation), the THF, water, and ammonia were removed by evaporation. The nanoparticles were mixed with PLLA in chloroform, followed by removing the solvent by evaporation. The resulting composite was extruded at 180°C for 10 minutes using a screw-type mixer and then molded by pressing while heating at 180°C for 10 minutes, followed by crystallization of the PLLA in the composite by heating at 100°C for 4 hours. It is likely that the condensation reaction to form the elastomeric silicone phase in the particle was mostly finished during the extruding and molding of the PLLA composite because the same heating condition as that for the solvent in the second step above resulted in the same condensation reaction.

As shown in Figure 5 (A, B), the three-phased nanoparticles greatly increased the PLLA's tenacity without degrading its high breaking strength. The elongation of the PLLA nanocomposite was more than twice that of PLLA while the elasticity modulus and breaking (maximum) strength were comparable to those of PLLA.

As references, PLLA composites containing commercial silica nanoparticles, two-phased nanoparticles (cores and elastomeric silicone phase), or the organosiloxane consisting of PMPS and CLO units were prepared using the method described above. The composite containing the commercial silica nanoparticles, average diameter 12 nm, showed a slight increase in the modulus, but its strength and elongation were less than those of PLLA (Figure 5A (c)). The two-phased nanoparticles, average diameter of 10 nm, were formed using the organosiloxane with IPTS and PMPS units through the same core formation step described above. The composite containing the nanoparticles showed only a slight increase in elongation compared with that of PLLA (Figure 5A (d)). The composite containing the organosiloxane with PMPS and CLO (PMPS might be cross-linked when mixing with PLLA during heating) showed substantial increase in elongation compared with that of PLLA, but its strength and modulus were less (Figure 5A (e)). The composite containing CLO alone decreased the elasticity modulus of the PLLA composite and insufficiently increased its elongation (Figure 5A (f)). The amount of CLO alone in the composite was 2 wt%, which is near to the amounts of the CLO parts of the three-phased nanoparticles and the organosiloxane with PMPS and CLO. Increasing the amount of CLO further decreased both the modulus and also the strength while it increasing the elongation. These mean that each phase in the three-phased nanoparticles is necessary to increase the PLLA's tenacity while maintaining its breaking strength and modulus.

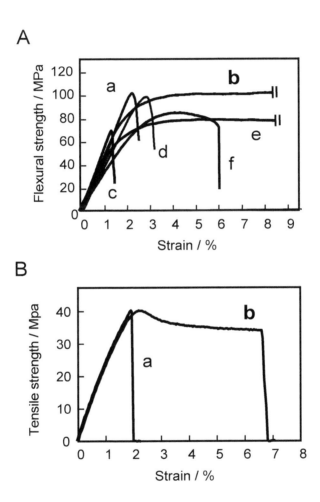

Fig. 5. Flexural (A) and tensile (B) testing of PLLA composites
(a) PLLA, **(b) PLLA nanocomposite with three-phased nanoparticles (5 wt%)**, (c) PLLA nanocomposite with silica nanoparticles (5 wt%), (d) PLLA nanocomposite with two-phased nanoparticles consisting of cores and elastomeric silicone phase (5 wt%), (e) PLLA composite with organosiloxane consisting of PMPS and CLO (5 wt%), and (f) PLLA composite with CLO (2 wt%).

These results suggest that the mechanism of the improved tenacity due to the use of the three-phased nanoparticles is as follows. During the initial period of the PLLA nanocomposite deformation, the nanoparticles create a high elasticity modulus and maximize the strength because of their core rigidity and the high affinity of the outside CLO phase for the PLLA matrix. This idea is supported by the results that the reference organosiloxane with PMPS and CLO, not forming cores and also, the nano-silica, which aggregated in the PLLA composites due to its low affinity with PLLA, did not maintain such a high modulus and strength, simultaneously. In the middle and final periods of the deformation, the three-phased nanoparticles elongated the composite due to the rubber-like elasticity of the elastomeric silicone phase derived from PMPS and the plasticity of the outside CLO phase. While the nano-silica, the reference two-phased particles without the CLO phase, and CLO alone did not perform such elongation of the PLLA composites, the organosiloxane with PMPS and CLO elongated the composite, which can supports the proposed mechanism.

Furthermore, we have cleared that the influence of adding the three-phased nanoparticles on the heat resistance of PLLA (Table 1). The glass transition temperature and heat distortion temperature of the PLLA composite with the nanoparticles (5wt%) slightly decreased comparing with PLLA, but these levels were fairly kept. The decomposition temperature measured by TGA increased. Adding typical elastomers such as rubber or plasticizer reduces heat resiatance of the composites. However, the nanoparticles maintained the heat resistance because of its core rigidity and its higher thermo-degradation resistance as above mentioned.

	Glass transition temp. (°C)	Heat distortion temp.(°C) Load: 0.45MPa / 1.80MPa	Decomposition temp.(°C) / 10% weight loss
PLLA	63	124 / 66	337
PLLA with three-phased nanoparticle (5wt%)	59	119 / 64	344

Table 1. Heat resistance of PLLA and PLLA composites with three-phased nanoparticle

3. Conclusion

In conclusion, we developed self-assembling siloxane nanoparticles with three phases: a high-density cross-linked siloxane phase (plural cores), an elastomeric silicone phase around the cores, and an outside CLO phase with a high affinity for the PLLA matrix. These nanoparticles self-assemble by aggregation and condensation of the organosiloxane with three units, IPTS, PMPS, and CLO that respectively form each phase. Adding these nanoparticles to PLLA increases the tenacity of the PLLA while maintaining its high breaking strength. Their use will expand the use of PLLA in durable product applications and other new applications. These nanoparticles can also be applied to various other brittle polymers by modifying the structure of the outside phase to achieve a high affinity with these polymers.

4. References

Hussain, F., Hojjati, M., Okamoto, M., & Gorga, R. E. (2006). Polymer-Matrix Nanocomposites, Processing, Manufacturing, and Application. *J. Compos. Mat.* Vol.40, pp.1511–1575.

Jordan, J., Jacob, K. I., Tannenhaum, R., Sharif, M. A., & Jasiuk, I. (2005). Experimental Trends in Polymer Nanocomposites- A Review. *Mat. Sci. Eng.* Vol.A 393, pp.1–11.

Usuki, A., Hasegawa, N., & Kato, M. (2005).Polymer-Clay Nonocomposites. *Adv. Polym. Sci.* Vol.179, pp.135–195.

Rosso, P., Ye, L., Friedrich, K., & Sprenger, S. (2006). A Toughened Epoxy Resin by Silica Nanoparticle Reinforcement. *J. Appl. Polym. Sci.* Vol. 100, pp.1849–1855.

Zelikman, E., Tchoudakov, R., & Moshe, N. (2006).Particulate Multi-Phase Polymeric Nanocomposites. *Polym. Comp.* Vol. 27, No. 4, pp. 425–430.

Moniruzzaman, M., & Winey, K. I. (2006). Polymer Nanocomposites Containing Carbon Nanotubes. *Macromolecules.* Vol. 39, pp.5194–5205.

Li, T., & Turng, L. (2006). Polylactide, Nanoclay, and Core-Shell Rubber. *Polym. Eng. Sci.* pp.1419–1427 .

Shibata, M, Someya, Y., Orihara, M., & Miyoshi, M. (2006). Thermal and Mechanical Properties of Plasticized Poly (L-lactide) Nanocomposites with Organo-Modified Montmorillonites. *J. Appl. Polym. Sci.* Vol.99, pp.2594–2602.

Ha, C. S. & Cho, W. (2000). Microstructure and Interface in Organic/Inorganic Hybrid Composites. *J. Polym. Adv. Technol.* Vol. 11, pp.145–150.

Chujo, Y. & Saegusa T. (1992). Organic Polymer Hybrids with Silica Gel Formed by Means of the Sol-Gel Method. *Adv. Polym.* Sci. Vol. 100, pp.11–29.

Tamaki, R. & Chujo, Y. (1999). Synthesis of Polystyrene and Silica Gel Polymer Hybrids Utilizing Ionic Interactions. *Chem. Mater.* Vol.11, pp.1719–1726.

Li, G. Z., Wang, L., Toghiani, H., Daulton, T. L., Koyama, K., & Pittman, C. U. (2001).Viscoelastic and Mechanical Properties of Epoxy/Multifunctional Polyhedral Oligomeric Silsesquioxane Nanocomposites and Epoxy/Ladderlike Polyphenylsilsesquioxane Blends. *Macromolecules.* Vol.34, pp.8686–8693.

Kim, G. M., Qin, H., Fang, X., Sun, F.C., & Mather, P.T. (2003).Hybrid Epoxy-Based Thermosets Based on Polyhedral Oligosilsesquioxane: Cure Behavior and Toughening Mechanisms. *J. Polym. Sci. B, Polym. Phys.* Vol.41, No.24, pp. 3299–3313.

Yamano, A., & Kozuka, H. (2009).Preparation of Silica Coatings Hevily Doped with Spiropyran Using Perhydropolysilazane as the Silica Source and Their Photochromic Properties. *J. Phys. Chem. B,* Vol. 113, pp. 5769–5776.

Petersson, L., & Oksman, K. (2006). Biopolymer Based Nanocomposites: Comparing Layered Silicates and Microcrystalline Cellulose as Nanoreinforcement. *Com. Sci, Tech.* Vol.66, pp.2187–2196.

Jiang, L., Zhang, J., & Wolcott, M. P. (2007). Comparison of Polylactide/Nano-Sized Calcium Carbonate and Polylactide /Montmorillonite Composites: Reinforcing Effects and Toughening Mechanism. *Polymer,* Vol.48, pp.7632–7644.

Iji, M., Morishita, N. & Kai,H.(2011). Self-assembling siloxane nanoparticles with three phases that increase tenacity of poly L-lactic acid", *Polymer Journal,* Vol. 43, pp.101-104.

11

One-Step Synthesis of Oval Shaped Silica/Epoxy Nanocomposite: Process, Formation Mechanism and Properties

Nopphawan Phonthammachai[1], Hongling Chia[1] and Chaobin He[1,2]
[1]Institute of Materials Research and Engineering, Singapore,
A*STAR (Agency for Science, Technology and Research),
[2]Department of Materials Science & Engineering, National University of Singapore,
Singapore

1. Introduction

Silica/epoxy nanocomposites have been widely employed as high strength material for aerospace, automobile, electronic and sporting equipment industries (Deng et al., 2007; Preghenella et al., 2005; Hsiue et al., 2001; Fu et al., 2008). The thermal mechanical properties of nanocomposites were reported to be influenced by the shape and size of nano-filler, volume fraction, quality of dispersion, and the interaction between filler and matrix (Ragosta et al., 2005; Kwon et al., 2008; Adachi et al., 2008; Zhang et al., 2006). These parameters determine the molecular mobility of matrix and the amount of energy dissipation at crack initiation and propagation as the stress transfer from matrix can be promoted by these high specific surface area particles.

To obtain silica/epoxy nanocomposites with good properties, many approaches have been devised to improving the dispersion of silica and the interaction between silica and epoxy (Fu et al., 2008; Kwon et al., 2008; Zhang et al., 2006; Chen et al., 2008; Wang et al., 2005; Zhang et al., 2008). The solution blending process has been reported as a method providing good silica dispersion and silica-epoxy bonding (Hsiue et al., 2001; Zhang et al., 2008; Deng et al., 2007; Liu et al., 2003; Mascia et al., 2006; Huang et al., 2005; Araki et al., 2008). However, the multiple steps are involved in this process. The spherical shaped silica synthesized by sol-gel process is normally used as a silica source due to its availability in solvents. The further step to functionalize silica surfaces by amine-terminated coupling agents is required to improve the dispersion of silica nanoparticles and their adhesion with matrix (Mascia et al., 2006). Then, the high pressure and temperature mixing with epoxy compositions is required after the above process. As the solvents are involved in this process, the steps to remove, recycle and dispose solvents are needed. This poses the environmental, health and safety issues, in addition to the additional costs involved in solvent removal/disposal. Moreover, large amounts of silica at 5-30 % by weight of total composite composition are required to obtain nanocomposite with good mechanical and thermal properties (Deng et al., 2007; Preghenella et al., 2005; Zhang et al., 2008). The high percentages of silica significantly increase the viscosity of compositions and influence many intrinsic properties of epoxy matrix such as weight, ductility, processability and transparency. Therefore, there is a need to develop an effective, convenient and low-cost process to prepare high performance silica/epoxy nanocomposite that exhibits uniform dispersion of silica with great silica- epoxy adhesion, to target a wide range of applications.

In the present work, a "Solvent-Free One-Pot Synthesis" method was developed for the preparation of high performance nanocomposites with uniform dispersion of oval shaped silica in epoxy and strong silica-epoxy bonding. The silica formation, surface functionalization and dispersion in epoxy compositions are combined into one step at 50 °C under mechanical stirring. In this process, the solvent was not involved in the nanocomposite preparation. Thus, it is friendly to the environment and benefits to the formation of oval shaped silica because high shear rate was applied for the mixing of viscous mixture. The details on the synthesis process, the chemical composition and morphology, the thermal mechanical properties of silanized silica/epoxy nanocomposites were studied. The properties of prepared nanocomposite were compared with those of neat epoxy, non-functionalized silica/epoxy and commercial available silica/epoxy systems.

2. Experimental

2.1 Materials

Diglycidyl ether of bisphenol A (DGEBA, D.E.R.™ 332) was supplied by Dow Chemicals. Diethyltoluenediamine (Ethacure 100-LC) was obtained from Albemarle. Tetraethylorthosilicate (TEOS, ≥ 99%) and (3-aminopropyl)trimethoxysilane (APTMS, 97%) were purchased from Sigma-Aldrich. Ammonia solution (25 wt-%) was supplied by Merck. The commercial available silica/epoxy (Nanopox F400) was supplied by Nanoresins AG.

2.2 Preparation of silanized silica/epoxy nanocomposite

A mixture of epoxy, Ethacure 100-LC, TEOS and APTMS was stirred vigorously at 50 °C. The weight ratio of epoxy: Ethacure 100-LC was fixed at 3.8:1. The amount of TEOS was varied to obtain 1-4 wt-% silica in epoxy composition. The amount of APTMS was fixed at 10 wt-% APTMS to silica amount. An ammonia solution with NH_3:TEOS molar ratio of 2.3:1 was injected into the above solution and aged for 60 min. The mixture was degassed under vacuum at 75 °C and poured into a mold coated with releasing agent. The sample was then cured in an air purged oven at 130 °C for 1 h, 160 °C for 2 h and 270 °C for 4 h.

2.3 Preparation of comparative samples

The neat epoxy sample was prepared by mixing an epoxy resin with Ethacure 100-LC at weight ratio of 3.8:1. A commercial available silica/epoxy sample (Nanopox F-400, 2 wt-% silica in epoxy composition) was formed by mixing epoxy, Ethacure 100-LC and Nanopox F-400 at the weight ratio of 3.6:1:0.24. The non-functionalized silica/epoxy (2 wt-% silica in epoxy composition) was prepared following the process shown in 2.2 without adding APTMS. These mixtures were degassed under vacuum at 75 °C before transferred into a mold and cured following the same curing process as that for silanized silica/epoxy nanocomposite.

2.4 Material characterization

2.4.1 Morphology and chemical composition

The morphology, size and dispersion of silica were investigated using transmission electron microscope (TEM) that was conducted in high resolution mode using a JEOL 2100F instrument and operated at 200 kV. The samples were cut using a Leica Ultracut UCT ultramicrotome and placed on 200 mesh copper grids. The chemical compositions of nanocomposites were analyzed using an energy dispersive X-ray spectroscopy in a transmission electron microscope (EDX-equipped TEM). The chemical state of elements in nanocomposite was determined by X-ray photoelectron spectroscopy (XPS) that was conducted using a VG Escalab 220i instrument with monochromatic Al radiation and spot size of 700 μm.

2.4.2 Thermal properties

Single-cantilever mode of the dynamic mechanical analyzer (DMA Q800, TA Instruments) was used to measure the dynamic modulus (E') and glass transition temperature (T_g) of materials by heating the samples from 25 to 250 °C with a ramping rate of 3 °C/min, frequency of 1 Hz and oscillation amplitude of 20 μm. Thermogravimetric analysis (TGA) was performed with a TA instrument Q500 thermogravimetric analyzer. The degradation temperature (T_d) of materials was measured under nitrogen atmosphere by heating the samples to 800 °C at a ramping rate of 5 °C/min. The temperature at the middle of thermal transition of composites was defined as the degradation temperature (T_d).

2.4.3 Mechanical properties

The flexural strength and modulus of nanocomposites were determined by 3-point bending test according to the ASTM Standard D 790-96, with specimens of $55 \times 13 \times 2.2$ mm³. The tests were conducted with crosshead speed of 1 mm/min, at a span length of 40 mm. The tensile tests were carried out according to the ASTM Standard D 638-03 using an Instron 5569 testing machine at tensile speed of 1 mm/min. The specimens were cut into dog-bone shape with dimension of $55 \times 3 \times 2.2$ mm³. The fracture toughness was measured using the Single-Edge-Notch 3-Point-Bend (SEN-3PB) Tests. The Mode-I critical stress intensity factor (K_{Ic}) was measured using SEN-3PB geometry (span = 50.8 mm) and single-edge-notched (SEN) specimens of $60 \times 12.7 \times 3.0$ mm³, which meets the plane strain condition requirements. A sharp notch was introduced by pressing afresh razor blade at the bottom of a saw-slot in the middle of the rectangular bar with the Instron 5569 at a crosshead speed of 0.5 mm/min. The tests were conducted on the same Instron 5569 at a crosshead speed of 1 mm/min.

3. Results and discussion

3.1 Synthesis of silica/epoxy nanocomposite

In "Solvent-Free One-Pot Synthesis" method, silanized silica nanoparticles were synthesized through the sol-gel process of TEOS, APTMS and ammonia solution (25 wt-% ammonia in water) in epoxy resin (Scheme 1).

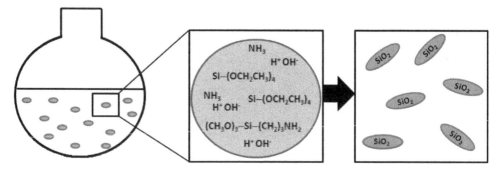

Scheme 1. The formation of silanized silica/epoxy nanocomposite prepared using the "Solvent-Free One-Pot Synthesis" method.

In this process, the water molecules conduct the hydrolysis of TEOS and APTMS, while NH_3 is a basic catalyst that accelerates the reaction (reaction a and b, Scheme 2). The condensation of hydrolyzed TEOS was subsequently occurred to form silica nanoparticles (reaction c, Scheme 2). The hydrolyzed APTMS was functionalized onto the silica surfaces through the reaction between silanol groups of silica and hydroxyl groups of hydrolyzed APTMS to form silanized silica with $-NH_2$ functional groups those can be reacted with epoxy resin to form strong silica-epoxy bonding (reaction d and e, Scheme 2).

The formation of silanized silica as proposed mechanism was confirmed by the transmission electron microscope (TEM) with an energy dispersive X-ray spectroscopy (EDX) and the X-ray photoelectron spectroscopy (XPS). From the TEM image of 4 wt-% silanized silica/epoxy nanocomposite (Fig. 1a), the oval-shaped nanoparticles with diameter of 65-140 nm were uniformly dispersed inside epoxy resin. The preferable oval-shape of these particles (aspect ratio > 1) is possibly occurred through a high shear rate of viscous epoxy\TEOS\APTMS\$NH_3$\$H_2O$ mixture under vigorous stirring. These oval shaped silica nanoparticles could provide added benefit to the mechanical properties of the resulting nanocomposite system as it has higher aspect ratio than those spherical shaped silica. The chemical composition of dispersed particles was confirmed by EDX analysis to be Si, C and O atoms (Fig. 1b).

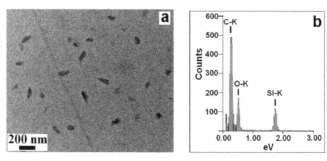

Fig. 1. Homogeneous dispersion of oval shaped silica nanoparticles in epoxy matrix prepared using the "Solvent-Free One-Pot Synthesis" method (a), the chemical composition of silanized silica/epoxy nanocomposite detected by EDX-equipped TEM (b).

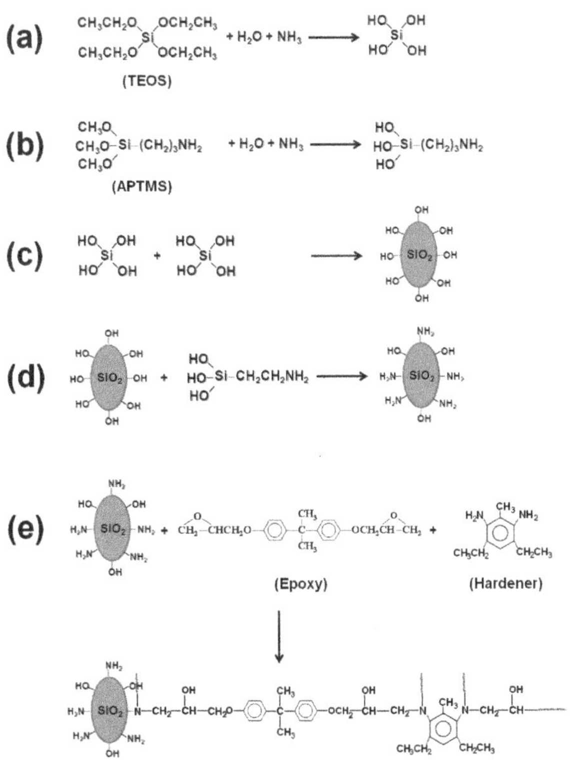

Scheme 2. Proposed mechanisms for the formation of silanized silica/epoxy nanocomposite prepared using the "Solvent-Free One-Pot Synthesis" method.

The XPS of silanized silica/epoxy nanocomposite showed broad peak of Si2p corresponding to three Si species (Fig. 2a). The part at highest binding energy (B.E.) of 103.8 eV represents the silanol groups (Si-OH) on silica surfaces. The Si-O-Si structure of silica particles and Si-alkylamine of functionalized APTMS was found at 102.5 and 101.7 eV respectively. The XPS result of non-functionalized silica/epoxy nanocomposite was compared with the above sample to confirm the present of APTMS functionalized on the silica surfaces (Fig. 2b). As expected, only two Si species were obtained from the non-functionalized silica/epoxy sample at 103.1 eV and 102.1 eV, respectively. The species at high B.E. refers to the silanol groups (Si-OH) on silica surfaces, while the one at lower B.E. is Si-O-Si structure of silica particles.

In accordance with the Si2p spectra, the N1s of silanzied silica/epoxy nanocomposite showed two Ni species represented the N1s of Ethacure 100-LC at B.E. 398.9 eV and APTMS at B.E. of 400.3 eV (Fig. 2c). In Etacure 100-LC, N atoms are connected with a stronger electron donor group (benzene ring) than N atoms of APTMS, therefore, they showed a signal at lower binding energy (Liu et al., 2006). However, only one N from C-N linkage between epoxy and Etacure 100-LC at 399.9 eV was observed from the non-functionalized sample (Fig. 2d).

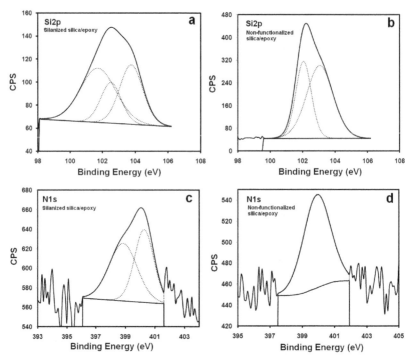

Fig. 2. XPS shows Si2p and N1s curve fit of 2 wt-% silanized silica/epoxy (a and c) and 2 wt% non-functionalized silica/epoxy nanocomposites (b and d).

Therefore, it can be concluded that the oval shaped silica nanoparticles were formed by the sol-gel process of TEOS in the presence of ammonia solution. The APTMS was functionalized on silica surfaces and acts as a linker to form a strong filler-matrix bonding during curing process at elevate temperature of 130-270 °C.

3.2 Thermal mechanical properties of silanized silica/epoxy nanocomposite

The thermal mechanical properties of silanized silica/epoxy nanocomposites were studied at various silica contents of 1-4 wt-% in epoxy composition. The thermal properties of nanocomposites were tested by DMA and TGA. The mechanical properties such as tensile modulus, flexural modulus and fracture toughness were done by tensile, 3-point bending and single-edge-notch 3-point-bend (SEN-3PB) tests.

From the DMA measurement, the glass transition temperature (T_g) of nanocomposites with 0-2 wt-% silanized silica were comparable at 205-210 °C (Fig. 3). The depletion of T_g to 185 °C and 161 °C was occurred when incorporated 3-4 wt-% silica into epoxy resin. Similar trend of result was achieved from the degradation temperature (T_d), in which the comparable T_d at 375-376 °C was shown in 0-2 wt-% silanized silica. However, the T_d of composite became lower to 374 and 360 °C at 3 and 4 wt-% silica contents. The depletion of T_g and T_d at high silica content may occurred from the retardant of cross-linked reaction of epoxy network by large amount of silica particles (Chen et al., 2008).

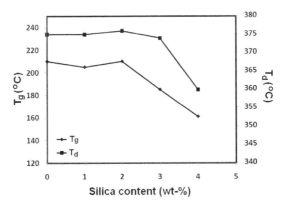

Fig. 3. T_g and T_d of neat epoxy and silanized silica/epoxy nanocomposites prepared at different silica contents of 1-4 wt-% in epoxy composition.

The storage modulus of nanocomposite was increased in accordance with the amount of silica incorporated into epoxy resin (Fig. 4). The similar trend of result was obtained from the fracture toughness, represented as the mode-I critical stress intensity factor (K_{Ic}) (Fig. 5). The K_{Ic} was dramatically improved for 54 % by incorporated 4 wt-% silanized silica into epoxy. The value was increased from 0.35±0.20 MPa.m$^{1/2}$ at 0 wt-% (neat epoxy) to 0.76±0.13 MPa.m$^{1/2}$ at 4 wt-%. It means that the solid silica nanoparticles those uniformly dispersed and formed strong bond with epoxy provide extra reinforcement to the composite structure.

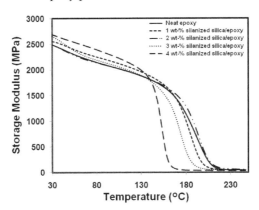

Fig. 4. Plot of storage modulus against temperature of neat epoxy and silanized silica/epoxy nanocomposites prepared at different silica contents of 1-4 wt-% in epoxy composition.

The enhancement of flexural and tensile modulus emphasizes the advantage of oval-shaped silica of the present method to the reinforcement of epoxy resin (Fig. 6). 20% improvement on the flexural and tensile modulus was achieved when introduced few percentages of oval shaped silica (2-3 wt-%) into epoxy, whereas the modulus became maintained at silica content greater than 3 wt-%. As the oval-shaped silica nanoparticles of the present method are uniformly dispersed in epoxy matrix with very low degree of aggregation between particles and form strong bond with epoxy, only few percentages of silica were required to improve the properties of neat resin. No different on the flexural property was observed when added too small quantity of silica (1 wt-%) into epoxy because the number of reinforced silica particles is not enough to prevent the fracture of brittle matrix. At the optimum ranges where the number of silica nanoparticles is enough to reinforce epoxy structure, the improvements of properties were achieved.

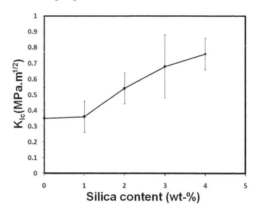

Fig. 5. Mode-I critical stress intensity factor (K_{Ic}) represents the fracture toughness of neat epoxy and silanized silica/epoxy nanocomposites prepared at different silica contents of 1-4 wt-% in epoxy composition. The K_{Ic} value is increased at higher amount of silica content in composite.

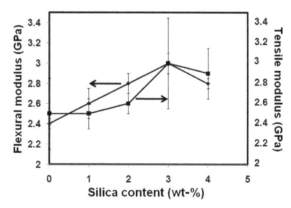

Fig. 6. Flexural and tensile modulus of neat epoxy and silanized silica/epoxy nanocomposites prepared at different silica contents of 1-4 wt-% in epoxy composition.

3.3 Comparison between the thermal mechanical properties of silanized silica/epoxy nanocomposite and comparative references

The silanized silica/epoxy nanocomposite prepared using the present method was compared with three comparative references, which are the neat epoxy resin, non-functionalized silica/epoxy and commercial silica/epoxy. The silica content of nanocomposite systems was fixed at 2 wt-% in epoxy composition.

From DMA and TGA results, it was shown that the T_g and T_d of neat epoxy resin and silanized silica/epoxy were comparable at T_g of 210 °C and T_d of 376 °C (Fig. 7). The commercial silica/epoxy showed slightly lower temperature at T_g of 209 °C and T_d of 374 °C. The lowest T_g was obtained from non-functionalized silica/epoxy nanocomposite at T_g of 207 °C and T_d of 372 °C. It was found from these results that the surface functionalization of silica particles is important to the strength of composite network, in which the thermal stability of nanocomposite structure can be promoted by strong silica-epoxy bonding.

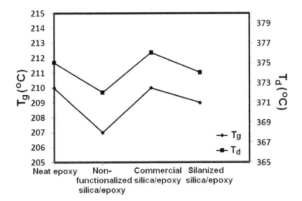

Fig. 7. Comparison between the T_g and T_d of 2 wt-% silanized silica/epoxy nanocomposites and comparative references (neat epoxy, 2 wt-% non-functionalized silica/epoxy and 2 wt-% commercial silica/epoxy).

The comparative results of storage modulus were correspondence with the T_g (Fig. 8.). The highest value was achieved from silanized silica/epoxy at 2510 MPa, while the lower modulus was obtained from non-functionalized silica/epoxy and commercial silica/epoxy nanocomposites at 2330 and 2370 MPa, respectively.

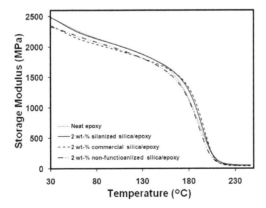

Fig. 8. Comparison between the storage modulus of 2 wt-% silanized silica/epoxy nanocomposites and comparative references (neat epoxy, 2 wt-% non-functionalized silica/epoxy and 2 wt-% commercial silica/epoxy).

The flexural and tensile modulus of neat epoxy resin were found to significantly be enhanced by incorporating silanized silica of the present method (Fig. 9.). In contrast, the poorest properties were achieved when mixed commercial silica with epoxy resin. This may occurred through the aggregation of high concentrated masterbatch of commercial silica (40 wt-%). The size and shape of silica also influence the mechanical properties of nanocomposite. As the silanized silica is oval in shape and its size is suitable for good composite reinforcement, only few percentages of silica are required for the property enhancement. Therefore, the special mixing process may require for the dispersion process of commercial silica in epoxy. Larger amount of commercial silica is also required to achieve a similar range of properties as our silanized silica/epoxy system. Therefore, it can clearly be seen from the comparison that the silanized silica/epoxy nanocomposite prepared by the present method is a competitive method where good thermal mechanical properties of epoxy can be enhanced through one-pot, solvent-free process and required only few percentages of silica.

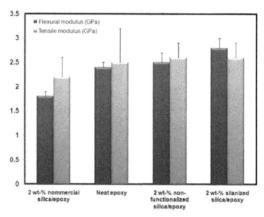

Fig. 9. Comparison between the flexural and tensile modulus of 2 wt-% silanized silica/epoxy nanocomposite and comparative references (neat epoxy, 2 wt-% non-functionalized silica/epoxy and 2 wt-% commercial silica/epoxy).

4. Conclusion

Oval shaped silica/epoxy nanocomposites with uniform silica dispersion and strong silica-epoxy adhesion were effectively and conveniently synthesized via the "Solvent-Free One-Pot Synthesis" method from TEOS, APTMS, ammonia solution, and epoxy compositions at 50°C. Small amount of silica incorporation in the nanocomposite (1-4 wt-% silica in epoxy composition) could enhance the property of silica/epoxy composite significantly. The lower T_g at higher silica loading is due to the decrease of the cross-linked density of epoxy network. The high performance silica/epoxy nanocomposites prepared using the present method exhibit better mechanical properties over neat epoxy (20% and 17% improvements on the flexural and tensile modulus) and commercial available silica/epoxy nanocomposite systems (36% improvement on the flexural modulus).

5. Acknowledgment

This work is funded by the Science and Engineering Research Council (SERC), A*STAR (Agency for Science, Technology and Research) under Grant No. 092 137 0013.

6. References

Deng, S., Ye, L., & Friedrich, K.J. (2007). Fracture Behaviours of Epoxy Nanocomposites with Nano-Silica at Low and Elevate Temperatures. *Journal of Materials Science*, Vol.42, No.8, (April 2007), pp. 2766-2774, ISSN 0022-2461

Preghenella, M., Pegoretti, A., & Migliaresi, C. (2005). Thermo-Mechanical Characterization of Fumed Silica-Epoxy Nanocomposites. *Polymer*, Vol.46, No.26, (December 2005), pp. 12065-12072, ISSN 0032-3896

Hsiue, G.H., Liu, Y.L., & Liao, H.H. (2001). Flame-Retardant Epoxy Resins: An Approach from Organic-Inorganic Hybrid Nanocomposites. *Journal of Polymer Science*, Vol.39, No.7, (April 2001), pp. 986-996, ISSN 1099-0518

Fu, S.-Y., Feng, X.-Q., Lauke, B., & Mai, Y.-W. (2008). Effects of Particle size, Particle/Matrix Interface Adhesion and Particle Loading on Mechanical Properties of Particulate-Polymer Composites. *Composites Part B*, Vol.39, No.6, (September 2008), pp. 933-961, ISSN 1359-8368

Ragosta, G., Abbate, M., Musto, P., Scarinzi, G., & Mascia, L. (2005). Epoxy-Silica Particulate Nanocomposites:Chemical Interactions, Reinforcement and Fracture Toughness. *Polymer*, Vol.46, No.23, (November 2005), pp. 10506-10516, ISSN 0032-3896

Kwon, S-C., Adachi, T., & Araki, W. (2008). Temperature Dependence of Fracture Toughness of Silica/Epoxy Composites: Related to Microstructure of Nano- and Micro-Particles Packing. *Composites Part B*, Vol.39, No.5, (July 2008), pp. 773-781, ISSN 1359-8368

Adachi, T., Osaki, M., Araki, W., & Kwon, S.-C. (2008). Fracture Toughness of Nano- and Micro-Spherical Silica-Particle-Filled Epoxy Composites. *Acta Materialia*, Vol.56, No.9, (May 2008), pp. 2101-2109, ISSN 1359-6454

Zhang, H., Zhang, Z., Friedrich, K., & Eger, C. (2006). Property Improvements of In Situ Epoxy Nanocomposites with Reduced Interparticle Distance at High Nanosilica Content. *Acta Materialia*, Vol.54, No.7, (April 2006), pp. 1833-1842, ISSN 1359-6454

Chen, Q., Chasiotis, I., Chen, C., & Roy, A. (2008). Nanoscale and Effective Mechanical Behavior and Fracture of Silica Nanocomposites. *Composites Science and Technology*, Vol.68, No.15-16, (December 2008), pp. 3137-3144, ISSN 0266-3538

Wang, K., Chen, L., Wu, J., Toh, M.L., He, C., & Yee, A.F. (2005). Epoxy Nanocomposites with Highly Exfoliated Clay: Mechanical Properties and Fracture Mechanisms. *Macromolecules*, Vol.38, No.3, (January 2005), pp. 788-800, ISSN 0024-9297

Zhang, H., Tang, L.C., Zhang, Z., Friedrich, K., & Sprenger, S. (2008). Fracture Behaviours of In Situ Silica Nanoparticle-Filled Epoxy at Different Temperatures. *Polymer*, Vol.49, No.17, (August 2008), pp. 3816-3825, ISSN 0032-3896

Deng, S., Hou, M., & Ye, L. (2007). Temperature-Dependent Elastic Moduli of Epoxies Measured by DMA and Their Correlations to Mechanical Testing Data. *Polymer Testing*, Vol.26, No.6, (September 2007), pp. 803-813, ISSN 0142-9418

Liu, Y.-L., Hsu, C.Y., Wei, W.-L., & Jeng, R.-J. (2003). Preparation and Thermal Properties of Epoxy-Silica Nanocomposites from Nanoscale Colloidal Silica. *Polymer*, Vol.44, No.18, (August 2003), pp. 5159-5167, ISSN 0032-3896

Mascia, L., Prezzi, L., & Haworth, B. (2006). Substantiating the Role of Phase Bicontinuity and Interfacial Bonding in Epoxy-Silica Nanocomposites. *Journal of Materials Science*, Vol.41, No.4, (February 2006), pp. 1145-1155, ISSN 0022-2461

Huang, C.J., Fu, S.Y., Zhang, Y.H., Lauke, B., Li, L.F., & Ye, L. (2005). Cryohenic Properties of SiO_2/Epoxy Nanocomposites. *Cryogenics*, Vol.45, No.6, (June 2005), pp. 450-454, ISSN 0011-2275

Araki, W., Wada, S., & Adachi, T. (2008). Viscoelasticity of Epoxy Resin/Silica Hybrid Materials with an Acid Anhydride Curing Agent. *Journal of Applied Polymer Science*, Vol.108L, No.4, (May 2008), pp. 2421-2427, ISSN 0021-8995

Li, X., Cao, Z., Zhang, Z., & Dang, H. (2006). Surface-Modification In Situ of Nano-SiO_2 and Its Structure and Tribological Properties. *Applied Surface Science*, Vol.252, No.22, (September 2006), pp. 7856-7861, ISSN 0169-4332

12

Nanoparticles in Ancient Materials: The Metallic Lustre Decorations of Medieval Ceramics

Philippe Sciau

CEMES-CNRS, Université de Toulouse
France

1. Introduction

The scientific community has begun to focus on optical properties of metallic colloids since the early twentieth century with Gustav Mie's works (Mie 1908). However the use of their outstanding properties is much older and dates back to several millennia ago (Colomban 2009; Garcia 2011).

Investigations using various techniques showed that red glasses of the late Bronze Age (1200-1000 BCE) from Frattesina di Rovigo (Italy) were coloured thanks to the excitation of phasmon surface modes of copper nanoparticles (Angelini *et al.* 2004; Artioli *et al.* 2008). The protohistoric community of this region developed advanced glass-manufacturing technology and was able to induce the exsolution of metallic copper crystals in the top layer of glass by exposing the material to reducing conditions. The presence of copper nanoparticles and cuprous oxide (cuprite Cu_2O) had already been reported in Celtic red enamels dated from 400 to 100 BCE (Brun *et al.* 1991). The use of metallic particles for colouring glass spread during the Roman period. Most of the red tesserae used in Roman mosaics were made of glass containing a dispersion of copper nanocrystals (Brun, Mazerolles *et al.* 1991; Colomban *et al.* 2003; Ricciardi *et al.* 2009). In addition to the copper crystals, gold nanoparticles were identified in some red tesserae showing that other metallic nanocrystals were used during Roman times (Colomban, March *et al.* 2003). It is precisely the case of the well-known Roman Lycurgus Cup in glass dated from the 4th century CE and currently exhibited in the British Museum (Freestone *et al.* 2007). The glass of this cup is dichroic and resembles jade with an opaque greenish-yellow tone, but when light shines through the glass (transmitted light) it turns into a translucent ruby colour. It has been demonstrated that the spectacular colour change is caused by colloidal metal and more precisely by nanocrystals of a silver-gold alloy dispersed throughout the glassy matrix (Barber & Freestone 1990). A handful of other Roman glasses showing a dichroic effect were also reported and although the colour change is not so spectacular, the Lycurgus Cup is obviously the result of a good technical mastery of Roman glass-workers (Freestone, Meeks *et al.* 2007). The Roman craftsmen knew that glass could be red coloured and that unusual colour change effect generated by the addition of noble metal bearing material when the glass was molten could be engineered. Nevertheless, the difficulties in controlling the coloration process meant that relatively few glasses of this type were produced, and even fewer have survived.

During the Middle Ages, glass manufacturing expanded considerably, especially to address the demand for stained glass (Kurmann-Schwarz & Lautier 2009). This development was accompanied by an increase in the type of colloidal metal used for colouring glass (Perez-Villar *et al.* 2008; Rubio *et al.* 2009; Gimeno *et al.* 2010). This age also saw the emergence of lusterware, a special type of glazed ceramics, with striking optical effects again obtained from metallic nanoparticles (Caiger-Smith 1991; Pérez-Arantegui *et al.* 2001). Then the progress in glass chemistry during the Renaissance period (Simmons & Mysak 2010) and especially in modern times allowed for better tuning of coloration effects based on the surface plasmons of metallic nanoparticles (Gil *et al.* 2006; Hartland 2011).

The manufacturing process of red glass was used worldwide. The famous Satsuma glasses produced in Japan in the mid-19th century were obtained using a similar technique and their ruby colour comes also from the absorption properties of copper nanocrystals (Nakai *et al.* 1999). It is also the case of the famed red *flambé* and mixed blue-red Jun glazed porcelains from Song and Ming to Qing Chinese Dynasties (Wood 1999).

2. Lustre decorations of medieval ceramics

The lustre is a variety of glaze decoration on ceramics, which appears in medieval times as mentioned in the introduction. Like the ruby glass, the colour of the lustre decorations has a physical basis coming from metallic nanoparticles (Bobin *et al.* 2003; Colomban 2009; Lafait *et al.* 2009). However, lustres possess the particularity of having a colour which can change depending on the angle from which it is observed. An example of these types of ceramic decorations is given in Figure 1 (Mirguet, Roucau *et al.* 2009). The colour change under specular reflection is often spectacular and produces a very intense coloured metallic shine, which can be golden-yellow, blue, green, pink, etc … The density of nanoparticles in the top layers of glaze is higher than that for ruby glass and shows a structuration in depth, which can be more or less complex, as in the lustre of figure 1. This multi-layer structuration on the scale of wavelengths of visible light gives rise to interference phenomena and scattering through rough interfaces, which adds to the surface plasmon effect and strongly contributes to the observed colour. As pointed out by Jacques Lafait *et al.* in their paper concerning the physical colours in cultural heritage, the colours with structural origin are particularly striking and very brilliant. The understanding of these structural effects on optical properties (photonic crystals) is very recent and it is fascinating to see that Islamic potters were able to create such complex structure through empirical chemical means in order to exploit their outstanding optical properties. However before going into detail about these various aspects, a word about the historical context in order to have a few chronological and geographic references.

2.1 Historical context

The earliest lustred potteries were found in Mesopotamia and most of them originate from the site of the Abbasid Caliphs' palace of Samarra in present-day Iraq (Caiger-Smith 1991). This monumental palace-complex whose building was begun by Caliph Mu'tasim in 836 CE, was abandoned in 883 CE.

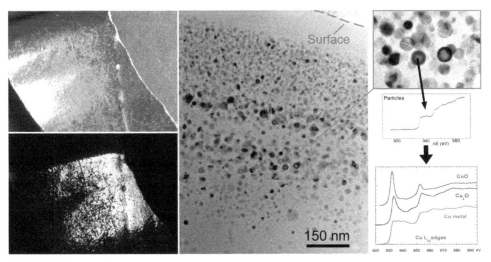

Fig. 1. 9th century CE lusterware from Mesopotamia (Susa). The colour changes from red (top left) to blue/green (bottom left) (Micrographs, courtesy D. Chabanne), nanoparticle distribution observed by TEM (in the centre) and on the right, the identification of particle nature (copper metal) by EELS. Details can be found in references (Chabanne 2005) and (Mirguet *et al.* 2009).

The early lustre manufacturing is therefore rather well dated, though it is currently assumed that first experiments may have occurred earlier, possibly in the time of Harun-al-Rashid (766-809 CE). Abbasid lustres were also found in some quantity in other Mesopotamian cities such as Baghdad, Basra, Kufa or Susa in present-day Iran. Theses cities are often presented as potential production centres, but it is yet an open question. Mesopotamian lustres were discovered outside this geographic area. Tiles with lustre decorations from Iraq were used in the partial reconstruction of the Kairouan Great Mosque (Tunisia), in the 9th century. Fragments have been found at Fustat, which was the main citadel of Lower Egypt in the 9th century. Shards have also been excavated from the site of the palace of Qal'a in Algeria, which was until 1052 CE the capital of the Hammamid princes. In fact, lustre decorations were certainly created in the early 9th century for courts and courtiers and seldom appeared in any other setting. For several centuries the lusterware kept its status of luxury tableware for princely courts.

The annexation of Egypt by the Fatimids (969 CE) led to profound modifications, not only on a governmental level, but also in the population. The Fatimid capital was transferred from Tunisia to al-Qahira, modern Cairo, and the old city of Fustat provided quarters for craftsman who worked for the new capital a few kilometres to the north. The demand for lustre by the new court led to the development of the local production. It is now attested that lustres were made in Fustat before the Fatimid period. However this production, often called pre-Fatimid, seems to have been very limited and of poor quality. The Egyptian production actually began with the arrival of the Fatimids, and during two centuries, a great deal of good quality lustre was being made reflecting the interests and cultural traditions of the new dynasty and its courtiers.

During the 12th century, lustre technique began to extend from Egypt to Syria and to Persia (present-day Iran). Craftsmen from Fustat allegedly brought the technique there during the decline of the Fatimid dynasty, which occurred in the middle 12th century. Concerning the diffusion in Persia, the subject is treated in detail in Oliver Watson's book (Watson 1985).

It seems that the technique appeared in the Occident (southern Spain) during the same period, as soon as the taifa emerged after the dissolution of the Spanish Umayyad caliphate. However, it is only under the Nasrid dynasty (1237-1492 CE) that the lustre technique really flourished in Spain. Its apogee, between the 14th–15th centuries, gave rise to the Hispano-Moorish ceramic, which was elaborated in the Valencia region up to the 18th century.

The technique found a new application during the Italian Renaissance (15th and 16th centuries) where Deruta and Gubbio became the most famous production centres of lustred glazed majolica (Padeletti *et al.* 2006). The main production centres with chronological data are summarized in the figure 2.

Fig. 2. Localization of the main centres of lustre productions.

2.2 Nanoparticle layer

Many studies were devoted to determining the elementary composition of glazes. The results obtained by the C2RMF lustre team, which analysed a significant corpus of specimens, are available online[1]. A review of the main results of the other studies can be found in Philippe Colomban's paper (Colomban 2009). The composition of glaze used in the lustre decoration is highly varied with alkaline and high lead glaze. There is not a specific composition. During the Abbasid times, the first productions used alkaline glazes but lead was then introduced with, in some cases, tin. The glazes used during the Fatimid epoch were mostly leaded alkaline. The Hispano-Moorish productions were characterized by high lead glaze containing a small amount of sodium and potassium. Renaissance decorations were also applied on leaded glazes. Significant composition variations were observed inside the same geographic area, which could be linked to a relative chronology, according to C2RMF's work. Several research groups are seeking to confirm this. The ceramic bodies are

[1]*Ceramics with metallic lustre decoration. A detailed knowledge of Islamic productions from 9th century until Renaissance.* D. Chabanne, M. Aucouturier, A. Bouquillon1, E. Darque-Ceretti, S. Makariou, X. Dectot, A. Faÿ-Hallé, D. Miroudot (2011), 2011arXiv1101.2321C

also various. Some lustre decorations were affixed on ceramic clay bodies, whereas others were deposited on siliceous pasta. The various associations of ceramic/glaze in relation to the different geographic areas and periods are listed in the online paper of the C2RMF lustre team (cf. note 1).

The composition of nanoparticles is much less diverse. Only copper and silver were used in all lustre decoration from the 9th century to now. On the other hand, the size, the shape and the spatial organisation can be very different (Fig. 3). One of the best tools for studying an organisation on a nanometre scale is the transmission electron microscopy (TEM). It is this technique which allowed for the first observation of nanocrystals in an archaeological lustre, a 13th century lustre of Hispano-Moorish period (Pérez-Arantegui, Molera *et al.* 2001). Although it is an abrasive technique, which is an obstacle for the investigation of museum pieces, several specimens from different periods and different geographic areas have been analysed by TEM (Borgia *et al.* 2002; Padeletti & Fermo 2003b; Pérez-Arantegui & Larrea 2003; Fredrickx *et al.* 2004; Padeletti & Fermo 2004; Roqué *et al.* 2007; Mirguet *et al.* 2008; Mirguet, Roucau *et al.* 2009; Sciau *et al.* 2009a). For some of them, sampling was limited to a few thousand cubic micrometres, using focused ion beam (FIB) techniques (Sciau *et al.* 2009b). However, since these ceramics are not conductors, a carbon deposit must be performed, which is not easy to carry out on museum pieces.

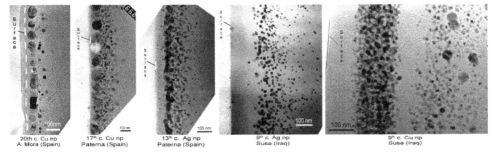

20th c. Cu np
A. Mora (Spain)   17th c. Cu np
Paterna (Spain)   13th c. Ag np
Paterna (Spain)   9th c. Ag np
Susa (Iraq)   9th c. Cu np
Susa (Iraq)

Fig. 3. A selection of bright field TEM images of various lustres from Mesopotamia (right) to Spain (left). More details can be found in the references (Chabanne 2005) and (Mirguet, Roucau *et al.* 2009).

Synchrotron radiation was used to obtain information on nanoparticles. In addition to the nature of nanoparticles, X-ray absorption fine structure (XAFS) measurements give interesting information on the presence of metallic ions (Ag^+, Cu^+, Cu^{2+}) in the glassy matrix (Padovani *et al.* 2006). In some cases, the size of the metallic oxide clusters can be estimated by fitting the extended X-ray absorption fine structure (EXAFS) spectra, whereas the average size of metallic nanoparticles can be deduced from the broadness of X-ray diffraction reflections. In addition, using glancing incidence X-ray diffraction (GIXRD) techniques, information can be obtained on structure and depth distribution of nanoparticles (Bontempi *et al.* 2006). Nevertheless, all these techniques give only partial information about the nanoparticle layers and require a synchrotron facility.

An alternative solution has been proposed using ion beam analysis (IBA) (Salomon *et al.* 2008; Pichon *et al.* 2010). The association of particle induced X-ray emission (PIXE) and elastic Rutherford backscattering spectrometry (RBS) allowed them to obtain significant data concerning the glassy matrix composition and the nature and the depth distribution of nanoparticles for a number of lustre ceramics, including valuable museum objects (Darque-Ceretti *et al.* 2005; Padeletti, Ingo *et al.* 2006; Chabanne *et al.* 2008; del Rio & Castaing 2010; del Rio *et al.* 2010). Whereas PIXE gives the chemical composition of the glassy matrix, RBS can provide detailed depth information. Thus, from a simulation of the experimental RBS spectrum, the nanoparticle distribution can be modelled. However, a calibration is necessary and for this, TEM observations are very useful (Chabanne, Bouquillon *et al.* 2008).

Several deductions can be made from all these investigations. The Abbasid and Fatimid samples observed by TEM showed a more complex structuration in depth than the Hispano-Moorish productions (Fig. 3). The particles of the first lustres are small with an average size of around 10-15 nm, and the thickness of the layer without particles below the glaze surface is superior to 100 nm (Chabanne 2005; Mirguet, Fredrickx *et al.* 2008). On the contrary, the Hispanic productions are characterized by a layer of big particles (50-100 nm) close to the glaze surface (Chabanne 2005; Mirguet, Roucau *et al.* 2009). This is even more pronounced for the 17th century lustres and modern replica made by Spanish artisans using a traditional process. Several Abbasid and Fatimid lustres present a partial structuring in depth such as shown in figure 3 and 4.

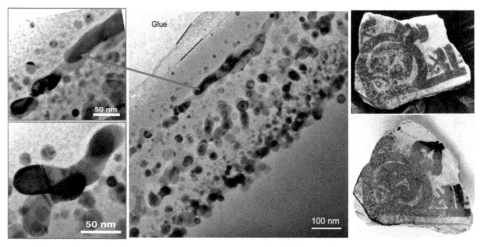

Fig. 4. Lustre from the Fatimid period (12th century CE) showing a partial multi layer silver nanoparticle organization (Mirguet, Roucau *et al.* 2009; Sciau, Mirguet *et al.* 2009a). On the right, the colour change from brown (scattering light) to pink (specular position) and on the left, magnifications of elongated silver particles ("metal worms").

TEM investigations even brought to light a lustre decoration from the Fatimid period with a very regular nanoparticle distribution in two well-separated layers (Fig. 5). Electron diffraction revealed that all particles were silver with a CFC structure (Sciau, Mirguet *et al.* 2009a). The electron energy loss spectroscopy (EELS) confirmed that silver is only present in the nanoparticle layers whereas copper is found everywhere in the glaze, however in the ionic form (Cu^{2+}). The green colour of the glaze indeed comes from the Cu^{2+} ions. The distance between the two layers is amazingly constant. With a value of around 430 nm, this distance is of the same order of magnitude as visible wavelengths. It results in that this double layer structure behaves as an optical network. A more in depth investigation of nanopaticles showed that the particles in the second layer are slightly larger and that their shapes are less spherical. Some particles have even coalesced forming larger particles. High resolution electron microscopy (HREM) showed that many of them have structural defects of a stacking fault type (Sciau, Mirguet *et al.* 2009a).

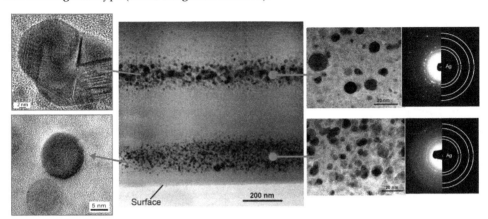

Fig. 5. Lustre from the Fatimid period (12th century CE) showing a well defined multi layer silver nanoparticle organization, with on the left, HREM images of nanoparticles of each layer and on the right, electron diffraction identification (Mirguet, Roucau *et al.* 2009; Sciau, Mirguet *et al.* 2009a).

2.3 Optical properties

Since the discovery of their astonishing colorimetric properties, several studies have sought to model the optical behaviour of lustre decorations.

Olivier Bobin was the first to carry out a theoretical investigation of optical properties of lusterware and thus to prove the role of the surface phasmon resonance of silver and/or copper nanocrystals in the colouring process (Bobin, Schvoerer *et al.* 2003). His modelling, using Mie's theory, and based on the copper-silver ratio, the particle size, the particle density and the nature of embedded glaze give rather good results for the colours observed in scattering light. In the modelling of the size effect on the surface plasmon resonance, two regimes are usually distinguished depending on the nanoparticle size range (Garcia 2011). For the small particles (smaller than light wavelength i.e. with a radius up to 50 nm), the particle can be properly described by a dielectric dipole. The size variation affects mainly the width and the intensity of the resonance band. On the other hand, the resonance wavelength is only slightly shifted. For the larger particles with a size comparable to the wavelength i.e. with a radius superior or equal to 50 nm, the dipole approximation is not sufficient and multipolar terms must be added leading to the splitting of the resonance band into several peaks: two peaks for quadrupole, three peaks for an octopole, etc ... (Kreibig *et al.* 1987). The metallic particles present in the lusterware are seldom superior to 50 nm in radius (cf. § 2.2); also the dipole approximation is sufficient to describe the surface plasmon resonance. Nevertheless, the size dispersion is large and that has as a consequence a significant broadening of the absorption band. Since the restoring force for surface plasmons is related to the charge accumulated at the surface, it is influenced as well by the particle shape. With elongated particles, the absorption band is split into two bands: the transversal and longitudinal bands. The frequency shift is proportional to the ratio between longitudinal and transversal lengths. While the resonant frequency of transversal plasmons falls at about the same position as for spherical particles (actually, at wavelengths slightly smaller), the resonance of longitudinal plasmons shifts towards larger wavelengths when the ratio increases. The intensity of the longitudinal plasmon band increases with the ratio while the one of the transversal band decreases. For the lustre of figure 4, the shape effect must be taken into account, but in this case other effects such as the interference phenomena must also be considered. It is obvious that the interferences have a significant influence on the colour of metallic shine (the specular position) for the lustres having a partial multi-layer structuration. The study of the lusterware with the double layer (Fig. 5) demonstrated that the interferences are at the origin of the colour variation of the metallic shine from blue to green (Sciau, Mirguet et al. 2009a).

The first model taking into account the interference phenomena was proposed by Vincent Reillon (Reillon & Berthier 2006). However, the modelling of such a complex system was not easy and it was only recently that a model integrating all phenomena (surface plasmon absorption, interference and scattering) was published (Reillon 2008; Reillon et al. 2010). From this model, it is now possible to correctly simulate the reflection spectra recorded as well in the specular direction as in the scattering directions. The evolution of the colour between the specular and the diffusion directions can be perfectly calculated. Thus, the key parameters determining colour behaviour are (Lafait, Berthier et al. 2009):

- in the specular direction, interference phenomena play a major role with the key parameters being the number of layers, the optical index and the thickness of each layer,
- in the scattering cone, plasmon absorption is predominant and the key parameters are the kind of metal, the metal volume fraction, the particle size and shape, and the glass matrix composition,
- in the intermediate cone, there is a transition between a coherent component (dominant close to the specular direction) and a scattered incoherent component (dominant close to the scattering cone).

The colour behaviour of lustre is schematized in figure 6. To simulate the experimental spectra, the modelling uses a schematic representation of the multilayer structure of lustre decoration. Hence inversely from a modelling of a set of experimental spectra collected from different directions, it is possible to obtain significant information on the nanoparticle distribution.

2.4 Manufacturing process

Several descriptions of the glazing technique were proposed (Pérez-Arantegui, Molera et al. 2001; Padeletti & Fermo 2003a; Colomban & Truong 2004; Pradell et al. 2005; Roqué et al. 2005; Pradell et al. 2006; Roqué et al. 2008; Colomban 2009) on the basis of experimental evidence and on information extracted from the transcription of ancient recipes (Abu al Qâsem 14th century, Picolpasso 16th century, Deck and Bertan 19th century and Artigas 20th century).

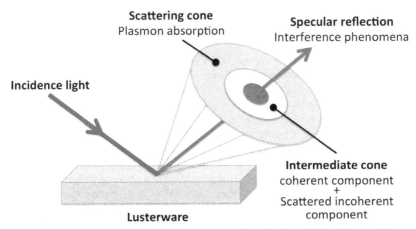

Fig. 6. Schematic representation of the light scattered and reflected by a lustre glaze decoration from Reillon's modelling (Reillon 2008; Reillon, Berthier *et al.* 2010).

Lustre decorations were obtained by applying a mixture of a paint, which contained copper and silver salts, water and more or less vinegar and lye, onto a glazed ceramic, which was subsequently annealed in a reducing atmosphere. Inside the kiln, the raw paint reacted with the glaze surface, and after firing, the remaining paint was washed off, revealing the lustre decoration beneath.

The role of ionic exchange was first identified in the lustre formation process (Smith *et al.* 2003; Pradell, Molera *et al.* 2005). From the analysis of medieval ceramics, some interesting trends and information were obtained concerning the driving force responsible for the diffusion of silver and copper ions into the glaze, consisting of an "ionic exchange" of Ag^+ and Cu^+ with Na^+ and K^+. This type of ionic exchange is a well-known mechanism in glasses and glazes (Pradell, Molera *et al.* 2006). When glasses are immersed in a solution containing copper and/or silver molten salt (typically sulphates or nitrates), atomic exchanges take place and the alkalis (K^+ and Na^+) of the glaze are replaced by Ag^+ and Cu^+ ions of the solution. Accurate chemical analysis of medieval ceramics showed a clear inverse correlation between the metal components of the lustre decorations (Cu and Ag) and the amount of Na and K in the glaze. This correlation has been found in early Islamic lusterwares from Iraq (9th century CE) as well as in late Hispano-Moorish lusterwares from Paterna (13th-17th centuries CE). In addition, reproductions and ancient lustre surfaces were observed by means of white light interferometry, atomic forces microscopy, X-ray diffraction and electron microprobe (Roqué, Pradell *et al.* 2005). These observations showed that lustre layers do not appear as superimposed layers on the top of the glaze, but rather as a surface roughness resulting in the nanocrystals growth inside the glassy matrix. The surface roughness increases during the formation process as a result of metal nanoparticle growth. The lustre formation process involves a two step process: ion exchange and crystallization (nucleation and crystal growth) of copper and silver metallic nanoparticles inside the glassy matrix (Roqué, Molera *et al.* 2008).

More recently, it has been assumed that the burning of organic residues could be used to control the surface temperature and embedded metal dispersion allowing one to set the final lustre colour in different places of the same item (Mirguet, Fredrickx *et al.* 2008). The strong temperature gradient arising from the combustion of surface acetate residues could control the self-organization of the metal particles leading to light diffraction. The multilayer particle distribution of lustre could be explained by special firing cycles where repeated heat flashes provoked by surface organic residue combustion make it possible to control the size, the shape and the distribution of nanoparticles. In several cases, very elongated silver particles ("metal worms") were observed by TEM. It is particularly well marked in the Fatimid lustre of Figure 4, which shows 3 layers of metal worm particles separated by areas of lower density in nanoparticles. The particles of these intermediate zones are smaller and spherical. The nanoparticle coalescences forming larger particles are also observed in the second layer of the double layer lustre (Fig. 5). High-resolution electron microscopy (HREM) showed that many of them have structural defects of a stacking fault type. It is not the case of nanoparticles of the first layer, which are smaller and quite spherical. The nanoparticles of the two layers are different enough to conclude that they were not formed under the same thermal conditions. Nanoparticle coalescences were also found in the two other lustres of figure 3 with silver particles (Abbasid from Susa and 13th century from Paterna).

The elongated shape of silver particles proves that the temperatures close to that of the silver melting point (~960°C) were reached. Diffusion-controlled phenomena cannot lead to such worm-shapes. The Melting of metallic silver demonstrates that glaze surface temperatures close to 1000°C were achieved at the peak temperature cycle. This is consistent with an increase in temperature obtained thanks to the combustion of organic residue. Except for some modern replica (Mora productions, Fig. 3), worm-shapes were not observed for copper particles, which indicated that the melting point of copper (~1080°C) was not reached.

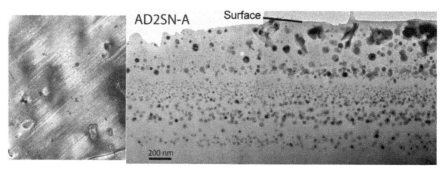

Fig. 7. Lustre creation made by Eva Haudum (Colomban 2009).

3. Conclusion

The review of present knowledge on ancient materials, in which the size and distribution as well as the reduction of metallic nanoprecipitates were organized, shows the high level of empirical control carried out by ancient potters and glass-makers. The ancient potters were certainly not aware of material nanostructuration, since they did not have any nanostructure checking facilities. However, their know-how allowed for the realization of two perfectly well separated layers (Sciau, Mirguet *et al.* 2009a). This special nanostructuration has been up to now observed on only one sample. It is important to keep in mind that this sample might have been accidently obtained or that the first layer (uppermost one) was created because the first visual aspect did not suit the craftsman. However, other lustres of the Fatimid (12th century CE) and Abbasid (9th - 10th centuries CE) periods show a partial organisation in multi-layered nanoparticles. So there is a good chance that this sample is not an isolated case and that this technology was developed by the Islamic craftsmen to exploit the complex nano-optical properties of multi layered particles.

The optical modelling performed by Vincent Reillon confirmed the role of interference in the colour of metallic shine reflection (specular direction). It is obvious that the multi-layer organisation of nanoparticles, with distances between layers comparable to the visible wavelengths, strengthens the interference effects allowing for the obtainment of very bright iridescent colours from a large palette of hues. It is likely that it was the aim of the Abbasid and Fatimid potters to develop empirical processes for creating such multi-layered nanoparticles through a know-how, which has nowadays partially been lost. Modern artists (S. Çizer, E. Haudum ...) search to recreate lustre decorations with strong iridescent effects and bright colours using modern kilns permitting alternative oxidising (oxygen flux) and reducing (CO flux) phases during the firing (Colomban 2009). Their best recreations exhibit the partial multi-layer structure observed in the Abbasid and Fatimid artefacts, but maybe with a weaker organisation and sometimes residues of raw mixture as observed in figure 7.

4. Acknowledgments

I would like to thank A. Bouquillon and D. de Camaret for their help in the writing of this chapter as well as colleagues who allowed me to develop this activity in CEMES-CNRS lab (G. BenAssayag, B. Brunetti, R. Carles, Ph. Colomban, H. Dexpert, C. Roucau). I thank all under-graduate and PhD students who experimentally contributed to this paper, especially D. Chabanne, C. Dejoie, Y. Leon, C. Mirguet and J. Roqué-Rosell. Research supported by the *Conseil Régional de Midi-Pyrénées* under several contracts.

5. References

Angelini, I.; Artioli, G.; Bellintani, P.; Diella, V.; Gemmi, M.; Polla, A. & Rossi, A. (2004). Chemical analyses of bronze age glasses from Frattesina di Rovigo, northern Italy. *Journal of Archaeological Science*, Vol.31, No.8, pp. 1175-1184, ISSN 0305-4403

Artioli, G.; Angelini, I. & Polla, A. (2008). Crystals and phase transitions in protohistoric glass materials. *Phase Transitions*, Vol.81, No.2-3, pp. 233-252, ISSN 0141-1594

Barber, D. J. & Freestone, I. C. (1990). An investigation of the origin of the color of Lycurgus cup by analytical transmission electron-microscopy. *Archaeometry*, Vol.32, (February 1990), pp. 33-45, ISSN 0003-813X

Bobin, O.; Schvoerer, M.; Miane, J. L. & Fabre, J. F. (2003). Coloured metallic shine associated to lustre decoration of glazed ceramics: a theoretical analysis of the optical properties. *Journal of Non-Crystalline Solids*, Vol.332, No.1-3, (December 2003), pp. 28-34, ISSN 0022-3093

Bontempi, E.; Colombi, P.; Depero, L. E.; Cartechini, L.; Brunetti, B. G. & Sgamellotti, A. (2006). Glancing-incidence X-ray diffraction of Ag nanoparticles in gold lustre decoration of Italian Renaissance pottery. *Applied Physics A: Materials Science & Processing*, Vol.83, No.4, (Jun 2006), pp. 543-546, ISSN 0947-8396

Borgia, I.; Brunetti, B.; Mariani, I.; Sgamellotti, A.; Cariati, F.; Fermo, P.; Mellini, M.; Viti, C. & Padeletti, G. (2002). Heterogeneous distribution of metal nanocrystals in glazes of historical pottery. *Applied Surface Science*, Vol.185, No.3-4, (January 2002), pp. 206-216, ISSN 0169-4332

Brun, N.; Mazerolles, L. & Pernot, M. (1991). Microstructure of opaque red glass containing copper. *Journal of Materials Science Letters*, Vol.10, No.23, (December 1991), pp. 1418-1420, ISSN 0261-8028

Caiger-Smith, A. (1991). *Lustre pottery. Technique, tradition and innovation in Islam and the Western World*, New Amsterdam Books, ISBN 1-56131-030-1, New York, USA.

Chabanne, D. (2005). *Le décor de lustre métallique des céramiques glaçurées (IXème-XVIIème siècles). Matériaux, couleurs et techniques*. PhD thesis, University Bordeaux 3.

Chabanne, D.; Bouquillon, A.; Aucouturier, M.; Dectot, X. & Padeletti, G. (2008). Physico-chemical analyses of Hipano-Moresque lustred ceramic: a precursor for Italian majolica? *Applied Physics A: Materials Science & Processing*, Vol.92, No.1, (July), pp. 11-18, ISSN 0947-8396

Colomban, P. (2009). The use of metal nanoparticles to produce yellow, red and iridescent colour, from bronze age to present times in lustre pottery and glass: solid state chemistry, spectroscopy and nanostructure. *Journal of Nano Research*, Vol.8, pp. 109-132, ISSN 1662-5250

Colomban, P.; March, G.; Mazerolles, L.; Karmous, T.; Ayed, N.; Ennabli, A. & Slim, H. (2003). Raman identification of materials used for jewellery and mosaics in Ifriqiya. *Journal of Raman Spectroscopy*, Vol.34, No.3, (March 2003), pp. 205-213, ISSN 0377-0486

Colomban, P. & Truong, C. (2004). Non-destructive raman study of the glazing technique in lustre potteries and faience (9-14th centuries): silver ions, nanoclusters microstructure and processing. *Journal of Raman Spectroscopy*, Vol.35, No.3, (March 2004), pp. 195-207, ISSN 0377-0486

Darque-Ceretti, E.; Hélary, D.; Bouquillon, A. & Aucouturier, M. (2005). Gold like lustre: nanometric surface treatment for decoration of glazed ceramics in ancient Islam, Moresque Spain and Renaissance Italy. *Surface Engineering*, Vol.31, No.5-6, (December 2003), pp. 352-358, ISSN 0267-0844

del Rio, A. P. & Castaing, J. (2010). Lustre decorated ceramics from a 15th-16th century production in Seville. *Archaeometry*, Vol.52, (February 2010), pp. 83-98, ISSN 0003-813X

del Rio, A. P.; Roehrs, S.; Aucouturier, M.; Castaing, J. & Bouquillon, A. (2010). Medinal Al-Zahra lustre ceramics: 10th century local nanotechnology or importation from middle east. *Arabian Journal for Science and Engineering*, Vol.35, No.1C, (Jun 2010), pp. 157-168, ISSN 1319-8025

Fredrickx, P.; Hélary, D.; Schryvers, D. & Darque-Ceretti, E. (2004). A TEM study of nanoparticles in lustre glazes. *Applied Physics A: Materials Science & Processing*, Vol.79, No.2, (July 2004), pp. 283-288, ISSN 0947-8396

Freestone, I.; Meeks, N.; Sax, M. & Higgitt, C. (2007). The Lycurgus Cup - A Roman nanotechnology. *Gold Bulletin*, Vol.40, No.4, pp. 270-277, ISSN 0017-1557

Garcia, M. A. (2011). Surface plasmons in metallic nanoparticles: fundamentals and applications. *Journal of Physics D: Applied Physics*, Vol.44, No.28, (July 2011), pp. 283001, ISSN 0022-3727

Gil, C.; Villegas, M. A. & Navarro, J. M. F. (2006). TEM monitoring of silver nanoparticles formation on the surface of lead crystal glass. *Applied Surface Science*, Vol.253, No.4, (December 2006), pp. 1882-1888, ISSN 0169-4332

Gimeno, D.; Aulinas, M.; Bazzocchi, F.; Fernandez-Turiel, J. L.; Garcia-Valles, M.; Novembre, D.; Basso, E.; Messiga, B.; Riccardi, M. P.; Tarozzi, C. & Mendera, M. (2010). Chemical characterization of the stained glass window from the rose window, Siena Duomo (Italy, 1288-1289). *Boletin De La Sociedad Espanola De Ceramica Y Vidrio*, Vol.49, No.3, (May-Jun 2010), pp. 205-213, ISSN 0366-3175

Hartland, G. V. (2011). Optical Studies of Dynamics in Noble Metal Nanostructures. *Chemical Reviews*, Vol.111, No.6, (Jun 2011), pp. 3858-3887, ISSN 0009-2665

Kreibig, U.; Schmitz, B. & Breuer, H. D. (1987). Separation of plasmon-polariton modes of small metal particles *Physical Review B*, Vol.36, No.9, (September 1987), pp. 5027-5030, ISSN 0163-1829

Kurmann-Schwarz, B. & Lautier, C. (2009). The Medieval stained-glass window in Europe: 10 years of abundant research. *Perspective-La Revue De L Inha*, No.1, (March 2009), pp. 99-130, ISSN 1777-7852

Lafait, J.; Berthier, S.; Andraud, C.; Reillon, V. & Boulenguez, J. (2009). Physical colors in cultural heritage: surface plasmons in glass. *Comptes Rendus Physique*, Vol.10, No.7, (September 2009), pp. 649-659, ISSN 1631-0705

Mie, G. (1908). Beiträge zur optik trüber medien, speziell kolloidaler metallösungen. *Annalen der Physik*, Vol.25, No.3, (March 1908), pp. 377-445, ISSN 0003-3804

Mirguet, C.; Fredrickx, P.; Sciau, P. & Colomban, P. (2008). Origin of the self-organisation of $Cu°/Ag°$ nanoparticles in ancient lustre pottery. A TEM study. *Phase Transitions*, Vol.81, No.2-3, pp. 253-266, ISSN 0141-1594

Mirguet, C.; Roucau, C. & Sciau, P. (2009). Transmission electron microscopy a powderful means to investigate the glazed coating of ancient ceramics. *Journal of Nano Research*, Vol.8, pp. 141-146, ISSN 1662-5250

Nakai, I.; Numako, C.; Hosono, H. & Yamasaki, K. (1999). Origin of the red color of satsuma copper-ruby glass as determined by EXAFS and optical absorption spectroscopy. *Journal of the American Ceramic Society*, Vol.82, No.3, (March 1999), pp. 689-695, ISSN 0002-7820

Padeletti, G. & Fermo, P. (2003a). How the masters in Umbria, Italy, generated and used nanoparticles in art fabrication during the Renaissance period. *Applied Physics A: Solids and Surfaces*, Vol.76, pp. 515-525,

Padeletti, G. & Fermo, P. (2003b). How the masters in Umbria, Italy, generated and used nanoparticles in art fabrication during the Renaissance period. *Applied Physics a-*

Materials Science & Processing, Vol.76, No.4, (March 2003), pp. 515-525, ISSN 0947-8396

Padeletti, G. & Fermo, P. (2004). Production of gold and ruby-red lustres in Gubbio (Umbria, Italy) during the renaissance period. *Applied Physics A: Materials Science & Processing*, Vol.79, No.2, (July 2004), pp. 241-245, ISSN 0947-8396

Padeletti, G.; Ingo, G. M.; Bouquillon, A.; Aucouturier, M.; Roehrs, S. & Fermo, P. (2006). First-time observation of Mastro Giorgio materpieces by means of non-destructive techniques. *Applied Physics A: Materials Science & Processing*, Vol.83, No.4, (Jun 2006), pp. 475-483, ISSN 0947-8396

Padovani, S.; Puzzovio, D.; Mazzoldi, P.; Borgia, I.; Sgamellotti, A.; Brunetti, B. G.; Cartechini, L.; D'Acapito, F.; Maurizio, C.; Shokouhi, F.; Oliaiy, P.; Rahighi, J.; Lamehi-Rachti, M. & Pantos, E. (2006). XAFS study of copper and silver nanoparticles in glazes of medieval middle-east lustreware (10th-13th century). *Applied Physics A: Materials Science & Processing*, Vol.83, No.4, (Jun 2006), pp. 521-528, ISSN 0947-8396

Pérez-Arantegui, J. & Larrea, A. (2003). The secret of early nanomaterials is revealed, thanks to transmission electron microscopy. *Trends in Analytical Chemistry*, Vol.22, No.5, (May 2003), pp. 327-329, ISSN 0165-9936

Pérez-Arantegui, J.; Molera, J.; Larrea, A.; Pradell, T.; Vendrell-Saz, M.; Borgia, I.; Brunetti, B. G.; Cariati, F.; Fermo, P.; Mellini, M.; Sgamellotti, A. & Viti, C. (2001). Luster pottery from the thirteenth century to the sixteenth century: a nanostructured thin metallic film. *Journal of the American Ceramic Society*, Vol.84, No.2, (February 2001), pp. 442-446, ISSN 0002-7820

Perez-Villar, S.; Rubio, J. & Oteo, J. L. (2008). Study of color and structural changes in silver painted medieval glasses. *Journal of Non-Crystalline Solids*, Vol.354, No.17, (April 2008), pp. 1833-1844, ISSN 0022-3093

Pichon, L.; Beck, L.; Walter, P.; Moignard, B. & Guillou, T. (2010). A new mapping acquisition and processing system for simultaneous PIXE-RBS analysis external beam. *Nuclear Instruments & Methods in Physics Research, Section B: Beam Interactions with Materials and Atoms*, Vol.268, No.11-12, (Jun 2010), pp. 2028-2033, ISSN 0168-583X

Pradell, T.; Molera, J.; Bayes, C. & Roura, P. (2006). Luster decoration of ceramics: mechanisms of metallic luster formation. *Applied Physics A: Materials Science & Processing*, Vol.83, No.2, (May 2006), pp. 203-208, ISSN 0947-8396

Pradell, T.; Molera, J.; Roqué, J.; Vendrell-Saz, M.; Smith, A. D.; Pantos, E. & Crespo, D. (2005). Ionic-exchange mechanism in the formation of medieval luster decorations. *Journal of the American Ceramic Society*, Vol.88, No.5, (May 2005), pp. 1281-1289, ISSN 0002-7820

Reillon, V. (2008). *Caractérisation et modélisation des propriétés optiques des ceramiques lustrées.* PhD thesis, Université Pierre et Marie Curie – Paris 6.

Reillon, V. & Berthier, S. (2006). Modelization of the optical and colorimetric properties of lustred ceramics. *Applied Physics A: Materials Science & Processing*, Vol.83, No.2, (May 2006), pp. 257-265, ISSN 0947-8396

Reillon, V.; Berthier, S. & Andraud, C. (2010). Optical properties of lustred ceramics: complete modelling of the actual structure. *Applied Physics A: Materials Science & Processing*, Vol.100, No.3, (September 2010), pp. 901-910, ISSN 0947-8396

Ricciardi, P.; Colomban, P.; Tournié, A.; Macchiarola, M. & Ayed, N. (2009). A non-invasive study of Roman Age mosaic glass tesserae by means of Raman spectroscopy. *Journal of Archaeological Science*, Vol.36, No.11, (November 2009), pp. 2551-2559, ISSN 0305-4403

Roqué, J.; Molera, J.; Cepria, G.; Vendrell-Saz, M. & Perez-Arantegui, J. (2008). Analytical study of the behaviour of some ingredients used in lustre ceramic decorations following different recipes. *Phase Transitions*, Vol.81, No.2-3, pp. 267-282, ISSN 0141-1594

Roqué, J.; Molera, J.; Pérez-Arantegui, J.; Calabuig, C.; Portillo, J. & Vendrell-Saz, M. (2007). Lustre colour ans shine from the Olleries xiques workshop in Paterna (Spain), 13th century AD: Nanostructure, chemical composition and annealing conditions. *Archaeometry*, Vol.49, No.3, (August 2007), pp. 511-528, ISSN 0003-813X

Roqué, J.; Pradell, T.; Molera, J. & Vendrell-Saz, M. (2005). Evidence of the nucleation and growth of the metal Cu and Ag nanoparticles in lustre: AFM surface characterization. *Journal of Non-Crystalline Solids*, Vol.351, No.6-7, (March 2005), pp. 568-575, ISSN 0022-3093

Rubio, F.; Perez-Villar, S.; Garrido, M. A.; Rubio, J. & Oteo, J. L. (2009). Application of Gradient and Confocal Raman Spectroscopy to Analyze Silver Nanoparticle Diffusion in Medieval Glasses. *Journal of Nano Research*, Vol.8, pp. 89-97, ISSN 1662-5250

Salomon, J.; Dran, J.-C.; Guillou, T.; Moignard, B.; Pichon, L.; Walter, P. & Mathis, F. (2008). Ion-beam analysis for cultural heritage on the AGLAE facility: impact of PIXE/RBS combination. *Applied Physics A: Materials Science & Processing*, Vol.92, No.1, (July 2008), pp. 43-50, ISSN 0947-8396

Sciau, P.; Mirguet, C.; Roucau, C.; Chabanne, D. & Schvoerer, M. (2009a). Double nanoparticle layer in the 12th century lustreware decoration: accident or technological mastery. *Journal of Nano Research*, Vol.8, pp. 133-139, ISSN 1662-5250

Sciau, P.; Salles, P.; Roucau, C.; Mehta, A. & Benassayag, G. (2009b). Applications of focused ion beam for preparation of specimens of ancient ceramic for electron microscopy and synchrotron X-ray studies. *Micron*, Vol.40, No.5-6, (July-August 2009), pp. 597-604, ISSN 0968-4328

Simmons, C. T. & Mysak, L. A. (2010). Transmissive properties of Medieval and Renaissance stained glass in European churches. *Architectural Science Review*, Vol.53, No.2, (May 2010), pp. 251-274, ISSN 0003-8628

Smith, A. D.; Pradell, T.; Molera, J.; Vendrell-Saz, M.; Marcus, M. A. & Pantos, E. (2003). MicroEXAFS study into the oxidation states of copper coloured Hipano-Moresque Lustre decorations. *Journal de Physique IV*, Vol.104, (March 2003), pp. 519-522, ISSN 1155-4339

Watson, O. (1985). *Persian lustre ware*, Faber and Faber, ISBN 9780571132355, London, UK.

Wood, N. (1999). *Chinese glazes, their origins, chemistry and recreation*, University of Pennsylvania Press, ISBN 978-0-8122-3476-3, Philadelphia, USA.

13

Low Energy Emulsification Methods for Nanoparticles Synthesis

Veronique Sadtler[1], Johanna M. Galindo-Alvarez[1]
and Emmanuelle Marie –Bégué[2]
[1]Laboratoire Réactions et Génie des Procédés – GEMICO,
CNRS-Nancy Université, Nancy
[2]UMR 8640 CNRS-ENS-UPMC, Ecole Normale Supérieure,
Paris,
France

1. Introduction

Nanoparticles synthesis by miniemulsion polymerization produces materials that are not obtainable by means of other techniques such as conventional emulsion polymerization. The reason is that, in miniemulsion polymerization, particles are mainly formed by droplet nucleation (Asua, 2002). However, the high energy requirement for preparation of nano-emulsions by traditional methods (Mason et al., 2006; Solans et al., 2005) has precluded widespread use and commercialization.

Nanoemulsions, also referred as miniemulsions or ultrafine emulsions, compose a particular class of emulsions consisting of colloidal dispersions, transparent or bluish for the smallest droplet sizes between 20–100 nm, or milky for sizes up to 500 nm (Solans et al., 2002). In opposition to microemulsions, these systems are thermodynamically unstable, and the droplet size tends to increase with time before phase separation. Nevertheless, the very small initial droplet size makes them kinetically stable (Tadros et al., 2004).

As nanoemulsions are non-equilibrated systems, external energy is required for their preparation. Two generating processes are reported in the literature. In the first case, high mechanical energy is applied during emulsification, generally by using high shear stirring, high pressure homogenizers and/or ultrasound generators. On the contrary, the lower energy method, or condensation method, is based on the phase transitions taking place during the emulsification process (Lamaallam et al., 2005; Solans, et al., 2002; Tadros, et al., 2004). These phase transitions result from changes in the spontaneous curvature of the surfactant and can be achieved (i) at constant composition by changing the spontaneous curvature of non-ionic surfactants with temperature, the well-known Phase Inversion Temperature, PIT, widely used in industry (Izquierdo et al., 2005; K. Shinoda & Saito, 1968)or (ii) at constant temperature by varying the composition of the system by the Emulsion Inversion Point (EIP) method (Forgiarini et al., 2001; Pey et al., 2006; Porras et al., 2008).

Thus, nanoemulsions are specially formulated heterophase systems where stable nanodroplets (with a diameter lower than 500 nm) of one monomer phase are dispersed in a second continuous phase before polymerization takes place, often following a radical mechanism. Since its introduction, this approach has extended the classical emulsion radical polymerization, as ideally each nanodroplet could be regarded as an individual batch reactor, a nanoreactor. Indeed, when (oligo) radicals are generated in the continuous phase, nanodroplets compete with micelles for their capture. In addition, the amount of surfactant added in the feed is usually adjusted so as to minimize (or avoid) the presence of micelles in the continuous phase (Anton et al., 2008; Antonietti & Landfester, 2002).

In those conditions, it could be a good approximation to consider that each droplet behaves as an independent reaction vessel, a hypothetical bulk state where the continuous phase may still transport initiators, side products and heat. Thus, miniemulsion polymerization allows preparing water-based formulated polymers with high solid contents. Additionally this particular mode of design of nanoparticles becomes an advantage, since the chemical composition and colloidal characteristics of the initial nanoemulsion can be used to prepare polymer nanoparticles by "miniemulsion polymerization" of the monomer contained in the oil droplets. The nanoemulsions used for that purpose are mainly prepared by high-energy emulsification methods (Asua, 2002). The aim of this chapter is to show the EIP Method and the Near – PIT concept as a tool to produce miniemulsion templates for miniemulsion polymerization.

2. Nanoparticles by Emulsion Inversion Point (EIP) method

Studies showed that nanoemulsions with very small droplet sizes can be obtained through low-energy methods if, during the emulsification process, the oil is completely dissolved in a single phase, like a bicontinuous microemulsion or a lamellar crystalline phase (Mohlin et al., 2003; Rang & Miller, 1999).The further evolution of the system led to the dislocation of this continuous phase into small nanodroplets. For instance, in the EIP method, the addition of water to a system of water/oil/surfactant forming a lamellar phase increases the hydratation degree of the surfactant polar head thereby increasing its spontaneous curvature. The lamellar phase is disrupted, and the oil, which was initially dissolved, forms small droplets in the size order of the thickness of the hydrophobic layer (See Figure 1). Such methods of nanoemulsion preparation have received increasing attention (Maestro et al., 2008), since even active molecule (i.e. lidocaïne) encapsulation in emulsions is achievable by these protocols (Sadurní et al., 2005).

As already mentioned, nanoemulsions can be used to prepare polymer nanoparticles by miniemulsion polymerization of the monomer contained in the oil droplets. The nanoemulsions used for that purpose are mainly prepared by sonifiers and high-pressure homogenizers (Asua, 2002). Only a few studies described the preparation of nanoparticles from nanoemulsions obtained by condensation methods (Calderó et al., 2011; Isabel Solè et al., 2010; Liat Spernath & Magdassi, 2007; L. Spernath et al., 2009).

In this section the formation by EIP, of monomer-in-water nanoemulsions, followed by their conversion in polymer nanoparticles will be considered (Sadtler et al., 2010). For this purpose, the water/Brij 98/styrene system was chosen. Brij compounds are POE-based non ionic surfactants which are commonly used for biomedical applications. Water always

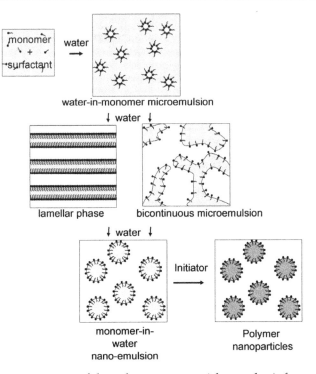

Fig. 1. Schematic representation of the polymer nanoparticles synthesis from nanoemulsions prepared by the EIP method (Adapted from Sadtler, et al., 2010)

contained 0.1 M of NaCl and a small quantity of hexadecane was always solubilized in styrene (in the 5:95 ratio) to avoid Ostwald ripening after nanoemulsion formation (Kabal'nov et al., 1987; Marie et al., 2007). Because lamellar liquid crystalline phase and/or bicontinuous microemulsions are necessary to generate nanoemulsions, the partial phase diagram of the system was determined prior to the nanoemulsion preparation and its miniemulsion polymerization. The phase diagram has been carried out at 50°C due to the suitable temperature to styrene polymerization (Figure 2).

The phase behaviour was found to be in good agreement with the one of other systems containing polyoxyethylene alkyl ether non-ionic surfactant of technical grade. A domain of liquid isotropic phase extends along the surfactant/styrene axis solubilizing up to 10 % water. According to the literature, the most probable structures are inverse micelles or W/O microemulsion (Om). Higher amounts of water (up to 20% approximately) led to the appearance of the lamellar crystalline phase (Lα) that coexists with the Om phase. A wide multiphasic region, with two or three phases comprising liquid crystalline phases (equilibrium not determined), occupies the centre of the diagram for water composition from 20% to 60%. The lamellar crystalline phase in equilibrium with water, (W + Lα) is observed at high-medium surfactant concentration. For the higher amounts of water (up to 90%), the two phase region is present: oil-in-water microemulsion (Wm) and free oil (O). Above this area, by increasing the surfactant concentration, the oil is completely incorporated in the oil-in-water microemulsion. Finally, at low Brij 98 concentrations (O:S ratio above 80:20), appears the three-phase region (Forgiarini, et al., 2001; I. Solè et al., 2006).

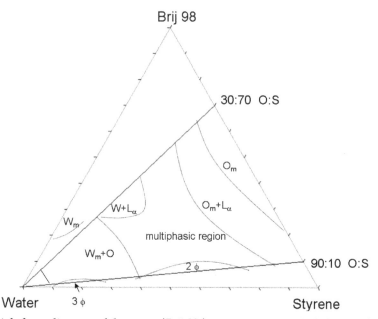

Fig. 2. Partial phase diagram of the water/Brij 98/styrene system, at 50° C. Wm: direct micellar solution or O/W microemulsion; Om: reverse micellar solution or W/O microemulsion); Lα: anisotropic phase (lamellar liquid crystalline phase); W: water phase; O: oil phase. 2ϕ: two isotropic liquid phases; 3ϕ: three isotropic liquid phases. (Adapted from Sadtler, et al., 2010)

The equilibrium phase diagram (figure 2), allows identifying a suitable region for nanoemulsion formation. This domain corresponds to the two-phase region, Wm + O, for O/S ratio between 30:70 and 80:20. The emulsification process path is schematically represented by an arrow on the phase diagram (figure 3). Thus, the addition of water at constant rate to different mixtures of Brij 98 and styrene (inside the suitable region), allowed the system to cross the multiphasic central region, with two or three phases comprising liquid crystalline phases, favouring nanoemulsion formation. The final water concentration was fixed at 80 wt. %, to keep a relatively high percentage of dispersed phase. These aqueous dispersions can be regarded as O/W nanoemulsions (and not microemulsions) because they are formed in the multiphase region (W_m + O). Bluish dispersions were obtained. After addition of KPS solution (the water soluble thermal initiator), miniemulsion polymerization was carried out at 50° C for 24 h.

Figure 4 shows the evolution of nanoparticle sizes as a function of O:S ratio for nanoemulsion containing 80 wt.% water. Results from Sadtler *et al.*, (2010), showed that nanoparticle sizes were clearly dependent on O:S ratio and increased with O:S ratio. After polymerization, nanoparticle sizes varied between 36 nm (O:S ratio of around 0.5) and

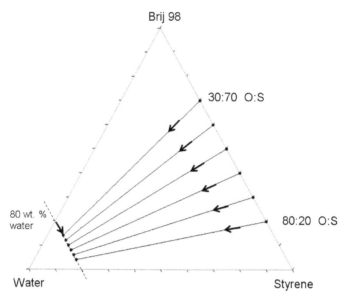

Fig. 3. Schematic representation of the emulsification paths: stepwise addition of 80 wt. % water to different oil:surfactant ratios mixtures

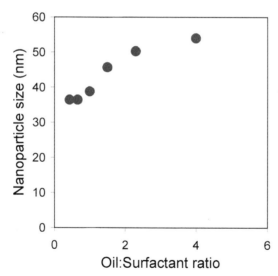

Fig. 4. Polystyrene nanoparticle diameter as function of O:S initial ratio (water addition rate = 4.6 ml/h). (Adapted from Sadtler, et al., 2010)

50 nm (O:S ratio = 4). The fact that the nanoparticle diameter progressively increased with the oil:surfactant ratio suggests that the styrene constituted the inner core of the nanodroplets, which was consistent with a direct O/W –type structure. It should be noticed that the polystyrene nanoparticles sizes obtained by this emulsification path, were

exceptionally small for the water/Brij 98/styrene system, compared to those reported in the literature from high-energy emulsification method (Antonietti & Landfester, 2002; Asua, 2002; Bouanani et al., 2008; Marie, et al., 2007).

Figure 5 shows the size of the nanoparticles synthesized from nanoemulsions prepared at different water flow rates, ranging from 4 ml/h to 150 ml/h, for a O:S ratio of 30:70 and final water composition of 80 wt. %. As predicted by Pey (2006), the polystyrene particle sizes increased with water addition rate. This could be related by the crossing rate of the phases along the emulsification paths, *i.e.* the kinetic of the whole emulsification process.

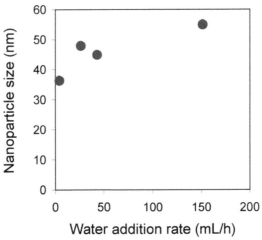

Fig. 5. Polystyrene nanoparticle size as function of the water addition rate (O:S ratio = 0.42). (Adapted from Sadtler, et al., 2010)

To confirm this point, the same path was followed as in the first experiment by adding water at once. In the second experiment, the order of the addition was modified: the oil was added at once to the mixture of water and surfactant (the sample compositions were identical in the two cases).

Figure 6 presents the particle sizes after polymerization obtained following these different pathways. Particles resulting from stepwise addition of water over the mixture of oil and surfactant were smaller than the one obtained by water addition at once (36 and 65 nm respectively).

Phase transitions that take place during the emulsification process (as result of the change in the spontaneous curvature of surfactant), allow to low energy emulsification methods make use of stored chemical energy to get a small drop size distribution. However, when styrene was added to the water and Brij 98 mixture (at once), milky emulsions were formed and the polymerization of the oil droplets did not produced small nanoparticles. Hence the resulting polystyrene dispersion presented an average size of 420 nm (figure 6). The polymerization process might even be totally different in this case, switching from real miniemulsion polymerization to a "simple" emulsion polymerization process. Indeed, miniemulsion polymerization is only possible if the droplets are nucleated thereby leading to polymerization inside the droplets. The smaller the droplet size, the higher the probability of radical entry into the droplets because of the higher interface area. When the emulsion droplets are bigger, the radical entry probability decreases while the micellar and/or homogeneous nucleation increased. Theses nucleation processes are found in the emulsion polymerization process.

Concerning the emulsification protocol, internal phase addition should favour a proper mixing, to assure to reach the equilibrium with all the oil dissolved into the critical phases (*i.e.*, cubic liquid crystal or lamellar phase) (Isabel Solè, et al., 2010).

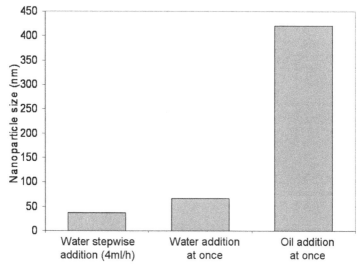

Fig. 6. Polystyrene nanoparticle diameter as function of the emulsification process. The nanoemulsions were prepared at 30:70 O:S ratio, with a final water concentration of 20 wt.%.

3. Nanoparticles by near-PIT method

The formulation-composition map is the graphical representation of the so-called generalized formulation (see Figure 7). The middle shaded zone corresponds to the three phase behavior at or near the optimum formulation. The formulation variable scale is such that the hydrophilicity increases from top to bottom and the stair like bold line is the standard inversion frontier. This line separates the regions in which O/W and W/O emulsions are formed as the result of the stirring of an equilibrate surfactant–oil–water system (Salager, 2000b; Salager et al., 1983). The crossing over through the inversion frontier represents a dynamic phase inversion, since the curvature of the liquid–liquid interface swaps its bending from one way to the other. This change is the consequence of the variation in one of formulation variables (*i.e.*, surfactant affinity) or composition variables (*i.e.*, oil/water ratio) during the stirring process.

If the change is rendered in the map as a vertical shift (crossing through the horizontal branch of inversion line), as for instance in the continuous change in temperature (in the case of non-ionic surfactant), the inversion will always take place under the same conditions (at so-called optimum formulation). Such a dynamic inversion, which is found to be reversible, has been called transitional because it is linked to a phase behavior transition (Salager, 2000a). On the other hand, when the inversion takes place by crossing through a vertical branch of the inversion line, it is called catastrophic because it may be modeled by using catastrophe theory (Salager, 1988; Salager, et al., 1983).

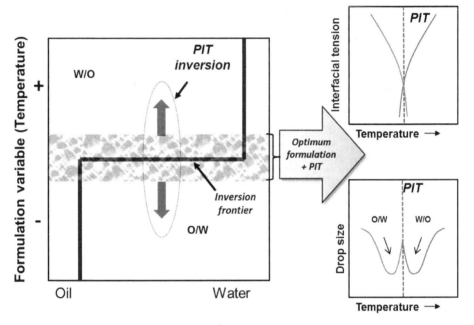

Fig. 7. Formulation – composition map. The bold line is the standard inversion frontier. At right side upper schema illustrates the minimum of interfacial tension obtained near to PIT value. Bottom schema show the regions near to PIT value where droplet size presents a minimum.

The transitional phase inversion is based on the particular ability of emulsions stabilized by poly(ethylene oxide) (PEO)-based non-ionic surfactants to undergo a phase inversion upon temperature variation (Kōzō Shinoda & Arai, 1964). A change of formulation (i.e. induced by temperature increase) along a vertical line, results in a minimum of both the interfacial tension and the emulsion stability at the optimum formulation (see Figure 5). The minimum of stability at optimum formulation has been attributed either to the percolation through liquid crystals located across the thin film, or to the trapping of all surfactant in the microemulsion (Antón et al., 1986). Thus as optimum formulation is approached (either from above or from below the standard inversion line) both the interfacial tension and the emulsion stability decrease. As far as the emulsion droplet size is concerned, the two resulting effects are opposite. The weakening of interfacial tension tends to enhance the efficiency of stirring-mixing process and thus produces smaller droplets, while the decrease in emulsion stability favors the occurrence of coalescence events, and thus results in larger droplets (Salager et al., 1996) (see Figure 7).

The use of the low-energy PIT method has been reported for miniemulsion polymerization by heating above PIT temperature to inverse the emulsion and then cooling to induce the re-inversion followed sometimes by a rapid cooling in an ice bath to set droplet size within the submicronic range (Jahanzad et al., 2007; Liat Spernath & Magdassi, 2007; L. Spernath & Magdassi, 2010). In this section the Near - PIT method (Galindo-Alvarez et al., 2011), for which the strong decrease in interfacial tension near to optimum formulation is used to form submicronic droplets, will be discussed. In contrast to other protocols, Near-PIT method does not reach and cross temperatures aboves PIT, thus temperature sensitive molecules can be use through a carefully match of surfactant system.

It has been reported that stable O/W nanoemulsions can be produced by the PIT method if the dispersed system is rapidly cooled by about 30°C away from its temperature of transitional phase inversion (Solans, et al., 2005). In those conditions, droplet coalescence becomes negligible because the non-ionic surfactant molecules provide an efficient steric barrier. Therefore the miniemulsion templates should exhibit a PIT value about 30 °C higher than the targeted polymerization temperature. PIT value results from interaction between overall surfactant concentration, surfactant mixing ratio and weight fraction oil (K. Shinoda & Arai, 1967). Thus, the PIT value of the studied system was tuned by the appropriate selection of the constituents.

In the case of non-ionic surfactant mixtures, it is well-known that increasing the length of the poly(ethylene oxide) chain results in higher HLB numbers and thus the increase in PIT. Two non-ionic surfactants, PEO stearyl ethers (Brij 78 and Brij 700), differing by the length of the PEO chain (20 and 100 repeat units, respectively) were used for formulating the nanoemulsions allowing a certain adjustment of the PIT value within the convenient range. Figure 8 shows the influence of surfactant mixing ratio and weight fraction of dispersed phase over PIT value. Thus a water/Brij 78 + Brij 700/styrene system containing 1%w/v of NaCl, a surfactant mixing ratio of 0.35/0.65 Brij 700/Brij 78 and 35 wt% of dispersed phase with PIT value around 80°C has been chosen to carry out the miniemulsion polymerization at 50°C using potassium persulfate (KPS) as water-soluble initiator.

On the basis of the previously selected formulation, the Near-PIT emulsification procedure is designed and compared to classical sonification and emulsion polymerization with regard to the final nanoparticle size obtained after reaction completion. For used conditions, a polymerization temperature of 50 °C ensures fast enough initiator decomposition so that no limitation by the polymerization reaction is considered. About particle nucleation mechanism, droplet size distribution of miniemulsion polymerization templates was similar to droplet size distribution of latex particles, suggesting predominance of droplet nucleation mechanism.

Near-PIT protocol, as discussed at the beginning of this section, is based on the effect that droplet size decreases when PIT temperature is approached as the result of an enhanced stirring efficiency due to the very low interfacial tension. Nevertheless, in that temperature range, close to the PIT, resulting emulsions turn out to be very unstable; and no theoretical relationships are available to discriminate zones of minimum droplet size from unstable emulsion. In a general way, in Near-PIT protocol the system is heated until a temperature close to PIT value, equivalent to: -5°C or – 10°C below PIT temperature. As this value is particular for each system formulation, Figure 9 illustrates the final average particle diameters obtained for suspensions resulting from miniemulsion polymerizations after Near-PIT protocols carried out at PIT-10°C, PIT-5°C and PIT-0.5 °C values. The used formulation was ϕ_{oil} = 0.2, $X_{Brij700}$ = 0.35 and the overall surfactant concentration equal to 5.2 wt%. For the three examined conditions, particles with diameters lower than 100 nm were obtained after polymerization. In addition, polydispersity indices were relatively low (<0.22) indicating reasonably narrow size distributions. Even if the average diameters were similar for the different temperatures, the better compromise was obtained for the experience carried out at PIT – 5°C, since its polydispersity index was the lowest.

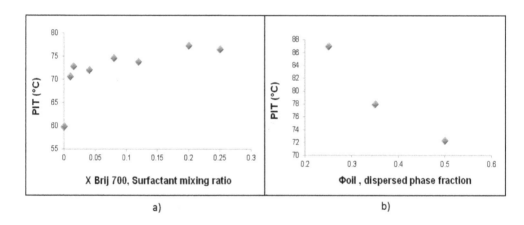

a)   b)

Fig. 8. a) Increase of PIT value as function of Brij® 700 content, XBrij700, ϕ_{oil} = 0.5 and 8wt% of surfactant concentration. b) Variation of PIT value with the weight fraction of dispersed phase, ϕ_{oil}, $X_{Brij\,700}$ = 0.25 and 8wt% of surfactant mixture.
(Adapted from Galindo-Alvarez, et al., 2011)

In miniemulsion polymerization, the use of an effective surfactant system may give very small (20–300 nm) monomer droplets with very large surface area and almost all the surfactant adsorbed at the droplet surface (the concept of critically "stabilized miniemulsion"). Particle nucleation occurs primarily via radical (primary or oligomeric) entry into monomer droplets, since little or no surfactant is present in the form of micelles. The reaction proceeds by polymerization of the monomer in these small droplets, since the loci of polymerization become the monomer droplets and ends when all monomer in droplet is consumed (Schork et al., 2005).

In contrast for macroemulsion polymerization, polymerization starts with large monomer droplets (diameters higher than 10 µm) stabilized by surfactant and coexisting with empty or monomer-swollen surfactant micelles. The water-soluble initiator forms oligoradicals with the slightly water-soluble monomer molecules and these oligoradicals go inside the micelles (heterogeneous nucleation) or start nucleate particles in the continuous phase after reaching a critical degree of polymerization (homogeneous nucleation). During polymerization, the monomer diffuses from the large monomer droplets through the continuous phase to the polymer particles and sustain polymer particle growth until the monomer droplets have vanished (Antonietti & Landfester, 2002). Thus in miniemulsion polymerization latex particles size distribution are expected to correspond to the primary emulsion droplets. On the contrary, in macroemulsion polymerization the particle size distribution is established by the contribution of several nucleation processes leading to average diameters usually larger than 100nm and sometimes to the formation of several populations within the final sample.

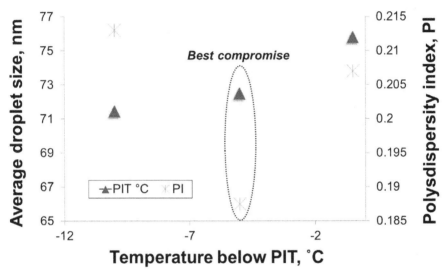

Fig. 9. Influence of polymerization temperature below PIT on final average particle size. Nanoemulsion formulation: $\phi_{oil} = 0.2$, $X_{Brij700} = 0.35$ and 5.2 wt% surfactant concentration. (Adapted from Galindo-Alvarez, et al., 2011)

Figure 10 illustrates the particle size obtained from two miniemulsion polymerization methods (low energy Near-PIT and ultrasound emulsification) and one coarse–emulsion polymerization (standard mechanical emulsification) with various surfactant to oil weight ratios and composition of surfactant mixture. In macroemulsion polymerization protocol, coarse–emulsion is agitated at 800rpm as in Near-PIT method, but the system is heated only until polymerization temperature and not 25 °C beyond as in Near-PIT. As expected, macroemulsion polymerization from coarse–emulsion gave the highest particle diameters and polydispersity indices which indicate a large and probably multimodal particle size distribution (see figures. 10 and 11). Thus, the viability of Near-PIT method to produce submicronic droplets as templates for miniemulsion polymerization, with slightly better efficiency than that found for ultrasonic emulsification method, has been confirmed.

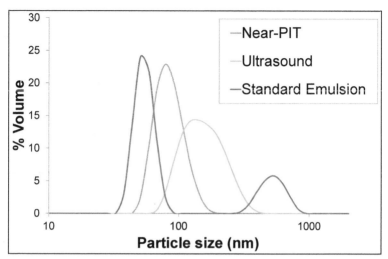

Fig. 10. Influence of emulsification process in particle size distribution, nanoemulsion formulation: $°oil = 0.35$, $X_{Brij\ 700} = 0.4$ and 5.2 wt% surfactant concentration. (Adapted from Galindo-Alvarez, et al., 2011)

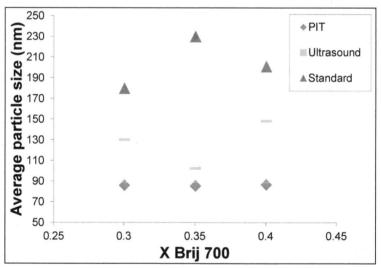

Fig. 11. Influence of emulsification process in average particle size, miniémulsion formulation: $°oil = 0.35$ and 5.2 wt% surfactant concentration. (Adapted from Galindo-Alvarez, et al., 2011)

4. Conclusion

This chapter has showed the viability to produce polystyrene nanoparticles by two different types of low energy emulsification methods: EIP, emulsion inversion point and Near-PIT, near phase inversion temperature.

The Emulsion Inversion Point technique was used on the water/Brij 98/styrene system to the formation of direct styrene-in-water nanoemulsions. After miniemulsion polymerization, particle sizes as low as 36 nm were obtained. These values are much lower than the one classically reached by high-energy emulsification methods. Thus Emulsion Inversion Point method is very attractive in industrial applications for nanoparticle synthesis, since the nano-emulsion formation does not require high concentration of surfactant, as in the case of microemulsion, or special high-shear equipments as in the case of most reported miniemulsion polymerization

The Phase Inversion Temperature concept as a tool to produce miniemulsion templates for miniemulsion polymerization is a promising methodology in polymerization field to obtain monodisperse and aggregate-free nanoparticle suspensions. In this review a low-energy emulsification method has been designed and showed to allow the preparation of polymeric particles smaller than usual ultrasound miniemulsification methods (about 75 nm) in a water/Brij 78 + Brij 700/styrene system. The operating conditions were adjusted so as to conciliate particle size distribution, colloidal stability and polymerization kinetics. Contrary to usual PIT methods, the Near-PIT procedure did not imply heating the samples at temperatures higher than PIT. Final particles had comparable characteristics to those obtained by traditional PIT methods. In addition, we showed the relevance of temperature control (PIT or PIT-ΔT) over nanoparticle size to obtain even slightly smaller particles than those obtained after ultrasound emulsification. Finally it is possible to vary the composition of the surfactant mixture, within a certain range, without strongly modifying nanoparticles final characteristics, but in way to control the thickness of the hydrophilic superficial layer.

5. Acknowledgments

This work has been funded by the BLAN-06-0174 ANR (National Research Agency) program. The authors thank Prof. Alain Durand, Dr. Marianna Rondon Gonzales, Mrs. Audrey Acrement, Mrs. Ioulia Habipi and Mr. David Boyd for their contributions to this study.

6. References

Anton, N., Benoit, J. P., & Saulnier, P. (2008). Design and production of nanoparticles formulated from nano-emulsion templates-A review. *Journal of Controlled Release*, *128*(3), 185-199.

Antón, R. E., Castillo, P., & Salager, J.-L. (1986). Surfactant-oil-water systems near the affinity inversion part IV: emulsion inversion temperature. *Journal of Dispersion Science and Technology 7*(3), 319-329.

Antonietti, M., & Landfester, K. (2002). Polyreactions in miniemulsions. *Progress in Polymer Science (Oxford)*, *27*(4), 689-757.

Asua, J. M. (2002). Miniemulsion polymerization. *Progress in Polymer Science (Oxford)*, *27*(7), 1283-1346.

Bouanani, F., Bendedouch, D., Hemery, P., & Bounaceur, B. (2008). Encapsulation of montmorillonite in nanoparticles by miniemulsion polymerization. *Colloids and Surfaces A: Physicochemical and Engineering Aspects, 317*(1-3), 751-755.

Calderó, G., García-Celma, M. J., & Solans, C. (2011). Formation of polymeric nano-emulsions by a low-energy method and their use for nanoparticle preparation. *Journal of Colloid And Interface Science, 353*(2), 406-411.

Forgiarini, A., Esquena, J., Gonzalez, C., & Solans, C. (2001). Formation of nano-emulsions by low-energy emulsification methods at constant temperature. *Langmuir, 17* 2076-2083.

Galindo-Alvarez, J., Boyd, D., Marchal, P., Tribet, C., Perrin, P., Marie-Bégué, E., et al. (2011). Miniemulsion polymerization templates: A systematic comparison between low energy emulsification (Near-PIT) and ultrasound emulsification methods. *Colloids and Surfaces A: Physicochemical and Engineering Aspects, 374*(1-3), 134-141.

Izquierdo, P., Feng, J., Esquena, J., Tadros, T. F., Dederen, J. C., Garcia, M. J., et al. (2005). The influence of surfactant mixing ratio on nano-emulsion formation by the pit method. *Journal of Colloid And Interface Science, 285*(1), 388-394.

Jahanzad, F., Chauhan, G., Mustafa, S., Saha, B., Sajjadi, S., & Brooks, B. W. (2007). Composite Polymer Nanoparticles via Transitional Phase Inversion Emulsification and Polymerisation. *Macromolecular Symposia, 259*(1), 145-150.

Kabal'nov, A. S., Pertzov, A. V., & Shchukin, E. D. (1987). Ostwald ripening in two-component disperse phase systems: Application to emulsion stability. *Colloids and Surfaces, 24*(1), 19-32.

Lamaallam, S., Bataller, H., Dicharry, C., & Lachaise, J. (2005). Formation and stability of miniemulsions produced by dispersion of water/oil/surfactants concentrates in a large amount of water. *Colloids and Surfaces A: Physicochemical and Engineering Aspects, 270-271*(1-3), 44-51.

Maestro, A., Solè, I., González, C., Solans, C., & Gutiérrez, J. M. (2008). Influence of the phase behavior on the properties of ionic nanoemulsions prepared by the phase inversion composition method. *Journal of Colloid And Interface Science, 327*(2), 433-439.

Marie, E., Rotureau, E., Dellacherie, E., & Durand, A. (2007). From polymeric surfactants to colloidal systems. 4. Neutral and anionic amphiphilic polysaccharides for miniemulsion stabilization and polymerization. *Colloids and Surfaces A: Physicochemical and Engineering Aspects, 308*(1-3), 25-32.

Mason, T. G., Graves, S. M., Wilking, J. N., & Lin, M. Y. (2006). Extreme emulsification: Formation and structure of nanoemulsions. *Condensed Matter Physics, 9*(1), 193-199.

Mohlin, K., Holmberg, K., Esquena, J., & Solans, C. (2003). Study of low energy emulsification of alkyl ketene dimer related to the phase behavior of the system. *Colloids and Surfaces A: Physicochemical and Engineering Aspects, 218*(1-3), 189-200.

Pey, C. M., Maestro, A., Solé, I., González, C., Solans, C., & Gutiérrez, J. M. (2006). Optimization of nano-emulsions prepared by low-energy emulsification methods at constant temperature using a factorial design study. *Colloids and Surfaces A: Physicochemical and Engineering Aspects, 288*(1-3), 144-150.

Porras, M., Solans, C., González, C., & Gutiérrez, J. M. (2008). Properties of water-in-oil (W/O) nano-emulsions prepared by a low-energy emulsification method. *Colloids and Surfaces A: Physicochemical and Engineering Aspects, 324*(1-3), 181-188.

Rang, M. J., & Miller, C. A. (1999). Spontaneous emulsification of oils containing hydrocarbon, nonionic surfactant, and oleyl alcohol. *Journal of Colloid And Interface Science, 209*(1), 179-192.

Sadtler, V., Rondon-Gonzalez, M., Acrement, A., Choplin, L., & Marie, E. (2010). PEO-Covered Nanoparticles by Emulsion Inversion Point (EIP) Method. *Macromolecular Rapid Communications, 31*(11), 998-1002.

Sadurní, N., Solans, C., Azemar, N., & García-Celma, M. J. (2005). Studies on the formation of O/W nano-emulsions, by low-energy emulsification methods, suitable for pharmaceutical applications. *European Journal of Pharmaceutical Sciences, 26*(5), 438-445.

Salager, J.-L. (1988). 2 Phase transformation and emulsion inversion on the basis of catastrophe theory. In P. Becher (Ed.), *Encyclopedia of Emulsion Technology: Basic Theory, Measurement, Applications* (Vol. 3, pp. 79 - 134). New York - Basel: Marcel Dekker.

Salager, J.-L. (2000a). Emulsion properties and related know-how to attain them. In F. Nielloud & G. Marti-Mestres (Eds.), *Pharmaceutical Emulsions and Suspensions* (pp. 73 - 125). New York: Marcel Dekker.

Salager, J.-L. (2000b). Formulation concepts for the emulsion maker. In F. Nielloud & G. Marti-Mestres (Eds.), *Pharmaceutical Emulsions and Suspensions* (pp. 19 - 72). New York: Marcel Dekker.

Salager, J.-L., Miñana-Perez, M., M. Pérez-Sánchez, Ramirez-Gouveia, M., & Rojas, C. I. (1983). Surfactant-oil-water systems near the affinity inversion part III: the two kinds of emulsion inversion. *Journal of Dispersion Science and Technology 4*(3), 313 - 329.

Salager, J.-L., Perez-Sanchez, M., & Garcia, Y. (1996). Physicochemical parameters influencing the emulsion drop size. *Colloid and Polymer Science, 274*(1), 81-84.

Schork, F. J., Luo, Y., Smulders, W., Russum, J. P., Butté, A., & Fontenot, K. (2005). Miniemulsion Polymerization *Polymer Particles* (pp. 129-255).

Shinoda, K., & Arai, H. (1964). The correlation between phase inversion temperature in emulsion and cloud point in solution of nonionic emulsifier. *Journal of Physical Chemistry, 68*(12), 3485-3490.

Shinoda, K., & Arai, H. (1967). The effect of phase volume on the phase inversion temperature of emulsions stabilized with nonionic surfactants. *Journal of Colloid And Interface Science, 25*(3), 429-431.

Shinoda, K., & Saito, H. (1968). The effect of temperature on the phase equilibria and the types of dispersions of the ternary system composed of water, cyclohexane, and nonionic surfactant. *Journal of Colloid and Interface Science, 26*(1), 70-74.

Solans, C., Esquena, J., Forgiarini, A. M., Uson, N., Morales, D., Izquierdo, P., et al. (2002). Nanoemulsions: Formation and Properties. In K. L. Mittal & D. O. Shah (Eds.), *Surfactants in Solution: Fundamentals and Applications* (pp. 525). New York: Marcel Dekker.

Solans, C., Izquierdo, P., Nolla, J., Azemar, N., & Garcia-Celma, M. J. (2005). Nano-emulsions. *Current Opinion in Colloid & Interface Science, 10*(3-4), 102-110.

Solè, I., Maestro, A., González, C., Solans, C., & Gutiérrez, J. M. (2006). Optimization of nano-emulsion preparation by low-energy methods in an ionic surfactant system. *Langmuir, 22*(20), 8326-8332.

Solè, I., Pey, C. M., Maestro, A., González, C., Porras, M., Solans, C., et al. (2010). Nano-emulsions prepared by the phase inversion composition method: Preparation variables and scale up. *Journal of Colloid And Interface Science, 344*(2), 417-423.

Spernath, L., & Magdassi, S. (2007). A new method for preparation of poly-lauryl acrylate nanoparticles from nanoemulsions obtained by the phase inversion temperature process. *Polymers for Advanced Technologies, 18*(9), 705-711.

Spernath, L., & Magdassi, S. (2010). Formation of silica nanocapsules from nanoemulsions obtained by the phase inversion temperature method. *Micro and Nano Letters, 5*(1), 28-36.

Spernath, L., Regev, O., Levi-Kalisman, Y., & Magdassi, S. (2009). Phase transitions in O/W lauryl acrylate emulsions during phase inversion, studied by light microscopy and cryo-TEM. *Colloids and Surfaces A: Physicochemical and Engineering Aspects, 332*(1), 19-25.

Tadros, T., Izquierdo, P., Esquena, J., & Solans, C. (2004). Formation and stability of nano-emulsions. *Advances in Colloid and Interface Science, 108-109*, 303-318.

Permissions

The contributors of this book come from diverse backgrounds, making this book a truly international effort. This book will bring forth new frontiers with its revolutionizing research information and detailed analysis of the nascent developments around the world.

We would like to thank Dr. Abbass A. Hashim, for lending his expertise to make the book truly unique. He has played a crucial role in the development of this book. Without his invaluable contribution this book wouldn't have been possible. He has made vital efforts to compile up to date information on the varied aspects of this subject to make this book a valuable addition to the collection of many professionals and students.

This book was conceptualized with the vision of imparting up-to-date information and advanced data in this field. To ensure the same, a matchless editorial board was set up. Every individual on the board went through rigorous rounds of assessment to prove their worth. After which they invested a large part of their time researching and compiling the most relevant data for our readers. Conferences and sessions were held from time to time between the editorial board and the contributing authors to present the data in the most comprehensible form. The editorial team has worked tirelessly to provide valuable and valid information to help people across the globe.

Every chapter published in this book has been scrutinized by our experts. Their significance has been extensively debated. The topics covered herein carry significant findings which will fuel the growth of the discipline. They may even be implemented as practical applications or may be referred to as a beginning point for another development. Chapters in this book were first published by InTech; hereby published with permission under the Creative Commons Attribution License or equivalent.

The editorial board has been involved in producing this book since its inception. They have spent rigorous hours researching and exploring the diverse topics which have resulted in the successful publishing of this book. They have passed on their knowledge of decades through this book. To expedite this challenging task, the publisher supported the team at every step. A small team of assistant editors was also appointed to further simplify the editing procedure and attain best results for the readers.

Our editorial team has been hand-picked from every corner of the world. Their multi-ethnicity adds dynamic inputs to the discussions which result in innovative outcomes. These outcomes are then further discussed with the researchers and contributors who give their valuable feedback and opinion regarding the same. The feedback is then collaborated with the researches and they are edited in a comprehensive manner to aid the understanding of the subject.

Apart from the editorial board, the designing team has also invested a significant amount of their time in understanding the subject and creating the most relevant covers. They scrutinized every image to scout for the most suitable representation of the subject and create an appropriate cover for the book.

The publishing team has been involved in this book since its early stages. They were actively engaged in every process, be it collecting the data, connecting with the contributors or procuring relevant information. The team has been an ardent support to the editorial, designing and production team. Their endless efforts to recruit the best for this project, has resulted in the accomplishment of this book. They are a veteran in the field of academics and their pool of knowledge is as vast as their experience in printing. Their expertise and guidance has proved useful at every step. Their uncompromising quality standards have made this book an exceptional effort. Their encouragement from time to time has been an inspiration for everyone.

The publisher and the editorial board hope that this book will prove to be a valuable piece of knowledge for researchers, students, practitioners and scholars across the globe.

List of Contributors

Anurag Mishra
National Institute for Occupational Safety and Health, HELD/PPRB, Morgantown, WV, USA
West Virginia University, Department of Pharmaceutical Sciences, Morgantown, WV, USA

Yon Rojanasakul
West Virginia University, Department of Pharmaceutical Sciences, Morgantown, WV, USA

Liying Wang
National Institute for Occupational Safety and Health, HELD/PPRB, Morgantown, WV, USA

Enzo Di Fabrizio
Nanostructures Department, Italian Institute of Technology, Genova, Italy
BioNEM lab., Departement of Clinical and Experimental Medicine, Italy
Magna Graecia University, Viale Europa, Catanzaro, Italy

Manohar Chirumamilla Chowdary, Ermanno Miele, Rosanna La Rocca, Roman Krahne, Gobind Das, Francesco De Angelis, Carlo Liberale, Andrea Toma, Luca Razzari, and Remo Proietti Zaccaria
Nanostructures Department, Italian Institute of Technology, Genova, Italy

Francesco Gentile and Maria Laura Coluccio
BioNEM Lab., Department of Clinical and Experimental Medicine, Magna Graecia University, Viale Europa, Catanzaro, Italy
Nanostructures Department, Italian Institute of Technology, Genova, Italy

Michela Perrone Donnorso
Nanophysics Department, Italian Institute of Technology, Genova, Italy
Nanostructures Department, Italian Institute of Technology, Genova, Italy

Rosaria Brescia and Liberato Manna
Nanochemistry Department, Italian Institute of Technology, Genova, Italy

Leonard F. Pease III
Department of Chemical Engineering, University of Utah, Salt Lake City, UT, USA
Department of Internal Medicine, Division of Gastroenterology, Hepatology and Nutrition, University of Utah, Salt Lake City, UT, USA
Department of Pharmaceutics and Pharmaceutical Chemistry, University of Utah, Salt Lake City, UT, USA

Rajasekhar Anumolu
Department of Chemical Engineering, University of Utah, Salt Lake City, UT, USA

Ken Kokubo
Division of Applied Chemistry, Graduate School of Engineering, Osaka University, Japan

S.V. Valueva and L.N. Borovikova
The Institution of the Russian Academy of Science, The Institute of High-Molecular Compounds, Saint-Petersburg, Russia

Neus G. Bastús, Eudald Casals, Isaac Ojea and Miriam Varon
Institut Català de Nanotecnologia, Barcelona, Spain

Victor Puntes
Institut Català de Nanotecnologia, Barcelona, Spain
Institut Català de Recerca i Estudis Avançats (ICREA) Barcelona, Spain

E.C. Jung and H.R. Cho
Nuclear Chemistry Research Division, Korea Atomic Energy Research Institute, Republic of Korea

C. Coutanceau, S. Baranton and T.W. Napporn
e-Lyse, Laboratoire de Catalyse en Chimie Organique, UMR CNRS-Université de Poitiers, France

L.P. Bulat
National Research University ITMO, St. Petersburg, Russia

D.A. Pshenai-Severin
Ioffe Physical Technical Institute, St Petersburg, Russia

V.V. Karatayev, V.B. Osvenskii , M. Lavrentev and A. Sorokin
GIREDMET Ltd., Moscow, Russia

V.D. Blank and G.I. Pivovarov
Technological Institute of Superhard and New Carbon Materials, Troitsk, Russia

V.T. Bublik and N.Yu. Tabachkova
National University of Science and Technology "MISIS", Moscow, Russia

Yu.N. Parkhomenko
GIREDMET Ltd., Moscow, Russia
National University of Science and Technology "MISIS", Moscow, Russia

Masatoshi Iji
NEC Corporation, Japan

Nopphawan Phonthammachai and Hongling Chia
Institute of Materials Research and Engineering, Singapore
A*STAR (Agency for Science, Technology and Research), Singapore

Chaobin He
Department of Materials Science & Engineering, National University of Singapore, Singapore
Institute of Materials Research and Engineering, Singapore
A*STAR (Agency for Science, Technology and Research), Singapore

Philippe Sciau
CEMES-CNRS, Université de Toulouse, France

Veronique Sadtler and Johanna M. Galindo-Alvarez
Laboratoire Réactions et Génie des Procédés – GEMICO, CNRS-Nancy Université, Nancy, France

Emmanuelle Marie –Bégué
UMR 8640 CNRS-ENS-UPMC, Ecole Normale Supérieure, Paris, France

Printed in the USA
CPSIA information can be obtained
at www.ICGtesting.com
JSHW011449221024
72173JS00004B/997